Student Solutions Manual

to accompany

Introduction to Chemistry

Third Edition

Richard C. Bauer
Arizona State University

James P. Birk
Arizona State University

Pamela S. Marks
Arizona State University

Prepared by
Kirk Kawagoe
Fresno City College

The McGraw·Hill Companies

Mc Graw Hill

Connect
Learn
Succeed™

Student Solutions Manual to accompany
INTRODUCTION TO CHEMISTRY, THIRD EDITION
RICHARD C. BAUER, JAMES P. BIRK, PAMELA S. MARKS

Published by McGraw-Hill Higher Education, an imprint of The McGraw-Hill Companies, Inc., 1221 Avenue of the Americas, New York, NY 10020. Copyright © 2013, 2010 and 2007 by The McGraw-Hill Companies, Inc. All rights reserved. Printed in the United States of America.

♻ This book is printed on recycled, acid-free paper containing 10% post consumer waste.

1 2 3 4 5 6 7 8 9 0 QDB/QDB 1 0 9 8 7 6 5 4 3 2

ISBN: 978-0-07-737817-2
MHID: 0-07-737817-2

Student's Solutions Manual
to accompany
Introduction to Chemistry

Table of Contents

Chapter 1 – Matter and Energy

1.1 (a) mass; (b) chemical property; (c) mixture; (d) element; (e) energy; (f) physical property; (g) liquid; (h) density; (i) homogeneous mixture; (j) solid state

1.3 When converting to scientific notation, count the number of places you need to move the decimal point. Zeros to the left of the number are always dropped. For example the number 0.002030 becomes 2.030×10^{-3} and the zeros to the left of 2030 are dropped. The zero to the right is only kept if it is significant (covered later in this chapter). If the decimal point moves Right, the Exponent Decreases (RED). The opposite is true if the decimal moves left (the exponent increases, LEI).
(a) 2.95×10^4; (b) 8.2×10^{-5}; (c) 6.5×10^8; (d) 1.00×10^{-2}

1.5 When converting from scientific notation to standard notation you may need to add place holder zeros so that the magnitude of the number is correct. Otherwise, RED and LEI still apply (**see problem 1.3**). For example, to get 1.86×10^{-5} into standard notation, you need to increase the power by five, so the decimal moves to the left (LEI). In addition, you'll need four placeholder zeros to show the magnitude of the number.
(a) 0.0000186; (b) 10000000; (c) 453000; (d) 0.0061

1.7 (a) 6.2×10^3; (b) 3.5×10^7; (c) 2.9×10^{-3}; (d) 2.5×10^{-7}; (e) 8.20×10^5; (f) 1.6×10^{-6}

1.9 Non zero digits and zeros between nonzero digits are significant. Zeros to the right of the number are significant only if it is written with a decimal. Zeros to the left of the number and exponentials (i.e. $\times 10^3$) are not counted. The number 0.0950 has three significant digits. The digits "950" are all significant because (1) the 9 and 5 are nonzero and (2) the zero is significant because the number was written with a decimal (0.0950).
(a) 3; (b) 2; (c) 4; (d) 2; (e) 3

1.11 For operations involving multiplication, division, and powers the answer will have the same number of significant figures as the number with the fewest significant figures. The number 1.201×10^3 has four significant figures and the number 1.2×10^{-2} has two significant figures. The answer (c) 14.412 will be rounded to two significant figures (14).
(a) 1.5; (b) 1.5; (c) 14; (d) 1.20

1.13 For operations involving addition and subtraction the answer will have the same *absolute precision* as the *least precise number*. A number that has its last significant digit in the tenths place has less absolute precision than a number that ends in the hundredths place (hundredths are smaller than tenths). If you add these two number together, you would have to round the answer to the tenths place. For example,
(a) 1.6 + 1.15 = 2.75. The number 2.75 will have to be rounded to the tenths place (2.8).
(a) 2.8; (b) 0.28; (c) 2.8; (d) 0.049

1.15 (a) 1.21; (b) 0.204; (c) 1.84; (d) 42.2; (e) 0.00710

1.17 When you are converting between a unit and the same base unit with a prefix (i.e. mm to m or visa versa) you can find the conversion factors in Math Toolbox 1.3. Suppose you want to convert between millimeters and meters. There are several ways you can do this. First, by definition milli is = 10^{-3}. So 1 mm = 10^{-3} m. This equation says, "one millimeter is **one thousandth** of a meter." You might also already know that there are "one thousand millimeters in a meter" (1000 mm = 1 m). Either conversion factor is correct. Next you set up your calculation so that the appropriate units cancel. The English-Metric conversions are also found in Math Toolbox 1.3 (i.e. 1 lb = 453.6 g).

(a) Map: Length in mm $\xrightarrow{\text{1 mm} = 10^{-3}\text{ m}}$ Length in m

       Problem solution:

$$\text{Length in m} = 36\text{ mm} \times \frac{10^{-3}\text{ m}}{1\text{ mm}} = 0.036\text{ m}$$

(b) Map: Mass in kg $\xrightarrow{\text{1 kg} = 10^{3}\text{ g}}$ Mass in g

       Problem solution:

$$\text{Mass in g} = 357\text{ kg} \times \frac{10^{3}\text{ g}}{1\text{ kg}} = 3.57 \times 10^{5}\text{ g}$$

(c) Map: Volume in mL $\xrightarrow{\text{1 mL} = 10^{-3}\text{ L}}$ Volume in L

       Problem solution:

$$\text{Volume in L} = 76.50\text{ mL} \times \frac{10^{-3}\text{ L}}{1\text{ mL}} = 0.07650\text{ L}$$

(d) Map: Length in m $\xrightarrow{\text{1 cm} = 10^{-2}\text{ m}}$ Length in cm

       Problem solution:

$$\text{Length in cm} = 0.0084670\text{ m} \times \frac{\text{cm}}{10^{-2}\text{ m}} = 0.84670\text{ cm}$$

(e) Map: Length in nm $\xrightarrow{\text{1 nm} = 10^{-9}\text{ m}}$ Length in m

       Problem solution:

$$\text{Length in m} = 597\text{ nm} \times \frac{10^{-9}\text{ m}}{1\text{ nm}} = 5.97 \times 10^{-7}\text{ m}$$

(f) This is the first metric-English conversion, but the process is exactly the same. Note that the in-cm conversion factor is exact, so it is not a factor in determining significant figures:

       Map: Length in in $\xrightarrow{\text{1 in} = 2.54\text{ cm (exact)}}$ Length in cm

       Problem solution:

$$\text{Length in cm} = 36.5\text{ in} \times \frac{2.54\text{ cm}}{1\text{ in}} = 92.7\text{ cm}$$

(g) Map: Mass in lb $\xrightarrow{\text{1 lb} = 453.6\text{ g}}$ Mass in g

       Problem solution:

$$\text{Mass in g} = 168\text{ lb} \times \frac{453.6\text{ g}}{1\text{ lb}} = 7.62 \times 10^{4}\text{ g}$$

(h) Map: Volume in qt $\xrightarrow{\text{1 qt} = 0.9464\text{ L}}$ Volume in L

       Problem solution:

$$\text{Volume in L} = 914\text{ qt} \times \frac{0.9464\text{ L}}{1\text{ qt}} = 865\text{ L}$$

(i) Map: Length in cm $\xrightarrow{\text{1 in} = 2.54\text{ cm (exact)}}$ Length in in

       Problem solution:

$$\text{Length in in} = 44.5\text{ cm} \times \frac{\text{in}}{2.54\text{ cm}} = 17.5\text{ in}$$

(j) Map: Mass in g $\xrightarrow{\text{1 lb} = 453.6\text{ g}}$ Mass in lb

       Problem solution:

$$\text{Mass in lb} = 236.504\text{ g} \times \frac{\text{lb}}{453.6\text{ g}} = 0.5214\text{ lb}$$

(k) Map: Volume in L $\xrightarrow{\text{1 qt} = 0.9464\ \text{L}}$ Volume in qt

Problem solution:

$$\text{Volume in qt} = 2.0\ \cancel{L} \times \frac{1\ \text{qt}}{0.9464\ \cancel{L}} = 2.1\ \text{qt}$$

1.19 For all conversion problems, you need to find the conversion factors which connect the starting units to the final units. In (a) for example we need to convert from meters to miles. In Math Toolbox 1.3, we find that 1 mile is 1.609 km and we also know that 1 km is 1000 m. Once you establish this connection, miles to kilometers to meters, you have the necessary information to do the calculation. *It is very important to recognize that there are always many different paths in unit conversion problems (it depends on which conversion factors you have handy), but they will all lead to the same answer.*

(a) Map: length in m $\xrightarrow{\text{1 km} = 10^{3}\text{m}}$ length in km $\xrightarrow{\text{1 mi} = 1.609\ \text{km}}$ length in mi

Problem solution:

$$\text{length in mi} = 947\ \cancel{m} \times \frac{1\ \cancel{km}}{10^{3}\ \cancel{m}} \times \frac{1\ \text{mi}}{1.609\ \cancel{km}} = 0.589\ \text{mi}$$

(b) Map: mass in kg $\xrightarrow{\text{1 kg} = 10^{3}\text{g}}$ mass in g $\xrightarrow{\text{1 lb} = 453.6\ \text{g}}$ mass in lb

Problem solution:

$$\text{mass in lb} = 6.74\ \cancel{kg} \times \frac{10^{3}\ \cancel{g}}{1\ \cancel{kg}} \times \frac{1\ \text{lb}}{453.6\ \cancel{g}} = 14.9\ \text{lb}$$

(c) Map: volume in mL $\xrightarrow{\text{1 mL} = 10^{-3}\text{L}}$ volume in L $\xrightarrow{\text{1 gal} = 3.785\ \text{L}}$ volume in gal

Problem solution:

$$\text{volume in gal} = 250.4\ \cancel{mL} \times \frac{10^{-3}\ \cancel{L}}{1\ \cancel{mL}} \times \frac{1\ \text{gal}}{3.785\ \cancel{L}} = 0.06616\ \text{gal}$$

(d) Map: volume in dL $\xrightarrow{\text{1 dL} = 10^{-1}\ \text{L}}$ volume in L $\xrightarrow{\text{1 mL} = 10^{-3}\ \text{L}}$ volume in mL

Problem solution:

$$\text{Volume in mL} = 2.30\ \cancel{dL} \times \frac{10^{-1}\ \cancel{L}}{1\ \cancel{dL}} \times \frac{1\ \text{mL}}{10^{-3}\ \cancel{L}} = 2.30 \times 10^{2}\ \text{mL}$$

(e) Map: length in cm $\xrightarrow{\text{1 cm} = 10^{-2}\ \text{m}}$ length in m $\xrightarrow{\text{1 nm} = 10^{-9}\ \text{m}}$ length in nm

Problem solution:

$$\text{length in nm} = 0.000450\ \cancel{cm} \times \frac{10^{-2}\ \cancel{m}}{1\ \cancel{cm}} \times \frac{1\ \text{nm}}{10^{-9}\ \cancel{m}} = 4.50 \times 10^{3}\ \text{nm}$$

(f) Map: length in in $\xrightarrow{\text{1 in} = 2.54\ \text{cm}}$ length in cm $\xrightarrow{\text{1 cm} = 10^{-2}\ \text{m}}$ length in m

Problem solution:

$$\text{length in m} = 37.5\ \cancel{in} \times \frac{2.54\ \cancel{cm}}{1\ \cancel{in}} \times \frac{10^{-2}\ \text{m}}{1\ \cancel{cm}} = 0.952\ \text{m}$$

(g) Map: mass in lb $\xrightarrow{\text{1 lb} = 453.6\ \text{g}}$ mass in g $\xrightarrow{\text{1 kg} = 10^{3}\ \text{g}}$ mass in kg

Problem solution:

$$\text{mass in kg} = 689\ \cancel{lb} \times \frac{453.6\ \cancel{g}}{1\ \cancel{lb}} \times \frac{1\ \text{kg}}{10^{3}\ \cancel{g}} = 312\ \text{kg}$$

(h) Map: volume in qt $\xrightarrow{\text{1 qt} = 0.9464\,\text{L}}$ volume in L $\xrightarrow{\text{1 mL} = 10^{-3}\,\text{L}}$ volume in mL

Problem solution:

$$\text{volume in mL} = 0.5\,\cancel{\text{qt}} \times \frac{0.9464\,\cancel{\text{L}}}{1\,\cancel{\text{qt}}} \times \frac{1\,\text{mL}}{10^{-2}\,\cancel{\text{L}}} = 5 \times 10^{2}\,\text{mL}$$

(i) Map: length in cm $\xrightarrow{\text{1 in} = 2.54\,\text{cm}}$ length in in $\xrightarrow{\text{1 ft} = 12\,\text{in}}$ length in ft

Problem solution:

$$\text{length in ft} = 125\,\cancel{\text{cm}} \times \frac{1\,\cancel{\text{in}}}{2.54\,\cancel{\text{cm}}} \times \frac{1\,\text{ft}}{12\,\cancel{\text{in}}} = 4.10\,\text{ft}$$

(j) Map: mass in mg $\xrightarrow{\text{1 mg} = 10^{-3}\text{g}}$ mass in g $\xrightarrow{\text{1 lb} = 453.6\,\text{g}}$ mass in lb

Problem solution:

$$\text{mass in lb} = 542\,\cancel{\text{mg}} \times \frac{10^{-3}\,\cancel{\text{g}}}{1\,\cancel{\text{mg}}} \times \frac{1\,\text{lb}}{453.6\,\cancel{\text{g}}} = 1.19 \times 10^{-3}\,\text{lb}$$

(k) Map: volume in nL $\xrightarrow{\text{1 nL} = 10^{-9}\,\text{L}}$ volume in L $\xrightarrow{\text{1 gal} = 3.785\,\text{L}}$ volume in gal

Problem solution:

$$\text{volume in gal} = 25\,\cancel{\text{nL}} \times \frac{10^{-3}\,\cancel{\text{L}}}{1\,\cancel{\text{mL}}} \times \frac{1\,\text{gal}}{3.785\,\cancel{\text{L}}} = 6.6 \times 10^{-9}\,\text{gal}$$

1.21   (a) There are actually two different conversions in this problem. It helps to consider these separately before setting up the problem. For example, the two conversions are meters into feet and seconds into minutes. As with other conversions, the conversion factors are set up so that units cancel properly. Whether you do the "meter to feet" or the "seconds to minutes" conversion first, the answer will be the same.

Map: $\dfrac{\text{m}}{\text{s}}$ $\xrightarrow{\text{1 min} = 60\,\text{s}}$ $\dfrac{\text{m}}{\text{min}}$ $\xrightarrow{\text{1 yd} = 0.9144\,\text{m}}$ $\dfrac{\text{yd}}{\text{min}}$ $\xrightarrow{\text{1 yd} = 3\,\text{ft}}$ $\dfrac{\text{ft}}{\text{min}}$

Problem solution:

$$\frac{\text{ft}}{\text{min}} = 375\,\frac{\cancel{\text{m}}}{\cancel{\text{s}}} \times \frac{60\,\cancel{\text{s}}}{1\,\text{min}} \times \frac{1\,\cancel{\text{yd}}}{0.9144\,\cancel{\text{m}}} \times \frac{4\,\text{ft}}{1\,\cancel{\text{yd}}} = 7.38 \times 10^{4}\,\frac{\text{ft}}{\text{min}}$$

(b) For conversions of units with exponents, you will have to apply the conversion factor the same number of times as the magnitude of the exponent. It helps if you remind yourself that cm^3 is actually cm × cm × cm. When you convert cm^3 to in^3, the conversion factor is actually applied three times so that each cm factor in the unit is cancelled.

Map: cm^3 $\xrightarrow{\text{1 in} = 2.54\,\text{cm}}$ $\xrightarrow{\text{1 in} = 2.54\,\text{cm}}$ $\xrightarrow{\text{1 in} = 2.54\,\text{cm}}$ in^3

Problem solution:

$$\text{Volume in}^3 = 24.5\,\cancel{\text{cm}} \times \cancel{\text{cm}} \times \cancel{\text{cm}} \times \frac{1\,\text{in}}{2.54\,\cancel{\text{cm}}} \times \frac{1\,\text{in}}{2.54\,\cancel{\text{cm}}} \times \frac{1\,\text{in}}{2.54\,\cancel{\text{cm}}} = 1.50\,\text{in}^3$$

Most likely when you get accustomed to this process you will find it easier to write:

$$\text{Volume in}^3 = 24.5\,\cancel{\text{cm}^3} \times \left(\frac{1\,\text{in}}{2.54\,\cancel{\text{cm}}}\right)^3 = 1.50\,\text{in}^3$$

(c) Make sure that you understand conversions with exponents given in part (b) of the problem. You also need to recall that 1 mL = 1 cm^3.

Map: $\dfrac{\text{g}}{\text{mL}}$ $\xrightarrow{\text{1 lb} = 453.6\,\text{g}}$ $\dfrac{\text{lb}}{\text{mL}}$ $\xrightarrow{\text{1 mL} = \text{cm}^3}$ $\dfrac{\text{lb}}{\text{cm}^3}$ $\xrightarrow{(\text{1 in} = 2.54\,\text{cm})^3}$ $\dfrac{\text{lb}}{\text{in}^3}$

Problem solution:

$$\frac{lb}{in^3} = 19.3\ \frac{g}{mL} \times \frac{1\ lb}{453.6\ g} \times \frac{1\ mL}{1\ cm^3} \times \frac{2.54\ cm}{1\ in} \times \frac{2.54\ cm}{1\ in} \times \frac{2.54\ cm}{1\ in} = 0.697\ \frac{lb}{in^3}$$

1.23   When you are trying to classify matter, it helps to carefully read the description. If it contains two or more pure substances, it is some type of mixture. If it only contains one type of substance, you have to consider that it might be an element or compound. Remember, compounds are also called pure substances since each unit of the compound is the same. Water (H_2O) is a pure substance.
   (a)   Water and dye is a mixture. It is a homogeneous mixture if the dye is evenly mixed into the water.
   (b)   The pipe is made of copper and nothing else is mentioned. That makes it a pure substance. Since it only contains one type of atom, it is an element.
   (c)   Air is made up of several different kinds of gases. That means it is a mixture. Also, if you blow up a balloon, you are adding moisture (water vapor) to the mixture. Because the composition is most likely uniform throughout (gases mix quickly), it is a homogenous mixture.
   (d)   Pizza is not an element even though you might think it is essential to life. Pizza is made (at the very least) of cheese, bread, and anchovies. That makes it a mixture. Since each slice is not the same, it is a heterogeneous mixture.

1.25   Matter has mass and occupies space. Any object or substance is matter. It might also be helpful to remember that if a substance has a smell or taste, it is a form of matter because your body has to interact with it for you to sense it. If something makes you hot or cold, it may be some form of energy (for example sunlight). Only (a) is not a form of matter. Any type of light or heat, although it occupies space, does not have mass.
   (a)   Not matter. Light or heat are forms of energy and do not have mass. They still occupy space.
   (b)   Gasoline occupies space and has mass. Also, you can smell it so it is some form of matter.
   (c)   Even though you might not be able to see it, automobile exhaust has mass and occupies space. Also, since you can feel the pressure of the exhaust as it leaves the engine, you can assume it has mass.
   (d)   Oxygen gas occupies space and is made of oxygen molecules that have mass.
   (e)   Any object is matter.

1.27   Elements are composed of only one type of atom. Compounds are made up of two or more different elements in some fixed proportion. Natural gas, CH_4, also called methane is an example of a compound. Any sample of methane is composed of one part carbon and four parts hydrogen.

1.29   Metals are lustrous (shiny) and conduct heat and electricity. In addition, you can form wires with metals (ductile) and you can make foil out of them by hitting them with a hammer (malleable).

1.31   (a) titanium; (b) tantalum; (c) thorium; (d) technetium; (e) thallium

1.33   (a) boron; (b) barium; (c) beryllium; (d) bromine; (e) bismuth

1.35   (a) nitrogen; (b) iron; (c) manganese; (d) magnesium; (e) aluminum; (f) chlorine

1.37   (a) Fe; (b) Pb; (c) Ag; (d) Au; (e) Sb

1.39   Ir is the symbol for the element iridium. While many elements have symbols that start with the same letter as the name of the element, some do not. Iron's symbol is Fe which comes from the Latin word for iron, ferrum.

1.41   Only the first letter of an element symbol is capitalized. No is the correct way to write the element symbol. NO is a compound formed from nitrogen and oxygen.

1.43   The hamburger is a heterogeneous mixture. The salt is a pure substance (NaCl). The soft drink is a heterogeneous mixture until it goes flat. The ketchup is also a heterogeneous mixture; after sitting for awhile, liquid collects on the top.

1.45   The chemical formula for hydrogen gas would be H_2. Hydrogen is normally represented by white colored spheres. There are different ways to draw H_2. The spheres represent the atoms and the line or "stick" represents the bond that holds the atoms together.

| Ball and Stick | Space Filled |

1.47   There are four oxygen atoms (drawn as red spheres) and the two nitrogen atoms (colored blue). We write the chemical formula as N_2O_4. Subscripts following each atom type are used to indicate the number of each type of atom.

1.49   Elements and compounds are types of pure substances. A mixture of elements and compounds would contain two different substances. Image A represents a compound; each molecule is composed of two types of atoms. B represents an element; each molecule is exactly the same and only one kind of atom is present in each. What about C? The substances in C represent a mixture of elements. Compounds are not present. In D you have a compound and an element mixed together. E represents a mixture of two compounds.

1.51   The term *element* can refer to species of atoms all of one type (N or O) or pure substances composed of only one type of atom (N_2 or O_2). O_2 is what is called an elementary substance or elemental form and is one way we would find the element in nature. This means that O_2, P_4, and He are elements. Fe_2O_3, NaCl, and H_2O are compounds.

1.53   The atoms or molecules in the liquid state are close together but do not have a rigid form. In the solid state, the atoms or molecules are close together and are not free to move. This often means they will take on some sort of ordered structure (which we call the crystal lattice).

| Liquid State | Solid State |

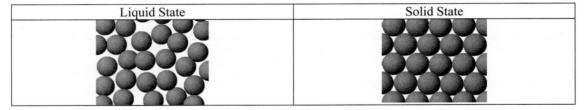

1.55   Gas. In the gas state, molecules are spaced far apart. As a result, they can easily be compressed. The molecular attractions of gas molecules tend to be weak compared to the amount of kinetic energy they have. If the attractions were strong, the molecules would prefer to be in a liquid or solid state. Since the attractive forces are weak, the molecules easily separate from each other (expand).

1.57   There are three states of matter; solid (*s*), liquid (*l*), and gas (*g*). (a) gas; (b) liquid; (c) solid

1.59   The diagram is an illustration of the solid state. You can make this conclusion since there is very little space between the atoms, and the atoms are in a very organized arrangement. If the distances between the atoms were large, you would conclude that it represents the gas state. Liquids do not show long range organized structure like that seen in solids.

1.61   When a substance is dissolved in water we say that it is an aqueous solution. This is abbreviated by the notation (*aq*). Oxygen gas can be dissolved in water to make a homogenous mixture (which is why fish can survive in water). This solution is represented as $O_2(aq)$.

1.63   These are physical properties. You can observe physical properties without changing the substance. Chemical properties are only observed when new substances are formed.

1.65   When you are converting a unit with a prefix to the same base unit but with a different prefix (i.e. milligrams to micrograms) you can find the conversion factors in Math Toolbox 1.3. Suppose you want to convert between grams and micrograms (μg). There are several ways you can do this. First, by definition micro = 10^{-6}. So 10^{-6} g = μg. This equation says, "one millionth of a gram is one microgram". We could also use 1 g = 10^6 μg =1000000 μg. This equation says, "1 gram is one million micrograms". In part (b), you can see both ways of doing this conversion.

(a) Map: Sodium mass in mg $\xrightarrow{\text{1000 mg = 1 g}}$ Sodium mass in g
     Problem solution:

$$\text{Mass in g} = 45 \text{ mg} \times \frac{1 \text{ g}}{1000 \text{ mg}} = 0.045 \text{ g}$$

(b) Map: Sodium mass in g $\xrightarrow{\text{16 oz = 453.6 g}}$ Sodium mass in oz
     Problem solution:

$$\text{Mass in oz} = 0.045 \text{ g} \times \frac{16 \text{ oz}}{453.6 \text{ g}} = 1.6 \times 10^{-3} \text{ oz}$$

(c) Map: Sodium mass in g $\xrightarrow{\text{1 lb = 453.6 g}}$ Sodium mass in lb

$$\text{Mass in lb} = 0.045 \text{ g} \times \frac{1 \text{ lb}}{453.6 \text{ g}} = 9.9 \times 10^{-5} \text{ lb}$$

1.67   (a) Map: Salt mass in g $\xrightarrow{\text{1 g = 1000 mg}}$ Salt mass in mg
             Problem solution:

$$\text{Mass in mg} = 1.0 \times 10^{-6} \text{ g} \times \frac{1000 \text{ mg}}{1 \text{ g}} = 0.10 \text{ mg}$$

(b) Map: Salt mass in g $\xrightarrow{10^{-6} \text{ g = 1 } \mu g}$ Salt mass in μg
     Problem solution:

$$\text{Mass in } \mu g = 1.0 \times 10^{-4} \text{ g} \times \frac{1 \text{ } \mu g}{10^{-6} \text{ g}} = 1.0 \times 10^2 \text{ } \mu g$$

Alternate

Map: Salt mass in g $\xrightarrow{\text{1 g = 1000000 } \mu g}$ Salt mass in μg
Problem solution:

$$\text{Mass in } \mu g = 1.0 \times 10^{-4} \text{ g} \times \frac{1000000 \text{ } \mu g}{1 \text{ g}} = 1.0 \times 10^2 \text{ } \mu g$$

(c) Map: Salt mass in g $\xrightarrow{\text{1 kg = 1000 g}}$ Salt mass in kg
     Problem solution:

$$\text{Mass in kg} = 1.0 \times 10^{-4} \text{ g} \times \frac{1 \text{ kg}}{1000 \text{ g}} = 1.0 \times 10^{-7} \text{ kg}$$

1.69   (a) Map: Drink volume in L $\xrightarrow{\text{1000 mL = 1 L}}$ Drink volume in mL

        Problem solution:

$$\text{Volume in mL} = 1.2\,\cancel{L} \times \frac{1000\ \text{mL}}{1\,\cancel{L}} = 1.2 \times 10^3\ \text{mL}$$

        (b) Map: Drink volume in L $\xrightarrow{\text{1000 cm}^3 \text{ = 1 L}}$ Drink volume in cm^3

        Problem solution:

$$\text{Volume in cm}^3 = 1.2\,\cancel{L} \times \frac{1000\ \text{cm}^3}{1\,\cancel{L}} = 1.2 \times 10^3\ \text{cm}^3$$

        (c) Map: Drink volume in L $\xrightarrow{\text{1000 L = 1 m}^3}$ Drink volume in m^3

        Problem solution:

$$\text{Volume in m}^3 = 1.2\,\cancel{L} \times \frac{1\ \text{m}^3}{1000\,\cancel{L}} = 1.2 \times 10^{-3}\ \text{m}^3$$

1.71   You are given the length, width, and height of the box and asked to calculate the volume in milliliters and liters. Notice that length and volume are *different types of units*. When the type of unit given and the unit in the answer are different (i.e. length and volume units), this often means you will need to use an equation. The key equation is the volume equation:

$$\text{Volume} = \text{length} \times \text{width} \times \text{height}$$

$$\text{Volume in cm}^3 = 8.0\ \text{cm} \times 5.0\ \text{cm} \times 4.0\ \text{cm} = 1.6 \times 10^2\ \text{cm}^3$$

Notice that the answer of the equation is a volume unit, but not the units you want (mL or L)! You can solve this problem through two separate unit conversions:

    Problem map: Volume in cm^3 $\xrightarrow{\text{1 mL = 1 cm}^3}$ Volume in mL

    Problem solution:

$$\text{Volume in mL} = 1.6 \times 10^2\ \cancel{\text{cm}^3} \times \frac{1\ \text{mL}}{1\,\cancel{\text{cm}^3}} = 1.6 \times 10^2\ \text{mL}$$

    Problem map: Volume in cm^3 $\xrightarrow{\text{1 L = 1000 cm}^3}$ Volume in L

    Problem solution:

$$\text{Volume in mL} = 1.6 \times 10^2\ \cancel{\text{cm}^3} \times \frac{1\ \text{L}}{1000\,\cancel{\text{cm}^3}} = 0.16\ \text{L}$$

1.73   The density is defined as density $= \dfrac{\text{mass}}{\text{volume}}$.

$$\text{Density} = \frac{28\ \text{g}}{21\ \text{mL}} = 1.333\ \text{g/mL} = 1.3\ \text{g/mL}$$

Since 1 mL = 1 cm^3, we find that 21 mL = 21 cm^3

$$\text{Density} = \frac{28\ \text{g}}{21\ \text{cm}^3} = 1.3\ \text{g/cm}^3$$

Alternately, we can convert the density as follows (using the conversion 1 mL = 1 cm^3):

$$\text{Density in } \frac{\text{g}}{\text{cm}^3} = \frac{28\ \text{g}}{21\,\cancel{\text{mL}}} \times \frac{1\,\cancel{\text{mL}}}{1\ \text{cm}^3} = 1.3\ \text{g/cm}^3$$

Note that the units of g/cm^3 and g/mL are equivalent.

1.75　You are given the mass and density and asked to calculate the volume. These are related by the density equation: Density = mass/volume. The equation can be rearranged to solve for volume.

$$\text{Volume} = \frac{\text{mass}}{\text{density}} = \frac{50.0\,\cancel{g}}{1.30\,\dfrac{\cancel{g}}{\text{mL}}} = 38.5\,\text{mL}$$

1.77　Molecules in the liquid state are closer together than molecules in the gas state. This means that when a liquid or gas occupies the same size container, there will be more molecules of the liquid than the gas. Since density is $d = m/V$, the density of the liquid is higher than the density of the gas.

1.79　When the plastic is placed in the water, it floats. This means that its density is lower than that of water. However, it sinks when placed in the oil. From this information, we can order the substances according to increasing density.

　　　　　　oil (least density) < plastic < water (greatest density)

　　What happens when oil is placed in water? Based on our order, oil should float on water. If you make your own oil and vinegar (which is mostly water) salad dressing, you may have already noticed this yourself.

1.81　Temperature conversions are always done using equations. The following equation is used to convert between Celsius and kelvin scales: $T_K = T_{^\circ C} + 273.15$

　　　　$T_K = 56\,^\circ C + 273.15 = 329\ K$

1.83　In the Celsius scale, water freezes at 0°C and boils at 100°C. There are 100 degrees between the boiling and freezing temperature of water.

　　The freezing and boiling points of water from these values in Celsius using the equation: $T_K = T_{^\circ C} + 273.15$

　　　　Freezing point in K $= 0\,^\circ C + 273.15 = 273.15\ K$

　　　　Boiling point in K $= 100\,^\circ C + 273.15 = 373.15\ K$

　　In Fahrenheit we calculate the freezing and boiling points using the equation: $T_{^\circ F} = 1.8\left(T_{^\circ C}\right) + 32$

　　　　Freezing point in $^\circ F = 1.8 \times (0\,^\circ C) + 32 = 32\,^\circ F$

　　　　Boiling point in $^\circ F = 1.8 \times (100\,^\circ C) + 32 = 212\,^\circ F$

	Freezing Point	Boiling Point	Difference
Celsius	0°C	100°C	100°C
Kelvin	273.15 K	373.15 K	100 K
Fahrenheit	32°F	212°F	180°F

1.85　No. Boiling point is a property that depends on what the substance is, not how much is present. In a microwave oven, you can boil half a cup of water much faster than two cups of water. However, the temperature at which both boil is the same (100°C).

1.87   Physical properties are (a) mass, (b) density, and (e) melting point. In each case, you can measure the property without actually changing the substance. The mass of a penny can be measured without changing its composition. Similarly, when ice melts to form water, the chemical formula is still the same (H_2O). Chemical properties are (c) flammability, (d) resistance to corrosion, and (f) reactivity with water. Chemical properties are observed when new substances are formed. When substances burn, corrode, or react, new substances are formed.

1.89   Physical changes are (a) boiling acetone, (b) dissolving oxygen gas in water, and (e) screening rocks from sand. In each case the substance is not changed. Chemical changes are (c) combining hydrogen and oxygen to make *water*, (d) *burning* gasoline, and (f) conversion of ozone to *oxygen*. In each case, new substances are formed.

1.91   Symbolically the condensation of chlorine gas can be represented by: $Cl_2(g) \rightarrow Cl_2(l)$. This representation means that chlorine in the gas state is converted to chlorine in the liquid state. This is a phase change since the substance has not changed. At the molecular level, we know that molecules in the gas state are relatively far apart and that they are moving very rapidly. In the liquid state, the molecules are much closer together and are no longer able to move freely from each other. Note that vaporization is the opposite of condensation.

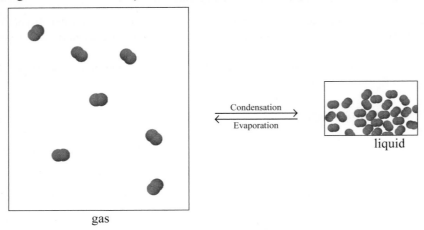

gas

1.93   Chemical change. During a physical change, all the molecules would remain the same. In this case, molecules are transformed from pure A and B_2 to AB_4. This is a chemical change since a new substance is formed.

1.95   When methane condenses, it goes from the gas state to the liquid state. One possible representation of this process would be the figure shown to the right. Condensation occurs when methane leaves the gas phase (top of the container) and goes to the bottom (liquid phase). Vaporization could be described by the same image if we reverse the process and imagine molecules leaving the liquid and going into the gas.

1.97   In each of the sections which expand the molecular structure, the iodine atoms are paired. In other words, since new substances are not formed, this must be a physical change.

1.99   The ability to understand if something possesses kinetic or potential energy really depends on how well we understand what is going on. On a large scale (macroscopic), kinetic energy is the easiest to observe. If it moves, it has kinetic energy. We assume that sparks, the welder's hands, and the smoke are all moving. These possess kinetic energy. On a macroscopic level, potential energy is also relatively easy to find. By definition, if an object is attracted to (e.g. pulled on by gravity of the earth) or repelled by an object (i.e. a spring) and is separated by some distance (e.g. it can fall) it possesses potential energy. On a molecular and atomic level (microscopic), kinetic and potential energy are much harder to classify. Can the energy be used to do work or to create? That would be potential energy. For example, the welder's fuel is probably acetylene and oxygen. It contains stored or potential energy. Electricity is also a form of potential energy (when it is not being used). However, if it is making something happen, then we are seeing kinetic energy.

1.101   The molecules in image A appear to be moving faster. Since kinetic energy increases with the velocity of the molecules, the molecules in A have greater kinetic energy.

1.103   The ability to understand if something possesses kinetic or potential energy really depends on how well we understand what is going on. Objects that possess potential energy will move if they are allowed. For example, a picture will fall if the nail that holds it up comes out of the wall. Because the picture starts moving when it is released, it had potential energy. When the picture reaches the ground, all the potential energy has been used. Your bed mattress expands when you get up. This means that the bed, when you were lying on it, had potential energy (the ability to move and do work).

1.105   If an object or substance is moving, it has kinetic energy. The people walking, the wheel chair rolling, and the suitcase being pushed all have kinetic energy. If an object or substance can fall, it has potential energy. Anything raised above the ground has potential energy. The people, the wall art, and objects on the tables all have potential energy. Many objects in this picture have both potential and kinetic energy.

1.107   Here is an example. Your car is on a hill and you have applied the brake. When you release the brake, what happens? The car rolls down the hill and begins to move faster and faster. As it moves down the hill, potential energy is lost, but kinetic energy is gained. When the car reaches the bottom of the hill, it begins to slow down. This happens because of friction, which produces heat energy. This heat energy is directly related to the motion of molecules (kinetic energy). When the car has reached the bottom of the hill, the potential energy has been used up.

1.109   When the batter swings the bat potential energy is converted into kinetic energy (moving bat). The bat strikes the ball and the kinetic energy of the bat is transferred to kinetic energy of the ball. As the ball leaves the bat, it rises against the gravity of the earth and some of the kinetic energy is converted to potential energy. When the ball starts dropping again, potential energy is converted to kinetic energy.

1.111   To calculate the BMI, you must first calculate the person's mass in kg. The height is converted from feet and inches into meters. These results are used to calculate the BMI according to the equation given.

Map: mass in lb $\xrightarrow{\text{1 lb} = 453.6 \text{ g}}$ mass in g $\xrightarrow{\text{1 kg} = 10^3 \text{g}}$ mass in kg

$$\text{mass in kg} = 169 \, \text{lb} \times \frac{453.6 \, \text{g}}{1 \, \text{lb}} \times \frac{1 \, \text{kg}}{10^3 \, \text{g}} = 76.7 \, \text{kg}$$

To calculate the height, we first calculate the height in inches. 6 feet is 72 inches, so the person's height is 74 inches.

Map: height in in $\xrightarrow{\text{1 in} = 2.54 \text{ cm}}$ height in cm $\xrightarrow{\text{1 cm} = 10^{-2} \text{ m}}$ height in m

$$\text{height in m} = 74 \, \text{in} \times \frac{2.54 \, \text{cm}}{1 \, \text{in}} \times \frac{10^{-2} \, \text{m}}{1 \, \text{cm}} = 1.88 \, \text{m}$$

$$\text{BMI} = \frac{\text{weight(kg)}}{\left(\text{height (m)}\right)^2} = \frac{76.7 \, \text{kg}}{\left(1.88 \, \text{m}\right)^2} = 21.7 \frac{\text{kg}}{\text{m}^2}$$

Based on the BMI, the person would be considered healthy.

1.113   As the water leaves the top of the fountain, it possesses kinetic energy (going up). However, its upward movement slows as it reaches the top of its arc. This happens because the kinetic energy it possessed is transformed into potential energy as it moves away from the earth. At some point, the kinetic energy is "used up" and the water starts moving in the downward direction. As it does so, its kinetic energy increases. When the water reaches the pool at the base of the fountain, it has used up its potential energy. As the water enters the pool, it causes the water in the pool to move (kinetic energy is distributed into the pool).

1.115   A hypothesis has three important features. It summarizes what you know, it allows predictions under certain conditions (i.e. experiments), and it is flexible or can be modified. In research, often an investigator has ideas about how something works. The ideas are based on the investigator's previous experiences and, to a large part, intuition. The hypothesis is formed from these ideas and guides the research that is done. Information gathered from the research allows the investigator to evaluate and modify the hypothesis. Thus a new hypothesis is formed. Rarely are complete answers found in research (otherwise we might simply call it "search" rather than "re-search").

1.117   To start with, none of these observations are laws since laws have no known exceptions. Also, the difference between a hypothesis and a theory is a little fuzzy. A hypothesis has many exceptions and a theory should have very few. Finally, an observation does not make any attempt to explain observations or predict outcomes of an experiment. This means (b) and (d) are observations. (a) is closer to a hypothesis since many people walk under ladders all day without bad luck (painters for example). (c) is similar to (b), but it attributes the floating of the oil to the density of the oil. This can be shown to be true in many different types of experiments. As a result, (c) is more similar to a theory than a hypothesis.

1.119   Quick, run to the library! A college librarian will be able to find data on many different types of woods (you could also do an internet search for wood densities). Included in this data will be data on types of woods and their densities (or specific gravities). Since water has a density of approximately 1.0 g/cm^3, you would look for woods that have densities both higher and lower than water. After obtaining samples of these woods, you could carefully measure and weigh samples of wood and see if they float on water. (Ironwood has a density of 63 lb/ft^3. Will that sink or float?)

ADDITIONAL QUESTIONS

1.121   At high altitude, the air pressure is lower. As a result, when a balloon rises, it expands. Since the mass of air in the balloon has not changed but the volume has increased, the density of the balloon is lower.

1.123   Volume depends on the density and mass. If we solve the density equation for volume we have:

$$\text{Volume} = \frac{\text{mass}}{\text{density}}$$

This means that if the density is smaller, the volume will be larger. Table 1.6 lists the densities of several common substances. The density of zinc and copper are listed as 7.14 g/mL and 8.92 g/mL respectively. Since the masses of the samples are the same and the density of zinc is lower, the block of zinc is larger.

1.125   (a) He; (b) Ne; (c) Ar; (d) Kr; (e) Xe; (f) Rn

1.127   Since the objects that are separated are not changed (i.e. they float or they don't float), this must be a process based on physical properties. If it was based on a chemical property, new substances would have to be formed.

1.129 Convert the person's body mass into kilograms and then use epinephrine's dosage information (1 kg = 0.1 mg) to calculate the milligrams of epinephrine (mg-epi).

$$\text{Mass-lb} \xrightarrow{\text{1 lb = 453.6 g}} \text{Mass-g} \xrightarrow{\text{1 kg = }10^3\text{ g}} \text{Mass-kg} \xrightarrow{\text{1 kg = 0.1 mg-epi}} \text{Mass-mg-epi}$$

$$\text{mg-epi} = 180\,\cancel{\text{lb}} \times \frac{453.6\,\cancel{\text{g}}}{1\,\cancel{\text{lb}}} \times \frac{1\text{ kg}}{10^3\,\cancel{\text{g}}} \times \frac{0.1\text{ mg-epi}}{1\,\cancel{\text{kg}}} = 8\text{ mg-epi}$$

1.131 The conversion of 240 mg/dL involves the conversion of mg to lbs and also dL to fluid ounces. Remember that these conversions can be applied in either order and that units must cancel properly.

$$\frac{\text{mg}}{\text{dL}} \xrightarrow{\text{1 mg = }10^{-3}\text{ g}} \frac{\text{g}}{\text{dL}} \xrightarrow{\text{1 lb = 453.6 g}} \frac{\text{lb}}{\text{dL}} \xrightarrow{\text{1 dL = }10^{-1}\text{ L}} \frac{\text{lb}}{\text{L}} \xrightarrow{\text{1 mL = }10^{-3}\text{ L}} \frac{\text{lb}}{\text{mL}} \xrightarrow{\text{1 fl oz = 29.57 mL}} \frac{\text{lb}}{\text{fl oz}}$$

$$\frac{\text{lb}}{\text{fl oz}} = 260\frac{\cancel{\text{mg}}}{1\,\cancel{\text{dL}}} \times \frac{10^{-3}\,\cancel{\text{g}}}{1\,\cancel{\text{mg}}} \times \frac{1\text{ lb}}{453.6\,\cancel{\text{g}}} \times \frac{1\,\cancel{\text{dL}}}{10^{-1}\,\cancel{\text{L}}} \times \frac{10^{-3}\,\cancel{\text{L}}}{1\,\cancel{\text{mL}}} \times \frac{29.57\,\cancel{\text{mL}}}{1\text{ fl oz}} = 1.7 \times 10^{-4}\frac{\text{lb}}{\text{fl oz}}$$

To determine the mass range of cholesterol, the masses must be calculated for both 4 and 6 liters of blood.

$$\text{Volume-L} \xrightarrow{\text{1 dL = }10^{-1}\text{ L}} \text{Volume-dL} \xrightarrow{\text{1 dL = 260 mg}} \text{Mass-mg} \xrightarrow{\text{1 mg = }10^{-3}\text{ g}} \text{Mass-g} \xrightarrow{\text{1 lb = 453.6 g}} \text{Mass-lb}$$

$$\text{Mass-kg} = 4\,\cancel{\text{L}} \times \frac{1\,\cancel{\text{dL}}}{10^{-1}\,\cancel{\text{L}}} \times \frac{260\,\cancel{\text{mg}}}{1\,\cancel{\text{dL}}} \times \frac{10^{-3}\text{ g}}{1\,\cancel{\text{mg}}} \times \frac{1\text{ lb}}{453.6\,\cancel{\text{g}}} = 0.02\text{ lb}$$

$$\text{Mass-kg} = 6\,\cancel{\text{L}} \times \frac{1\,\cancel{\text{dL}}}{10^{-1}\,\cancel{\text{L}}} \times \frac{260\,\cancel{\text{mg}}}{1\,\cancel{\text{dL}}} \times \frac{10^{-3}\text{ g}}{1\,\cancel{\text{mg}}} \times \frac{1\text{ lb}}{453.6\,\cancel{\text{g}}} = 0.03\text{ lb}$$

The mass range is 0.02 to 0.03 lb

1.133 The kelvin scale most directly describes the lowest possible temperature because it is absolute with no negative values and zero being the lowest possible temperature.
(a) 0.00 K (The number of significant figures here is arbitrary since this is an exact number.)
(b) $-273.15°C$ (Calculated from 0.00 K using the formula $T_{°C} = T_K - 273.15$)
(c) $-459.67°F$ (Calculated from $-273.15\ °C$ using the formula $T_{°F} = 1.8(T_{°C}) + 32$)

1.135
(a) $NaNO_3(s)$ represents a compound because the formula contains more than one element symbol (so it is not an element). The physical state symbol shows that it is a solid so it is a pure substance, not a mixture.
(b) $N_2(g)$ represents an element because the formula contains only the element symbol for one type of element (nitrogen). The physical state symbol shows that it is a gas so it is a pure substance, not a mixture.
(c) $NaCl(aq)$ represents a mixture because the physical state symbol (*aq*) shows that this compound is dissolved in water.

1.137 Blood is a mixture because it contains more than one substance. It is further classified as a heterogeneous mixture because the solids are suspended, not dissolved. If blood was homogeneous, it would be clear and transparent.

1.139 From the conversion factor 1 m = 100 cm we have to determine the relationship between m^3 and cm^3 so that units cancel properly. To cube the units, we have to cube the entire conversion factor ratio, including the numbers.

$$10.0 \text{ m}^3 \times \left(\frac{100 \text{ cm}}{1 \text{ m}}\right)^3 = 10.0 \text{ m}^3 \times \frac{(100)^3 \text{ cm}^3}{1^3 \text{ m}^3} = 10.0 \text{ m}^3 \times \frac{10^6 \text{ cm}^3}{1 \text{ m}^3} = 1.00 \times 10^7 \text{ cm}^3$$

1.141 Mass in kilograms (using density as conversion factor): $0.00500 \text{ m}^3 \times \dfrac{1060 \text{ kg}}{\text{m}^3} = 5.30 \text{ kg}$

Mass in pounds (converting kg to g to lb): $5.30 \text{ kg} \times \dfrac{1000 \text{ g}}{1 \text{ kg}} \times \dfrac{1 \text{ lb}}{453.6 \text{ g}} = 11.7 \text{ lb}$

Chapter 2 – Atoms, Ions, and the Periodic Table

2.1   (a) neutron; (b) law of conservation of mass; (c) proton; (d) main-group element; (e) relative atomic mass; (f) mass number; (g) isotope; (h) cation; (i) subatomic particle; (j) alkali metal; (k) periodic table

2.3   Dalton used the laws of conservation of mass (Lavoisier) and definite proportions (Proust). Dalton essentially reasoned that, because pure substances were always composed of elements in some fixed ratios, matter must be composed of discrete units (atoms).

2.5   The second postulate of Dalton's atomic theory, listed in section 2.1, states that atoms of different elements differ in their atomic masses and chemical properties.

2.7   Compounds contain discrete numbers of atoms of each element that form them. Because all the atoms of an element have the same relative atomic mass, the mass ratio of the elements in a compound is always the same (law of definite proportions).

2.9   No. On the left side of the diagram there are four white atoms and two blue atoms (represented as diatomic molecules). On the right, however, there are six white atoms. Since the numbers of each atom are not conserved, mass is not conserved. For mass to be conserved, another white molecule is needed on the left side of the diagram.

2.11   Thomson's cathode ray experiment.

2.13   The electron is the subatomic particle with a negative charge. A summary of the subatomic particles and their properties is given in Table 2.1.

2.15   The nucleus of helium has two protons and two neutrons. Two electrons can be found in the cloud surrounding the nucleus.

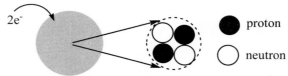

2.17   The neutron and proton (Table 2.1) have approximately the same atomic mass (1 amu).

2.19   Carbon atoms have 6 protons. The relative atomic mass of a carbon atom is 12.01 amu indicating the presence of 6 neutrons. Protons and neutrons have approximately equal masses so the nuclear mass is approximately two times the mass of the protons.

2.21   The atomic number (protons) is given on the periodic table. (a) 1 (element symbol: H); (b) 8 (element symbol: O); (c) 47 (element symbol: Ag)

2.23   The number of protons determines the identity of an element.

2.25   The atomic number of an atom is equal to the number of protons. If you know the name of the element, you can find the atomic number by finding the element on the periodic table. For example, for iron (Fe), you can find the atomic number, 26, listed with the element symbol in the fourth period of the periodic table.

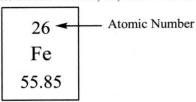

26 ← — Atomic Number

Fe

55.85

2.27   All atoms of an element have the same number of protons and electrons. Only (b), atomic number, is the same for different isotopes of an element. The mass number, neutron number, and mass of an atom are different for each isotope of an element.

2.29   The following table displays the atomic, neutron, and mass numbers for the isotopes of hydrogen:

	$_1^1H$	$_1^2H$	$_1^3H$
Atomic number	1	1	1
Neutron number	0	1	2
Mass number	1	2	3

2.31   The mass number (A) is defined as the number of protons (Z) plus the number of neutrons (N), $A = Z + N$. To find the number of protons (Z) in the nucleus (atomic number), we need to find the element on the periodic table. This allows us to calculate the neutron number (N) using: $N = A - Z$. For example, the element oxygen (Ar) has an atomic number of 18. For an oxygen isotope with a mass number of 36, the number of neutrons is: $N = 36 - 18 = 18$.

(a) $Z = 18$, $A = 36$, and $N = 18$
(b) $Z = 18$, $A = 38$, and $N = 20$
(c) $Z = 18$, $A = 40$, and $N = 22$

2.33   To find the number of protons (Z) in the nucleus (atomic number), we need to find the element on the periodic table. This allows us to calculate the number of neutrons (N) using $N = A - Z$, where A is the mass number. For example, the element oxygen (O) has an atomic number of 8. For an oxygen isotope with a mass number of 15, the number of neutrons is $N = 15 - 8 = 7$.

In an atom the number of protons is equal to the number of electrons.

	Protons	Neutrons	Electrons
(a) $_8^{15}O$	8	7	8
(b) $_{47}^{109}Ag$	47	62	47
(c) $_{17}^{35}Cl$	17	18	17

2.35   Isotope symbols have the general format $^{\text{mass number}}_{\text{protons}}X^{\text{charge}}$ or $^A_Z X$ where A is the mass number (neutrons plus protons), Z is the atomic number (number of protons), and X is the element symbol.

(a) From the periodic table you find that the element with an atomic number of 1 is hydrogen, H. Since the isotope has 2 neutrons, the mass number is 3. The isotope symbol for hydrogen-3 is ^3_1H.

(b) The element with an atomic number of 4 is beryllium, Be. Since the isotope has 5 neutrons, the mass number is 9. The isotope symbol for beryllium-9 is ^9_4Be.

(c) The element with an atomic number of 15 is phosphorus, P. Since the isotope has 16 neutrons, the mass number is 31. The isotope symbol for phosphorus-31 is $^{31}_{15}P$.

2.37   The atomic number is determined from the isotope symbol or by finding the element on the periodic table. For example, copper (Cu) has an atomic number of 29. The number of neutrons in an atom of copper-65 is $N = 65 - 29 = 36$.

	Protons (Z)	Neutrons (N) ($N = A - Z$)
(a) $^{56}_{26}Fe$	26	$N = 56 - 26 = 30$
(b) ^{39}K	19	$N = 39 - 19 = 20$
(c) copper-65	29	$N = 65 - 29 = 36$

2.39   The mass number is given by: $A = Z + N$, where Z is the number of protons and N is the number of neutrons. From the periodic table we find that nitrogen's atomic number is 7 so there are 7 protons. Since the mass number is 13, the number of neutrons is $Z = A - N = 13 - 7 = 6$.

2.41   The isotope symbol takes the form $^{\text{mass number}}_{\text{protons}}X^{\text{charge}}$ where the mass number is the neutrons plus protons and the charge is determined by the protons and electrons. X is the element symbol. All atoms are electrically neutral so the number of electrons and protons are the same.

Isotope Symbol	Number of Protons	Number of Neutrons	Number of Electrons
$^{23}_{11}Na$	11	12	11
$^{56}_{25}Mn$	25	31	25
$^{18}_8O$	8	10	8
$^{19}_9F$	9	10	9

2.43   They differ in the number of electrons. The identity of an atom or an ion is determined by the number of protons in the nucleus. However, ions have different numbers of electrons than protons. This is why ions are charged. For example, the ion N^{3-} is similar to the N atom because it has 7 protons, but the ion has 10 electrons. The three "extra" electrons give the ion the 3– charge.

2.45   (a) When an atom gains one electron, an anion with a 1– charge is formed. For example, when a fluorine atom, with 9 protons and 9 electrons, gains 1 electron, there are 10 negative charges and 9 positive charges. This means that the resulting ion will have a 1– charge (F^-).

(b) When an atom loses two electrons, a cation with a 2+ charge is formed. For example, when a magnesium atom, with 12 protons and 12 electrons, loses 2 electrons, there are 12 positive charges and 10 negative charges. This means that the resulting ion will have a 2+ charge (Mg^{2+}).

2.47   (a) Zinc atoms have 30 protons and 30 electrons. When two electrons are lost there are still 30 positive charges, but only 28 negative charges. The ion that results has a 2+ charge. The symbol for the ion is Zn^{2+}. Positive ions are called cations.

(b) Phosphorus atoms have 15 protons and 15 electrons. When a P atom gains three electrons there will be three more negative charges (18 electrons) than positive charges (15 protons). The resulting ion will have a 3– charge. The symbol for the ion is P^{3-}. Negative ions are called anions.

2.49   The number of protons is determined from the atomic symbol and the periodic table. For example, zinc (Zn) has 30 protons. The number of electrons is determined by looking at the charge on the ion. A Zn^{2+} ion has two fewer negative charges than positive charges. This means that there are 28 electrons (number of electrons $= 30 - 2 = 28$).

	Number of Protons	Number of Electrons
(a) Zn^{2+}	30	28
(b) F^-	9	10
(c) H^+	1	0

2.51   The completed table is shown below.

(a) The number of protons is 17 as indicated by the isotope symbol. Since the mass number, A, is 37, the number of neutrons is 20 ($N = A - Z$). Since the charge is 1–, there must be 18 electrons.

(b) The number of protons is 12, so the element is Mg. The mass number, 25, is the sum of the protons and neutrons. Since there are two more protons than electrons, the ion has a charge of 2+.

(c) The number of protons is 7, so the element is N. The mass number is 13 (sum of protons and neutrons). The charge is 3– because there are three more electrons than protons.

(d) Since the element is calcium, the number of protons is 20. Because the mass number is 40, the number of neutrons is also 20. Since the charge is 2+, there must be 18 electrons.

Isotope Symbol	Number of Protons	Number of Neutrons	Number of Electrons
$^{37}_{17}Cl^-$	17	20	18
$^{25}_{12}Mg^{2+}$	12	13	10
$^{13}_{7}N^{3-}$	7	6	10
$^{40}Ca^{2+}$	20	20	18

2.53   Since the ion has a charge of 1+, there must be one more proton than electrons. Since there are 18 electrons, there must be 19 protons. The element is potassium, K.

2.55   Since the charge is 2+, there must be two more protons than electrons. Since there are 27 electrons, there must be 29 protons. The element is copper, Cu.

2.57   Lithium-7, ^{7}Li, has three protons, three electrons, and four neutrons. ^{7}Li$^+$ has only two electrons, and ^{6}Li has only three neutrons. Otherwise they are the same as ^{7}Li. Lithium-6 differs the most in mass.

Isotope	Protons (Z)	Neutrons (N)	Mass number (A) $A = N + Z$	Electrons
^{7}Li	3	4	7	3
^{7}Li$^+$	3	4	7	2
^{6}Li	3	3	6	3

2.59   From the periodic table we find that potassium has 19 protons. Since it has a 1+ charge, there must be one more proton than electrons in the atom. There are 18 electrons.

2.61   The mass of carbon-12 is defined as exactly 12 amu. From this, the atomic mass unit is defined as $1/12^{th}$ the mass of one carbon-12 atom.

2.63   The approximate mass of an isotope is equivalent to its mass number. This is true since most of the mass of an atom comes from the protons and neutrons in the nucleus. (a) 2 amu; (b) 238 amu

2.65   Deuterium, an isotope of hydrogen with 1 neutron and 1 proton, has a mass number of 2. A molecule of D_2 has a mass of 4 amu and a molecule of H_2 has a mass of 2 amu. Therefore, the mass of D_2 is 2 amu greater (or two times greater) than the mass of H_2.

2.67   The mass of an atom is approximately equal to its mass number. The mass of a krypton-80 atom is about 40 amu greater **or twice the mass** of an argon-40 atom.

2.69   The numerical values of the masses of individual atoms are very small when measured on the gram scale. The size of the atomic mass unit allows us to make easier comparisons and calculations of masses of molecules.

2.71   The mass number is the sum of the number of protons and neutrons in the nucleus and is always an integer value. The mass number, which is not the actual mass, is usually close to the actual mass because the proton and neutron weigh approximately 1 amu each. The mass of an atom is the actual measurement of how much matter is in the atom and is never exactly an integer value (except carbon-12).

2.73   A mass spectrometer is used to determine the mass of an atom (Figure 2.15). The mass number of an atom is the sum of the number of protons and neutrons.

2.75   If there are only two isotopes, the relative mass will be closer to the mass of the isotope that is most abundant. Since the relative mass of calcium is 40.08 amu, we can assume that calcium-40 is the most abundant isotope.

2.77   To calculate the relative atomic mass we calculate the weighted average of the isotopes of X.

	Isotope mass $\times$ abundance	=	mass contribution from isotope
^{22}X	21.995 amu $\times$ 0.7500	=	16.50 amu
^{20}X	19.996 amu $\times$ 0.2500	=	5.00 amu
			21.50 amu (relative atomic mass of X)

2.79   (a) The tallest peak (nickel-58), with an abundance of 67.88%, is the most abundant isotope.
        (b) The shortest peak (nickel-64), with an abundance of 1.08% is the least abundant isotope.
        (c) The average mass will be closer to 58 because nickel-58 represents more than half of the stable isotopes.
        (d) Each isotope has the same number of protons (28). Since each isotope has a 1+ charge, they must each have 27 electrons. The neutrons are obtained using $A = Z + N$ and solving for N:
        nickel 58, $N = 30$; nickel-60, $N = 32$; nickel-61, $N = 33$; nickel-62, $N = 34$; nickel-64, $N = 36$.

2.81   The mass of 1000 boron atoms can be determined by multiplying the relative atomic mass by 1000.

$$\text{Total mass} = 1000 \text{ atom} \times \frac{10.81 \text{ amu}}{\text{atom}} = 10{,}810 \text{ amu or } 1.081 \times 10^4 \text{ amu}$$

2.83   Since the relative atomic mass of mercury (200.6 amu) is much higher than that of boron (10.81 amu) there will be more boron atoms in 2500 amu (i.e. it takes more of them to add up to 2500). This can be demonstrated by calculating the number of atoms:

$$\text{Boron atoms} = 2500 \text{ amu} \times \frac{1 \text{ atom}}{10.81 \text{ amu}} = 231 \text{ atoms}$$

$$\text{Mercury atoms} = 2500 \text{ amu} \times \frac{1 \text{ atom}}{200.6 \text{ amu}} = 12 \text{ atoms}$$

2.85   Most elements can be classified in multiple ways. Metals, metalloids, and nonmetals are distinguished in Figure 2.18. Many of the groups of the periodic tables have unique names.

Group	Name
"A" block	main group elements
"B" block	transition elements
IA (1)	alkali metals
IIA (2)	alkaline earth metals
VIIA (17)	halogens
VIIIA (18)	noble gases

(a) alkali metal: K
(b) halogen: Br
(c) transition metal: Mn
(d) alkaline earth metal: Mg
(e) noble gas: Ar
(f) main-group element: Br, K, Mg, Al, Ar

2.87   Halogens are found in Group VIIA (17). The halogen found in period 3 is chlorine, Cl.

2.89   Titanium, Ti, is found at the intersection of period 4 and group IVB (4).

2.91   The metals and nonmetals are shown in Figure 2.18. (a) Ca, metal; (b) C, nonmetal; (c) K, metal; (d) Si, metalloid

2.93   The major groups of the elements are shown in Figure 2.18. (a) O, main group; (b) Mg, main group; (c) Sn, main group; (d) U, actinide; (e) Cr, transition metal

2.95   Group VIIA (17), the halogen group, all occur as diatomic molecules (F_2, Cl_2, Br_2, I_2). Astatine has no stable isotopes, but is believed to form diatomic molecules. It is estimated that less than 1 oz of astatine exists in the earth's crust.

2.97   Neon does not occur as a diatomic molecule.

2.99   The noble gases, group VIIIA (18), all occur as gases of uncombined atoms.

2.101   electrons

2.103   Many main group elements gain or lose electrons to form ions with the same number of electrons as the closest noble gas. In general, metals lose electrons to form positively charged ions (cations) and nonmetals gain electrons to form negatively charged ions (anions).
(a) The atoms of the elements in group IA (1) lose one electron to form 1+ ions. For example, sodium atoms have 11 electrons. After losing one electron, a sodium ion will have a 1+ charge and the same number of electrons (10) as neon.
(b) The atoms of the elements of group IIA (2) lose two electrons to form 2+ ions. For example, the ion Ca^{2+} has the same number of electrons (18) as argon.
(c) The atoms of the elements of group VIIA (17) gain one electron to form 1– ions. For example, the ion I^- has the same number of electrons (54) as xenon.
(d) The atoms of the elements of group VIA (16) gain two electrons to form 2– ions. For example, the ion Se^{2-} has the same number of electrons (36) as krypton.

2.105   Metals will lose electrons to have the same number of electrons as the nearest noble gas. This means that they will form positively charged ions. Nonmetals gain electrons to have the same number of electrons as the nearest noble gas. Therefore, nonmetals form negatively charged ions.
(a) Sodium atoms, Na, lose one electron to match the number of electrons in neon. Since one electron is lost, the charge is 1+: Na^+.
(b) Oxygen atoms, O, gain two electrons to match the number of electrons in neon. Since two electrons are gained, the charge is 2–: O^{2-}.
(c) Sulfur atoms, S, gain two electrons to match the number of electrons in argon. Since two electrons are gained, the charge is 2–: S^{2-}.
(d) Chlorine atoms, Cl, gain one electron to match the number of electrons in argon. Since one electron is gained, the charge is 1–: Cl^-.
(e) Bromine atoms, Br, gain one electron to match the number of electrons in krypton. Since one electron is gained, the charge is 1–: Br^-.

2.107   Reactivity is a periodic property. Since sodium is in group IA, you might expect other elements in the same group (lithium, potassium, rubidium, cesium and francium) to react the same way.

2.109   The mass of oxygen added to form Fe_2O_3 causes an increase in the mass.

2.111   To show that the data is in agreement with the law of definite proportions, the mass ratio of Zn/S must be calculated for each substance. If the ratio is the same for both substances, the law of definite proportions is obeyed. In the first sample of zinc sulfide we have a ratio of
$$Zn/S = \frac{67.1 \text{ g zinc}}{32.9 \text{ g sulfur}} = 2.04 \text{ g Zn/g S}$$
For the second compound, we are given the mass of zinc and zinc sulfide, but must calculate the mass of the sulfur used by the reaction:
    Mass of sulfur = 2.00 g zinc sulfide – 1.34 g zinc = 0.66 g sulfur
The Zn/S ratio is calculated as
$$Zn/S = \frac{1.34 \text{ g zinc}}{0.66 \text{ g sulfur}} = 2.0 \text{ g Zn/g S}$$
The mass ratios of Zn/S are the same for the two samples (within the significant figures given).

2.113   Electrons were discovered first because they have charge and were readily studied in cathode ray tubes. Thomson put cathode rays (beams of electrons) in magnetic and electric fields to determine the charge and mass-to-charge ratio of the electron.

2.115   Only nickel (Ni) has an atomic number of 28. Since its mass number is 60, the isotope must be nickel-60.

2.117   Since potassium has 19 protons, potassium-39 must have 20 neutrons.

2.119 The most abundant isotopes of cobalt have masses greater than the masses of the most abundant isotopes of nickel. As a result, the relative atomic mass of cobalt is greater than that of nickel.

2.121 When there are many isotopes, some of the isotopes can be present in very low abundance. As a result, their masses cannot be determined as accurately, and their percentage contributions are also less well known. Both factors result in a decrease in the precision of the calculated relative atomic mass.

2.123 The mass number is 127. Since there is only one isotope, the mass number should be very close to the relative atomic mass.

2.125 Since each carbon atom has less relative atomic mass than an iodine atom, there will be more carbon atoms.

2.127 The relative atomic mass of boron is 10.81 amu. Since boron exists as either boron-10 or boron-11, the relative abundance of boron-11 will be much higher. Of the choices given only two make sense:

       20% boron-10 and 80% boron-11 (correct answer)
       5.0% boron-10 and 95.0% boron-11

       The best answer can be found by calculating the relative atomic mass based on these percentages:

	Isotope mass $\times$ abundance	=	mass contribution from isotope
^{11}B	11.009 amu $\times$ 0.800	=	8.81 amu
^{10}B	10.013 amu $\times$ 0.200	=	2.00 amu
			10.81 amu (relative atomic mass of B)

2.129 $Br_2(l)$

2.131 Hydrogen is a nonmetal.

2.133 Adding a proton changes the element. So by adding a proton you would form a cation of a different element.

2.135 The subatomic particle described in (a) corresponds to a neutron because it has a zero charge. The subatomic particle in (b) has a mass that is four orders of magnitude less than a neutron. This subatomic particle is an electron and carries a charge of 1−. The subatomic particle described in (c) has a mass very close to the mass of neutron. This corresponds to a proton which carries a charge of 1+.

	Particle	Mass (g)	Relative Charge
(a)	neutron	1.6749×10^{-24}	0
(b)	electron	9.1094×10^{-28}	1−
(c)	proton	1.6726×10^{-24}	1+

2.137 Fluorine-18 is the designation for an isotope of this element with a mass number of 18. The mass number (A) is the sum of the neutrons (N) and protons (Z): $A = N + Z$. Since the atomic number of fluorine is 9, the neutron number is $N = A − Z = 18 − 9 = 9$. A neutral atom of fluorine would have 9 electrons to balance the 9 protons in the nucleus.

2.139 Neutral Fe atoms differ from Fe^{2+} and Fe^{3+} ions in number of electrons. Each of these contains 26 protons. A neutral Fe atom has 26 electrons, an Fe^{2+} ion has 24 electrons, and an Fe^{3+} ion has 23 electrons.

2.141 Iodine is in group VIIA (17) and period 5 of the periodic table. An element in group VIIA (17) is also known as a halogen.

2.143 A relative atomic mass of 12.01 amu for carbon indicates that the predominant naturally occurring isotope must be carbon-12. That the relative atomic mass is just slightly greater than 12 indicates there is very little carbon-14 in natural abundance. The most likely abundance of carbon-14 from the possibilities listed is less than 0.1%.

Chapter 3 – Chemical Compounds

3.1   (a) formula unit; (b) strong electrolyte; (c) molecular compound; (d) acid; (e) nonelectrolyte; (f) oxoanion

3.3   Ionic compounds usually consist of a combination of metals and nonmetals. If the substance is composed of two or more nonmetals, it is usually molecular. For example, we identify the first two substances, K_2S and Na_2SO_4, as ionic because K and Na are both metals. In SO_2, both sulfur and oxygen are nonmetals, so SO_2 is molecular.

3.5   Whether a molecular or ionic compound is formed can usually be determined by classifying the elements involved as metals or nonmetals. Metals generally combine with nonmetals to form ionic compounds and when two or more nonmetals combine they usually produce molecular compounds. Molecular compounds would be formed from (a) and (c) because the elements involved are nonmetals. Ionic compounds would be formed from (b) and (d) because potassium (K) and magnesium (Mg) are metals and each is paired with a nonmetal (O and Cl). If you are having trouble identifying metals and nonmetals, refer to Figure 3.12.

3.7   Both (a) and (d) would be molecular since both elements involved are nonmetals. Both (b) and (c) are ionic because of the presence of metals (Li and Ba) and nonmetals (F and Cl).

3.9   The attraction between oppositely charged ions is very strong. As a result, the ionic compound LiF would be expected to have the highest melting point. The attraction between molecules in molecular compounds is generally much weaker and thus CO_2 and N_2O_5 would be expected to have lower melting points.

3.11   The position of an element on the periodic table can be used to predict the charges of monatomic ions. All three elements given are in Group 1A (1) and are expected to form 1+ ions. This gives them the same number of electrons as the nearest noble gas. (a) Na^+, sodium ion; (b) K^+, potassium ion; (c) Rb^+, rubidium ion

3.13   The charge of an ion can be predicted by the position of the element on the periodic table. Metals lose electrons to match the number of electrons in the nearest noble gas and nonmetals gain electrons to match the number of electrons in the nearest noble gas. Don't forget that as metals lose electrons, their charge becomes more positive because the electron carries a negative charge. (a) Ca^{2+}, calcium ion; (b) N^{3-}, nitride ion; (c) S^{2-}, sulfide ion

3.15   Table 3.4 lists the important polyatomic ions used in this chapter. From the image you should determine that there is one nitrogen (blue) and two oxygen atoms (red). Given that the charge is 1– and the formula is NO_2^-, the ion is the nitrite ion.

3.17   Table 3.4 lists the important polyatomic ions used in this chapter and Figure 3.17 can be used to quickly look up polyatomic oxoanions. (a) sulfate ion; (b) hydroxide ion; (c) perchlorate ion

3.19   The polyatomic and monatomic ions are listed in Figure 3.17 and Table 3.4. (a) N^{3-}; (b) NO_3^-; (c) NO_2^-

3.21   The polyatomic and monatomic ions are listed in Figure 3.17 and Table 3.4. (a) CO_3^{2-}; (b) NH_4^+; (c) OH^-; (d) MnO_4^-

3.23   Table 3.4 lists the important polyatomic ions used in this chapter. From the image you should determine that there is one sulfur (yellow) and three oxygen atoms (red). Given that the charge is 2– and the formula is SO_3^{2-}, the ion is the sulfite ion.

3.25 IO_3^-

3.27   A solution is electrically neutral. This means that the sum of the positive charges is equal to the sum of the negative charges. If there are five Al^{3+} ions, representing a total positive charge of 15+, an electrically neutral solution would have 15 Cl^- ions (larger spheres) to balance the 15+ charge.

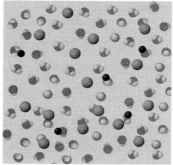

3.29   Compounds are electrically neutral. Here are two simple rules to follow: If the ions have the same charge, you only need one of each ion. If the charges are different, you can quickly determine the chemical formula by "crossing" the charges. The sign of the charge is not used (i.e. – or +), just the number. If there is only one of that element, the number "1" is omitted in the chemical formula.

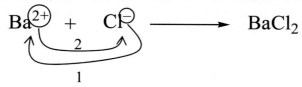

(a) $BaCl_2$; (b) $FeBr_3$; (c) $Ca_3(PO_4)_2$; (d) $Cr_2(SO_4)_3$

3.31   Ionic compounds are made up of cations and anions. The cation is always written first. Usually the cation is a monatomic metal ion. The common exceptions are Hg_2^{2+} and NH_4^+. The anion is either polyatomic or monatomic.

(a) KBr – Potassium and bromide ions are both monatomic and their charges are determined by their positions on the periodic table (see Figure 3.12). The ions in potassium bromide are K^+ and Br^-.

(b) $BaCl_2$ – Barium and chloride ions are both monatomic and their charges are determined by their positions on the periodic table. The ions in barium chloride are Ba^{2+} and Cl^-. A common error is to assume that the chloride ion is Cl_2^-. Barium chloride is a combination of one Ba^{2+} for every two Cl^-.

(c) $Mg_3(PO_4)_2$ – Magnesium ion is monatomic and phosphate ion is polyatomic. Magnesium ion's charge is determined by its position on the periodic table. The charges of polyatomic ions can be found in Figure 3.17 and Table 3.4. The ions in magnesium phosphate are Mg^{2+} and PO_4^{3-}.

(d) $Co(NO_3)_2$ – Cobalt ion is monatomic, but it can have several different charges. To identify the charge of cobalt ion, you need to figure out the total negative charge. Nitrate ion is a polyatomic ion with a charge of 1–. Since there are two nitrate ions, the total negative charge is 2–. Since all compounds are neutral (zero net charge) the charge on cobalt ion must be 2+. The ions in cobalt(II) nitrate are Co^{2+} and NO_3^-. Mathematically we can summarize the charge of any ion or compound as:

Total positive charge + total negative charge = overall charge

In the case of this compound we can write
(Charge from Cobalt)   +   (charge from nitrate) = 0
Co   +   2–   =   0
Cobalt ion's charge must be 2+.

3.33   (a) Fe^{2+} will form FeO. This balances the positive and negative charges. Fe^{3+} will form Fe_2O_3. If you have difficulty with this, you might consider "crossing" the charges as shown below. Notice that the least common multiple for these two ions is 6. That is $2 \times 3 = 6$. This means that if you combine two Fe^{3+} ions and three O^{2-} ions, the charge will balance to zero. This can be represented as follows:

   (b) Since the charge on a chloride ion is 1–, you need as many chloride ions as necessary to cancel the positive charge of the iron ions. Fe^{2+} will form $FeCl_2$, and Fe^{3+} will form $FeCl_3$.

3.35   (a) The overall charge is equal to the sum of the charges of the sodium ions and sulfate ions:

   Overall charge = (charge of sodium ions) + (charge of sulfate ion) = 0

   Since the charge of the two sodium ions must balance the charge of sulfate, sulfate must have a 2– charge.
   (b) Since sulfate ion has a charge of 2–, the charge of the strontium ion must be 2+.

3.37   For each "compound" written, the charges do not balance so the compounds would have a net charge. In compounds, the sum of the positive and negative charges balances to zero (i.e. all compounds are neutral).
   (a) Sodium ion is Na^+ and chloride ion is Cl^-. In $NaCl_2$ there are too many chloride ions. Only one chloride ion is needed to balance the charge of the sodium ion. The correct formula is NaCl.
   (b) Potassium ion is K^+ and sulfate ion is SO_4^{2-}. In the formula KSO_4 there are not enough potassium ions to balance the 2– charge on the sulfate ion. The correct formula is K_2SO_4.
   (c) Aluminum ion has a charge of 3+ and nitrate ion has a charge of 1–. In the formula Al_3NO_3 there are not enough nitrate ions for the number of aluminum ions. Three nitrate ions are needed to balance the charge of one aluminum ion. The correct formula is $Al(NO_3)_3$.

3.39   To name these compounds we state the name of the cation followed by the name of the anion. It is important to note that prefixes such as di– and tri– are not often used in naming ionic compounds because the number of each ion can be determined from the charges of the ions. The cations in these compounds are all metals that have charges predictable from their position on the periodic table so it is not necessary to write the charges of the ions.
   (a) magnesium chloride; (b) aluminum oxide; (c) sodium sulfide; (c) potassium bromide; (e) sodium nitrate; (f) sodium perchlorate

3.41   Charges which cannot be predicted by the position of the element on the periodic table must be specified. Figure 3.12 shows many of the common ions (those metals which do not need their charges specified). Manganese, Mn, and cobalt, Co, compounds should have the charges specified with the metal name.

3.43   When the charge of the metal (cation) can not be determined by its position on the periodic table, the charge is calculated from the chemical formula and the anion it is combined with. The charge can either be calculated (see (a)) or deduced by inspection. This is possible because the charge of the anion is almost always predictable.
   (a) To calculate the charge we state that the total charge is equal to the sum of the positive and negative charges.

   $$\text{Total charge} = \text{charge on copper} + \text{charge on oxygen}$$
   $$0 \quad = \quad 2x \quad + \quad 2-$$

   Solving for x:
   $$2x = 2$$
   $$x = 1$$

The charge on the copper ion is 1+. To name the compound, write the name of the metal with its charge (i.e. copper(I)) followed by the name of the anion. The compound is named copper(I) oxide or cuprous oxide.

(b) The charge on chloride ion is 1– and there are two chloride ions in the formula (a total of 2– charge). The charge on the chromium ion must be 2+ to balance the negative charge. The name of the compound is chromium(II) chloride.

(c) The charge on phosphate ion is 3–. The charge of the iron ion must be 3+ to balance the negative charge. The name of the compound is iron(III) phosphate or ferric phosphate.

(d) The charge on sulfide ion is 2–. The charge on the copper ion must be 2+ to balance the negative charge. The name of the compound is copper(II) sulfide or cupric sulfide.

3.45   To write the chemical formulas of ionic compounds, first determine the formula for each ion and combine the ions so that the positive and negative charges balance.

(a) Ca^{2+} and SO_4^{2-} (one of each ion is needed) $CaSO_4$

(b) Ba^{2+} and O^{2-} (one of each ion is needed) BaO

(c) NH_4^+ and SO_4^{2-} (two ammonium ions are needed to balance the charge of one sulfate ion) $(NH_4)_2SO_4$

(d) Ba^{2+} and CO_3^{2-} (one of each ion is needed) $BaCO_3$

(e) Na^+ and ClO_3^- (one of each ion is needed) $NaClO_3$

3.47   The charge of the metal is determined from the formula and the charge of the anion. First determine the total negative charge contributed by the anions and then use that information to determine the charge of the metal ion.

(a) The charge on chloride ion is 1– and there are two chloride ions in the formula (total charge 2–). The charge on the cobalt ion must be 2+ (Co^{2+}). The compound name is cobalt(II) chloride.

(b) The charge on oxide ion is 2– and there are two oxide ions in the formula (total charge 4–). The charge on the lead ion must be 4+ (Pb^{4+}). The compound name is lead(IV) oxide.

(c) The charge on nitrate ion is 1– and there are three nitrate ions in the formula (total charge 3–). The charge on the chromium ion must be 3+ (Cr^{3+}). The compound name is chromium(III) nitrate.

(d) The charge on sulfate ion is 2– and there are three sulfate ions in the formula (total charge 6–). Since there are two iron ions, the charge on each iron ion must be 3+ (Fe^{3+}). The compound name is iron(III) sulfate or ferric sulfate.

3.49   Use the name of the compound to determine the formulas for each ion. Then you can write the formula for the compound by balancing the total charge.

(a) From the name cobalt(II), we know that cobalt has a charge of 2+ (Co^{2+}). The formula for chloride ion is Cl^-. The compound formula is $CoCl_2$.

(b) From the name manganese(II) we know that manganese has a charge of 2+ (Mn^{2+}). The formula for nitrate ion is NO_3^-. The compound formula is $Mn(NO_3)_2$.

(c) From the name chromium(III) we know that chromium has a charge of 3+ (Cr^{3+}). The formula for oxide ion is O^{2-}. If the chemical formula is not immediately apparent based on the charges (as in this case), it is often useful to calculate the least common multiple. In this case the least common multiple is 6 (i.e. $3 \times 2 = 6$). It takes 2 Cr^{2+} to achieve a +6 charge and 3 O^{2-} to achieve a 6– charge. The compound formula is Cr_2O_3.

(d) From the name copper(II) we know that copper has a charge of 2+ (Cu^{2+}). The formula for phosphate ion is PO_4^{3-}. If the chemical formula is not immediately apparent based on the charges (as in this case), it is often useful to calculate the least common multiple. In this case the least common multiple is 6 (i.e. $2 \times 3 = 6$). It takes 3 Cu^{2+} to achieve a +6 charge and 2 PO_4^{3-} to achieve a 6– charge. The compound formula is $Cu_3(PO_4)_2$.

3.51   Table 3.7 lists the common names for the metal ions. Since nitrate ion has a charge of 1–, we determine that the charge of the iron ions in $Fe(NO_3)_2$ and $Fe(NO_3)_3$ are 2+ and 3+, respectively. The Fe^{2+} ion is called ferrous ion and the Fe^{3+} ion is called ferric ion. The common names for these compounds are ferrous nitrate and ferric nitrate.

3.53   To write the formulas for the compounds, the ions are combined so that the overall charge is zero. The easiest way to name the compounds in the chart is to first determine the name of each ion. The names of the cations and anions have been listed along with their formulas. The charges must be included in the names of compounds containing iron, nickel, and chromium because their charges cannot be determined by their position on the periodic table. The compound names are derived by combining the cation and anion names.

	Ca^{2+} calcium	Fe^{2+} iron(II)	K^+ Potassium	Mn^{2+} manganese(II)	Al^{3+} aluminum	NH_4^+ ammonium
Cl^- chloride	$CaCl_2$ calcium chloride	$FeCl_2$ iron(II) chloride	KCl potassium chloride	$MnCl_2$ manganese(II) chloride	$AlCl_3$ aluminum chloride	NH_4Cl ammonium chloride
O^{2-} oxide	CaO calcium oxide	FeO iron(II) oxide	K_2O potassium oxide	MnO manganese(II) oxide	Al_2O_3 aluminum oxide	$(NH_4)_2O$ ammonium oxide
NO_3^- nitrate	$Ca(NO_3)_2$ calcium nitrate	$Fe(NO_3)_2$ iron(II) nitrate	KNO_3 potassium nitrate	$Mn(NO_3)_2$ manganese(II) nitrate	$Al(NO_3)_3$ aluminum nitrate	NH_4NO_3 ammonium nitrate
SO_3^{2-} sulfite	$CaSO_3$ calcium sulfite	$FeSO_3$ iron(II) sulfite	K_2SO_3 potassium sulfite	$MnSO_3$ manganese(II) sulfite	$Al_2(SO_3)_3$ aluminum sulfite	$(NH_4)_2SO_3$ ammonium sulfite
OH^- hydroxide	$Ca(OH)_2$ calcium hydroxide	$Fe(OH)_2$ iron(II) hydroxide	KOH potassium hydroxide	$Mn(OH)_2$ manganese(II) hydroxide	$Al(OH)_3$ aluminum hydroxide	NH_4OH ammonium hydroxide
ClO_3^- chlorate	$Ca(ClO_3)_2$ calcium chlorate	$Fe(ClO_3)_2$ iron(II) chlorate	$KClO_3$ potassium chlorate	$Mn(ClO_3)_2$ manganese(II) chlorate	$Al(ClO_3)_3$ aluminum chlorate	NH_4ClO_3 ammonium chlorate

3.55   The formulas for the ions have been listed along with their names. To write the formulas for the compounds, the ions are combined so that the overall charge is zero.

	potassium K^+	iron(III) Fe^{3+}	strontium Sr^{2+}	aluminum Al^{3+}	cobalt(II) Co^{2+}	lead(IV) Pb^{4+}
iodide I^-	KI	FeI_3	SrI_2	AlI_3	CoI_2	PbI_4
oxide O^{2-}	K_2O	Fe_2O_3	SrO	Al_2O_3	CoO	PbO_2
sulfate SO_4^{2-}	K_2SO_4	$Fe_2(SO_4)_3$	$SrSO_4$	$Al_2(SO_4)_3$	$CoSO_4$	$Pb(SO_4)_2$
nitrite NO_2^-	KNO_2	$Fe(NO_2)_3$	$Sr(NO_2)_2$	$Al(NO_2)_3$	$Co(NO_2)_2$	$Pb(NO_2)_4$
acetate $CH_3CO_2^-$	KCH_3CO_2	$Fe(CH_3CO_2)_3$	$Sr(CH_3CO_2)_2$	$Al(CH_3CO_2)_3$	$Co(CH_3CO_2)_2$	$Pb(CH_3CO_2)_4$
hypochlorite ClO^-	KClO	$Fe(ClO)_3$	$Sr(ClO)_2$	$Al(ClO)_3$	$Co(ClO)_2$	$Pb(ClO)_4$

3.57   The formulas for silver and chloride ions are Ag^+ and Cl^-. The chemical formula for silver chloride is AgCl.

3.59   A molecular formula represents the exact numbers of each atom in the molecule. By counting the atoms in each molecule you can arrive at the following formulas: NF_3, P_4O_{10}, $C_2H_4Cl_2$.

3.61   To name molecular compounds, use the prefixes listed in Table 3.9 to indicate the number of each atom in the compound. A prefix is not needed if there is only one atom of the first element. The last element in the name of a diatomic molecular compound always ends in –ide. For example, CO is named carbon monoxide (not monocarbon monoxide).
(a) phosphorus pentafluoride; (b) phosphorus trifluoride; (c) carbon monoxide; (d) sulfur dioxide

3.63   The prefixes used in naming molecular compounds are listed in Table 3.9. If a prefix is not present, then only one of those atoms is present in the formula. (a) SF_4; (b) C_3O_2; (c) ClO_2; (d) SO_2

3.65   The central image represents phosphoric acid, H_3PO_4, with four oxygen atoms attached to one phosphorus atom and hydrogen atoms attached to three of the four oxygen atoms.

3.67   Acids are derived from the stem of the anion name. The following chart is useful in naming the acids:

suffix	new prefix	new suffix
-ide	hydro-	-ic acid
-ite		-ous acid
-ate		-ic acid

(a) HF(aq) – The anion is fluor**ide**. Since the new suffix is –ic, the acid name is **hydro**fluor**ic** acid.
(b) HNO₃(aq) – The anion is nit**rate**. Since the new suffix is –ic, the acid name is nit**ric** acid.
(c) H₃PO₃(aq) – The anion is phosph**ite**. Acids containing phosphorus (and sulfur) are tricky to name. In naming this acid, the stem becomes phosphor. Since the new suffix is -ous , the acid name is phosph**orous** acid.

3.69   To write the formula for an acid from its name, first determine the formula for the anion. One hydrogen is added to the formula for each negative charge on the anion to balance the charge. An "(aq)" is added to state that the substance is an acid when dissolved in water.

(a) hydrofluoric acid – The prefix *hydro*- with the suffix –ic mean that this is a binary acid (composed of only hydrogen and one other element). In this case the anion is fluoride (F⁻). Since hydrogen forms a 1+ ion and fluoride is a 1– ion, the acid formula is HF(aq).

(b) sulfurous acid – The –ous suffix means that the anion is sulfite (SO₃²⁻). Since hydrogen forms a 1+ ion, two hydrogens are needed to balance the charge. The formula for the acid is H₂SO₃(aq).

(c) perchloric acid – The –ic suffix means that the anion is perchlorate (ClO₄⁻). Since hydrogen forms a 1+ ion, one hydrogen is needed to balance the charge. The formula for the acid is HClO₄(aq).

3.71   The formula for nitric acid is HNO₃(aq). When HNO₃ is dissolved in water, the molecules ionize to form hydrogen ions and nitrate ions (H⁺ and NO₃⁻).

3.73   Many substances form crystalline solids, but only two types of substances are electrolytes: acids and ionic compounds (which include bases). Of these two types only ionic compounds form crystalline solids. Only KI, Mg(NO₃)₂, and NH₄NO₃ can be described by these properties. All ionic solids are also brittle because of the strong attraction of the opposite charges in the crystal structure.

3.75   These compounds contain a metal (Zn or Ti) and a nonmetal (O). They are ionic compounds. To arrive at the correct formula, make sure the positive and negative charges balance. Since titanium has a 4+ charge and oxide is 2-, the formula is TiO₂. The charge of zinc is always 2+. The correct formula for zinc oxide is ZnO because the charges are balanced.

3.77   Compounds containing only nonmetals are generally molecular (except ammonium compounds). Carbon dioxide and dinitrogen monoxide do not contain metals or ammonium ion, so they are classified as molecular compounds. Formulas for molecular compounds are derived from the prefix (Table 3.9). Absence of a prefix implies only one atom of an element in the formula. Carbon dioxide is CO₂ and dinitrogen monoxide is N₂O.

3.79   First classify each compound as either ionic or molecular. Both K₂S and Na₂SO₄ are ionic because of the presence of the metal ion. Since potassium and sodium have only one possible charge (1+) we do not specify the charge of the ion in the name. The names of the anions are written without modification so K₂S and Na₂SO₄ are named potassium sulfide and sodium sulfate, respectively.

SO₂ is a molecular compound and is named using prefixes to indicate the number of each element in the formula. Mono is not used with the first element so the name of SO₂ is sulfur dioxide.

3.81   The first step in naming a compound or substance is to identify whether it is ionic or molecular. Once that is determined, the rules for each type of substance can be applied.

(a) molecular, nitrogen trioxide; (b) ion, nitrate; (c) ionic, potassium nitrate; (d) ionic, sodium nitride; (e) ionic, aluminum chloride; (f) molecular, phosphorous trichloride; (g) ionic, titanium(II) oxide; (h) ionic, magnesium oxide

3.83   Once you have determined that a compound is ionic by the presence of a metal cation or ammonium ion, its formula is derived by balancing the charges of the cation and the anion. The ions can be found in Table 3.4 or Figure 3.17. Otherwise, the formula for molecular compounds is derived from the prefixes (Table 3.9). Note that acids are all molecular compounds, but the formula is derived from the anion it "contains" (the ion only forms when the substance ionizes in water). For example hydrofluoric acid contains fluoride. Fluoride has the formula F^- so the acid is HF (one hydrogen ion, H^+, is needed to balance the charge). The formulas of acids are followed by (aq) because they are ionized in water.
   (a) Ionic. Sodium has a charge of $1+$ and carbonate has a charge of $2-$. To balance the charge, the formula must be Na_2CO_3.
   (b) Ionic. Sodium has a charge of $1+$ and bicarbonate has a charge of $1-$. To balance the charge the formula must be $NaHCO_3$.
   (c) Acid. Carbonic acid contains carbonate, CO_3^{2-}. To balance the charge with hydrogen ion, H^+, the formula must be $H_2CO_3(aq)$.
   (d) Acid. Hydrofluoric acid contains chloride, F^-. To balance the charge with hydrogen ion, H^+, the formula must be $HF(aq)$.
   (e) Molecular. The prefixes indicate one sulfur and three oxygens; SO_3.
   (f) Ionic. The copper ion has a charge of $2+$ and sulfate has a charge of $2-$. To balance the charge the formula must be $CuSO_4$.
   (g) Acid. Sulfuric acid contains sulfate, SO_4^{2-}. To balance the charge with hydrogen ion, H^+, the formula must be $H_2SO_4(aq)$.
   (h) Acid. Hydrosulfuric acid contains sulfide, S^{2-}. To balance the charge with hydrogen ion, H^+, the formula must be $H_2S(aq)$.

3.85   When $HCl(g)$ is a molecular compound it is named hydrogen chloride. However, when it is dissolved in water to form $HCl(aq)$, it is named as an acid because it ionizes. To name a binary acid we replace the anions suffix, –ide, with –ic and add the prefix hydro-. So chloride is converted to hydrochloric acid.

3.87   (a) Prefixes are not used to name ionic compounds; sodium sulfate.
   (b) Charges are not specified in the name when the cation has only one charge state; calcium oxide.
   (c) Copper exists in two charge states; $2+$ and $1+$. The charge must be specified in the name; copper(II) oxide.
   (d) The prefix penta- is missing to indicate the presence of 5 chlorines; phosphorus pentachloride.

3.89   (a) Sulfide has the formula, S^{2-}, so two potassium atoms are required to balance the charge; K_2S.
   (b) The (II) represents the charge of the nickel ion, not the number of nickel ions in the formula. $NiCO_3$.
   (c) Nitride is the monatomic ion N^{3-}; Na_3N.
   (d) The prefix was applied to the wrong element; NI_3.

3.91   Charges on the ions are determined by looking at the positions of the atoms on the periodic table, from memory, or from the chemical formula.
   (a) one sodium ion, Na^+, and one chloride ion, Cl^-
   (b) one magnesium ion, Mg^{2+}, and two chloride ions, Cl^-
   (c) two sodium ions, Na^+, and one sulfate ion, SO_4^{2-}
   (d) one calcium ion, Ca^{2+}, and two nitrate ions, NO_3^-

3.93   Molecular compounds that are not acids (those that do not have hydrogen first in their formula) are nonelectrolytes. Acids, bases, and water-soluble ionic compounds all produce ions in solution and are, therefore, electrolytes when dissolved in water. (a) electrolyte (base); (b) electrolyte (acid); (c) electrolyte (ionic); (d) nonelectrolyte

3.95   The ions of silver, zinc, and cadmium can each have only one possible charge (see Figure 3.24). They are assumed to have these charges in the compounds they form.

3.97   The polyatomic ions are listed in Table 3.4 and Figure 3.17. (a) NO_3^-; (b) SO_3^{2-}; (c) NH_4^+; (d) CO_3^{2-}; (e) SO_4^{2-}; (f) NO_2^-; (g) ClO_4^-

3.99   Identify the type of compound and then name the compound according to its classification.
   (a) $MgBr_2(s)$ – This is an ionic compound. Since the charge of the cation is predictable from its position on the periodic table it is not specified in the name. The compound is named magnesium bromide.
   (b) $H_2S(g)$ – This is a molecular compound. It is not named as an acid because it is in the gas state. Prefixes are used to designate the numbers of each type of atom in the formulas of molecular compounds. The formula name is dihydrogen sulfide.
   (c) $H_2S(aq)$ – This is named as an acid since it is dissolved in water. In addition, the *–ide* suffix of sulfide indicates that the acid is named hydrosulfuric acid.
   (d) $CoCl_3(s)$ – This is an ionic compound. The charge of the cobalt ion must be determined from the chemical formula. Since chloride ion has a charge of 1–, the charge of the cobalt is 3+ to balance the contribution of the three chloride ions. The name of the compound is cobalt(III) chloride.
   (e) $KOH(aq)$ – This is an ionic compound (bases are ionic compounds). Since the charge of the potassium is predictable from its position on the periodic table it is not specified in the name. The compound is named potassium hydroxide.
   (f) $AgBr(s)$ – This is an ionic compound. Since the charge of the silver ion is predictable from its position on the periodic table it is not specified in the name. The compound is named silver bromide.

3.101   To write chemical formulas for ionic compounds, determine the formula for each ion and combine them so that the positive and negative charges balance. In molecular compounds, the formulas are written from the prefixes in the name.
   (a) Pb^{2+} and Cl^-, $PbCl_2$
   (b) Mg^{2+} and PO_4^{3-}, $Mg_3(PO_4)_2$
   (c) Since a prefix is not used on the nitrogen, there is only one nitrogen in the formula. The prefix *tri-* means three. The formula is NI_3.
   (d) Fe^{3+} and O^{2-}, Fe_2O_3
   (e) Ca^{2+} and N^{3-}, Ca_3N_2
   (f) Ba^{2+} and OH^-, $Ba(OH)_2$
   (g) The prefix *di-* means two and *pent-* means five. The formula is Cl_2O_5.
   (h) NH_4^+ and Cl^-, NH_4Cl

3.103   The formula for sodium ion is Na^+ and bicarbonate ion's formula is HCO_3^-. One of each ion is needed to balance the charge, so the formula is $NaHCO_3$.

3.105   Calcium ion is Ca^{2+} and the hypochlorite ion is ClO^-. Two hypochlorite ions are needed to balance the charge of calcium. The formula is $Ca(ClO)_2$.

3.107   Molecular compounds do not contain ions. If metal or ammonium ions are present in the formula, the substance is ionic. Water, $H_2O(l)$, is the only molecular substance in this equation.

3.109   The metals are solids, and the ionic compounds are both aqueous. The text can be interpreted as follows: "solid copper" is $Cu(s)$, "solution of silver nitrate" is $AgNO_3(aq)$, "solid silver" is $Ag(s)$, and copper(II) nitrate is $Cu(NO_3)_2$.

3.111   (a) NH_3; (b) $HNO_3(aq)$; (c) $HNO_2(aq)$

3.113   In acetic acid, the hydrogen atom attached to an oxygen atom is the one responsible for its acidic properties.

3.115   All three substances contain oxygen. Magnesium oxide, MgO, has metal and nonmetal ions which makes it an ionic compound. It is a solid at room temperature. Oxygen and carbon dioxide (O_2 and CO_2) are both molecular compounds and are gases at room temperature. In magnesium oxide, the oxide ions are attached to many different magnesium ions. This means the chemical formula represents the ratio of magnesium ions to oxygen ions. In oxygen and carbon dioxide, the chemical formulas represent how many atoms of each element are in one molecule of the substance.

3.117   Since sodium chloride contains a metal and a nonmetal, it is ionic. Sodium ion has a 1+ charge and chloride has a 1– charge, so the formula is NaCl. Sodium bicarbonate contains a metal and a group of nonmetals, the bicarbonate or hydrogen carbonate ion. Thus the compound is ionic. Sodium ion has a 1+ charge, and bicarbonate ion has a 1– charge (see Table 3.4), so the formula is $NaHCO_3$.

3.119   Zinc oxide contains the zinc and oxide ions. Zinc has only one ionic charge, 2+. Oxide has a 2– charge. The formula is ZnO.

3.121   An ionic compound has a name that includes a Roman numeral if the metal can have more than one charge, which is generally found with transition metals. Zinc, iron, and chromium are transition metals, but zinc has only one charge. A Roman numeral is found in the names for $FeCl_2$, iron(II) chloride, and $CrCl_3$, chromium(III) chloride.

3.123   The formula HF(g) represents a molecular compound, so it is named hydrogen fluoride. The formula HF(aq) represents HF dissolved in water where it is a binary acid, and is named hydrofluoric acid.

3.125   A high melting point is found for ionic compounds. Ionic compounds are generally formed from a metal and a nonmetal, so only $CrCl_3$ qualifies as an ionic compound and it has the highest melting point.

Chapter 4 – Chemical Composition

4.1   (a) mole; (b) Avogadro's number; (c) empirical formula; (d) solute; (e) molarity; (f) concentrated solution

4.3   The percent composition of calcium in calcium carbonate is set up as:

$$\% \text{ Ca} = \frac{\text{grams Ca}}{\text{grams CaCO}_3} \times 100\% = \frac{9.88 \text{ g Ca}}{24.7 \text{ g CaCO}_3} \times 100\% = 40.0\% \text{ Ca}$$

4.5   Mass percent is defined as the mass of the part divided by the mass of the whole times 100%. It does not matter what mass units are used, just so long as they are the same for the part and the whole. That allows the units to cancel properly.

$$\text{Mass \%} = \frac{\text{Mass of Part}}{\text{Mass of Whole}} \times 100\%$$

$$\text{Mass \%C} = \frac{30.0 \text{ μg}}{50.0 \text{ μg}} \times 100\% = 60.0\% \text{ C}$$

4.7   To calculate the actual mass of a substance in a sample, we rearrange the percent equation in the following way:

$$\text{Mass \%} = \frac{\text{Mass of Part}}{\text{Mass of Whole}} \times 100\% \text{ becomes Mass of Part} = \frac{\text{Mass \%} \times \text{Mass of Whole}}{100\%}$$

$$\text{Mass of lithium} = \frac{1.2 \text{ g} \times 18.8\%}{100\%} = 0.226 \text{ g lithium}$$

4.9   Chemical formulas must have whole number subscripts. (a) H_2S; (b) A ratio of 1.5 oxygen atoms to 1 nitrogen atom means that the number of each atom needs to be doubled to give whole-number coefficients for both atoms. Doubling the values works because it converts 1.5 to 3 (a whole number). This gives a ratio of 3 oxygen atoms to 2 nitrogen atoms (note that both quantities are doubled). The chemical formula is N_2O_3.
(c) A ratio of one-half calcium ion to one chloride ion means that the number of each ion needs to be doubled to give whole-number coefficients for both ions. This gives a ratio of 1 calcium ion to 2 chloride ions. The chemical formula for calcium chloride is $CaCl_2$.

4.11   The formulas are derived by counting the atoms of each element in the figures. (a) H_2SO_4; (b) SCl_4; (c) C_2H_4

4.13   Each molecule pictured contains 1 carbon atom and 2 oxygen atoms. The formula is CO_2.

4.15   (a) The formula unit for an ionic compound is described by its formula, which shows the ratio of ions in lowest possible whole numbers. Since the formula of sodium chloride is NaCl, one formula unit is NaCl. From the image, you can see a one-to-one correspondence of sodium ions and chloride ions in the nearest atoms (you should be able to count 14 chloride and 14 sodium ions). This also gives the formula NaCl (1:1 ratio of Na^+ and Cl^-).
(b) For molecular compounds consisting of discrete molecules, the formula unit is the same as its molecular formula. In this case there are two Cl atoms in each molecule, so the formula unit is Cl_2.
(c) For methane, another molecular compound, each molecule contains one carbon atom and four hydrogens so the molecular formula and the formula unit is CH_4.
(d) The image of silicon dioxide is more difficult to analyze because it is not composed of discrete molecules. As with ionic compounds, the formula unit is a ratio with lowest possible whole numbers. On average each silicon atom shares four oxygen atoms with its neighbors, so the formula unit is one silicon and two oxygens (½ × 4 oxygens); SiO_2. However, as with ionic compounds and molecules the formula unit can be derived from its name, silicon dioxide.

4.17 One mole of any object contains 6.022×10^{23} of those objects. One mole of NH_3 contains 6.022×10^{23} NH_3 molecules. One-half mole of NH_3 contains half of that:

$$\text{Molecules } NH_3 = 0.5 \text{ mol } NH_3 \times \frac{6.022 \times 10^{23} \; NH_3 \text{ molecules}}{1 \text{ mole } NH_3} = 3.0 \times 10^{23} \; NH_3 \text{ molecules}$$

Since there is one nitrogen atom for each NH_3, we can expect that there are 3.0×10^{23} N atoms as well. We can express that as follows:

$$N \text{ Atoms } = 3.0 \times 10^{23} \; NH_3 \times \frac{1 \text{ N atom}}{1 \; NH_3} = 3.0 \times 10^{23} \text{ N atoms}$$

In a similar fashion we can see that there are three hydrogen atoms in each NH_3.

$$H \text{ atoms} = 3.0 \times 10^{23} \; NH_3 \times \frac{3 \text{ H atoms}}{1 \; NH_3} = 9.0 \times 10^{23} \text{ H atoms}$$

4.19 One mole of Cu_2S contains 6.022×10^{23} Cu_2S formula units. One-half mole of Cu_2S contains half that value:

$$\text{Formula units of } Cu_2S = 0.50 \text{ mol } Cu_2S \times \frac{6.022 \times 10^{23} \; Cu_2S \text{ formula units}}{1 \text{ mole } Cu_2S} = 3.0 \times 10^{23} \; Cu_2S \text{ formula units}$$

4.21 To calculate the number of S atoms in 0.2 mol of SO_2, calculate the number of molecules in 0.2 mol and then multiply by the number of S atoms in each SO_2 molecule (1 S per molecule) as mapped below:

$$\text{Map: Moles} \xrightarrow{N_A} \text{Number of molecules} \xrightarrow{\text{formula ratio}} \text{Number of atoms}$$

$$\text{Atoms S} = 0.2 \text{ mol} \times \frac{6.022 \times 10^{23} \text{ molecule}}{\text{mol}} \times \frac{1 \text{ S atom}}{1 \text{ molecule}} = 1 \times 10^{23} \text{ S atoms}$$

4.23 To calculate the number of calcium ions in 1 mol of $CaCl_2$, calculate the number of formula units in 1 mol and then multiply by the number of calcium ions per formula unit (1 Ca^{2+} per $CaCl_2$) as mapped below:

$$\text{Map: Moles} \xrightarrow{N_A} \text{Number of formula units} \xrightarrow{\text{formula ratio}} \text{Number of ions}$$

$$Ca^{2+} \text{ ions} = 1 \text{ mol} \times \frac{6.022 \times 10^{23} \text{ formula unit}}{\text{mol}} \times \frac{1 \; Ca^{2+}}{1 \text{ formula unit}} = 6 \times 10^{23} \; Ca^{2+} \text{ ions}$$

4.25   The molar mass is the sum of the masses of the component elements in the chemical formula. The masses for one mole of each element are found on the periodic table.

NaCl

Mass of 1 mol of Na	= 1 mol × 22.99 g/mol =	22.99 g
Mass of 1 mol of Cl	= 1 mol × 35.45 g/mol =	35.45 g
Mass of 1 mol of NaCl =		58.44 g

The molar mass is 58.44 g/mol

Cl_2

Mass of 2 mol of Cl   = 2 mol × 35.45 g/mol = 70.90 g
The molar mass is 70.90 g/mol

CH_4

Mass of 1 mol of C	= 1 mol × 12.01 g/mol =	12.01 g
Mass of 4 mol of H	= 4 mol × 1.008 g/mol =	4.03 g
Mass of 1 mol of CH_4	=	16.04 g

The molar mass is 16.04 g/mol

SiO_2

Mass of 1 mol of Si=	1 mol × 28.09 g/mol =	28.09 g
Mass of 2 mol of O=	2 mol × 16.00 g/mol =	32.00 g
Mass of 1 mol of SiO_2	=	60.09 g

The molar mass is 60.09 g/mol

4.27   The molar mass is the sum of the masses of the component elements in the chemical formula. The masses for one mole of each element are found on the periodic table.

(a) Hg_2Cl_2

Mass of 2 mol of Hg	= 2 mol × 200.6 g/mol	= 401.2 g
Mass of 2 mol of Cl	= 2 mol × 35.45 g/mol	= 70.90 g
Mass of 1 mol of Hg_2Cl_2		472.1 g

Molar mass = 472.1 g/mol

(b) $CaSO_4 \cdot 2H_2O$ Note that the mass of 1 water molecule was calculated separately as 18.016 g/mol.

Mass of 1 mol of Ca	= 1 mol × 40.08 g/mol	= 40.08 g
Mass of 1 mol of S	= 1 mol × 32.06 g/mol	= 32.06 g
Mass of 4 mol of O	= 4 mol × 16.00 g/mol	= 64.00 g
Mass of 2 mol of H_2O	= 2 mol × 18.016 g/mol	= 36.032 g
Mass of 1 mol of $CaSO_4 \cdot 2H_2O$		= 172.17 g

Molar mass = 172.17 g/mol

(c) Cl_2O_5

Mass of 2 mol of Cl	= 2 mol × 35.45 g/mol	= 70.90 g
Mass of 5 mol of O	= 5 mol × 16.00 g/mol	= 80.00 g
Mass of 1 mol of Cl_2O_5		150.90 g

Molar mass = 150.90 g/mol

(d) $NaHSO_4$

Mass of 1 mol of Na	=	1 mol × 22.99 g/mol	=	22.99 g
Mass of 1 mol of H	=	1 mol × 1.008 g/mol	=	1.008 g
Mass of 1 mol of S	=	1 mol × 32.06 g/mol	=	32.06 g
Mass of 4 mol of O	=	4 mol × 16.00 g/mol	=	64.00 g
Mass of 1 mol of $NaHSO_4$			=	120.06 g

Molar mass = 120.06 g/mol

4.29   The molar mass is the sum of the masses of the component elements in the chemical formula. The masses for one mole of each element are found on the periodic table.

(a) I_2

Mass of 2 mol of I	=	2 mol ×126.9 g/mol	=	253.8 g

Molar mass = 253.8 g/mol

(b) $CrCl_3$

Mass of 1 mol of Cr	=	1 mol × 52.00 g/mol	=	52.00 g
Mass of 3 mol of Cl	=	3 mol × 35.45 g/mol	=	106.4 g
Mass of 1 mol of $CrCl_3$			=	158.4 g

Molar mass = 158.4 g/mol

(c) C_4H_8

Mass of 4 mol of C	=	4 mol × 12.01 g/mol	=	48.04 g
Mass of 8 mol of H	=	8 mol × 1.008 g/mol	=	8.064 g
Mass of 1 mol of C_4H_8				56.10 g

Molar mass = 56.10 g/mol

4.31   To measure out a useful number of atoms by counting would not be possible because atoms are too small for us to see and manipulate. If you could count 1000 water molecules per second for slightly over 20 billion years, you would have counted the molecules in one small drop of water (about 0.020 mL). Instead, because we know the mass of one mole of any substance, we can carefully weigh out an amount of any substance and determine the number of atoms our sample contains.

4.33   The molar mass and atomic mass have the same numerical value but different units. If the molar mass of LiCl is 42.39 g/mol, the mass of one formula unit of LiCl is 42.39 amu.

4.35   The units for molar mass are grams per mole:

$$Molar\ mass = \frac{g}{mol}$$

This means that if you have the numbers of grams and moles, you can calculate the molar mass by dividing the number of grams by the number of moles of substance this mass represents. In this case, you are given grams (12.0) and molecules of substance.

Map: Number of molecules $\xrightarrow{\ N_A\ }$ Moles

$$Molecules = 2.01 \times 10^{23}\ \text{molecules} \times \frac{mole}{6.022 \times 10^{23}\ \text{molecules}} = 0.334\ mol$$

$$Molar\ mass = \frac{12.0\ g}{0.334\ mol} = 35.9\ g/mol$$

4.37   The conversion from mass to number of moles uses the following problem solving map (MM is the molar mass):

Map: Mass in grams $\xrightarrow{MM}$ Moles

(a) The molar mass of $KHCO_3$ is 100.12 g/mol (calculated by adding the masses of the component elements of the chemical formula).

$$\text{Moles } KHCO_3 = 10.0 \text{ g } \cancel{KHCO_3} \times \frac{1 \text{ mol } KHCO_3}{100.12 \text{ g } \cancel{KHCO_3}} = 0.0999 \text{ mol } KHCO_3$$

(b) The molar mass of H_2S is 34.08 g/mol.

$$\text{Moles } H_2S = 10.0 \text{ g } \cancel{H_2S} \times \frac{1 \text{ mol } H_2S}{34.08 \text{ g } \cancel{H_2S}} = 0.293 \text{ mol } H_2S$$

(c) The molar mass of Se is 78.96.

$$\text{Moles } Se = 10.0 \text{ g } \cancel{Se} \times \frac{1 \text{ mol } Se}{78.96 \text{ g } \cancel{Se}} = 0.127 \text{ mol } Se$$

(d) The molar mass of $MgSO_4$ is 120.37 g/mol.

$$\text{Moles } MgSO_4 = 10.0 \text{ g } \cancel{MgSO_4} \times \frac{1 \text{ mol } MgSO_4}{120.37 \text{ g } \cancel{MgSO_4}} = 0.0831 \text{ mol } MgSO_4$$

4.39   The conversion from mass to number of moles uses the following problem solving map (MM is the molar mass). In these problems, if the mass is not given in grams, it must first be converted using an appropriate problem solving map:

Map: Mass in grams $\xrightarrow{MM}$ Moles

(a) The molar mass of NaCl is 58.44 g/mol.

$$\text{Moles } NaCl = 32.5 \text{ g } \cancel{NaCl} \times \frac{1 \text{ mol } NaCl}{58.44 \text{ g } \cancel{NaCl}} = 0.556 \text{ mol } NaCl$$

(b) The molar mass of $C_9H_8O_4$ is 180.15 g/mol. The mass is converted to grams using 1 mg = 10^{-3} g.

$$\text{Moles } C_9H_8O_4 = 250.0 \text{ mg } \cancel{C_9H_8O_4} \times \frac{10^{-3} \text{ g } \cancel{C_9H_8O_4}}{1 \text{ mg } \cancel{C_9H_8O_4}} \times \frac{1 \text{ mol } C_9H_8O_4}{180.15 \text{ g } \cancel{C_9H_8O_4}} = 1.388 \times 10^{-3} \text{ mol } C_9H_8O_4$$

(c) The molar mass of $CaCO_3$ is 100.09 g/mol. The mass is converted to grams using 1 kg = 10^3 g.

$$\text{Moles } CaCO_3 = 73.4 \text{ kg } \cancel{CaCO_3} \times \frac{10^3 \text{ g } \cancel{CaCO_3}}{1 \text{ kg } \cancel{CaCO_3}} \times \frac{1 \text{ mol } CaCO_3}{100.09 \text{ g } \cancel{CaCO_3}} = 733 \text{ mol } CaCO_3$$

(d) The molar mass of CuS is 95.61 g/mol. The mass is converted to grams using $1 \ \mu g = 10^{-6}$ g.

$$\text{Moles CuS} = 5.47 \ \mu g \ \cancel{CuS} \times \frac{10^{-6} \ g \ \cancel{CuS}}{1 \ \mu g \ \cancel{CuS}} \times \frac{1 \ \text{mol CuS}}{95.61 \ g \ \cancel{CuS}} = 5.72 \times 10^{-8} \ \text{mol CuS}$$

4.41   If the molar mass of a substance is relatively small, it will take more moles of that substance to equal 1 gram than it would take of a substance with a larger molar mass. This means that sodium, which has the smallest molar mass of the substances given, contains the most moles of atoms in a 1.0-g sample.

4.43   To convert moles to grams we use the following problem solving map (MM = molar mass)

Map: Moles $\xrightarrow{\ MM\ }$ Mass in grams

(a) The molar mass of $Ba(OH)_2$ is 171.32 g/mol.

$$\text{Mass } Ba(OH)_2 = 2.50 \ \text{mol} \ \cancel{BaSO_4} \times \frac{171.32 \ g \ BaSO_4}{1 \ \text{mol} \ \cancel{BaSO_4}} = 428 \ g \ Ba(OH)_2$$

(b) The molar mass of Cl_2 is 70.90 g/mol.

$$\text{Mass } Cl_2 = 2.50 \ \text{mol} \ \cancel{Cl_2} \times \frac{70.90 \ g \ Cl_2}{1 \ \text{mol} \ \cancel{Cl_2}} = 177 \ g \ Cl_2$$

(c) The molar mass of K_2SO_4 is 174.26 g/mol.

$$\text{Mass } K_2SO_4 = 2.50 \ \text{mol} \ \cancel{K_2SO_4} \times \frac{174.26 \ g \ K_2SO_4}{1 \ \text{mol} \ \cancel{K_2SO_4}} = 436 \ g \ K_2SO_4$$

(c) The molar mass of PF_3 is 87.97 g/mol.

$$\text{Mass } K_2SO_4 = 2.50 \ \text{mol} \ \cancel{PF_3} \times \frac{87.97 \ g \ PF_3}{1 \ \text{mol} \ \cancel{PF_3}} = 220. \ g \ K_2SO_4$$

4.45   The conversion of 2.7 moles of $Zn(CH_3CO_2)_2$ to an equivalent number of grams requires the molar mass (MM). We use the following problem solving map:

Map: Moles $\xrightarrow{\ MM\ }$ Mass in grams

The molar mass of $Zn(CH_3CO_2)_2$ is 183.47 g.

$$\text{Grams } Zn(CH_3CO_2)_2 = 2.7 \ \text{mol} \ \cancel{Zn(CH_3CO_2)_2} \times \frac{183.47 \ g \ Zn(CH_3CO_2)_2}{\text{mol} \ \cancel{Zn(CH_3CO_2)_2}} = 5.0 \times 10^2 \ g \ Zn(CH_3CO_2)_2$$

4.47   (a) The conversion of 30.0 g NH_3 to an equivalent number of moles requires the molar mass (MM). We use the following problem solving map:

Map: Mass in grams $\xrightarrow{\ MM\ }$ Moles

The molar mass of NH_3 is 17.03 g/mol.

$$\text{Moles } NH_3 = 30.0 \ g \ \cancel{NH_3} \times \frac{\text{mol} \ NH_3}{17.03 \ g \ \cancel{NH_3}} = 1.76 \ \text{mol} \ NH_3$$

(b) To calculate the number of NH_3 molecules in the sample, we use Avogadro's number ($N_A = 6.022 \times 10^{23}$) and the following problem solving map:

Map: Moles $\xrightarrow{N_A}$ Number of molecules

Molecules $NH_3 = 1.7612 \ \text{mol } NH_3 \times \dfrac{6.022 \times 10^{23} \ \text{molecules}}{\text{mol } NH_3} = 1.06 \times 10^{24}$ molecules NH_3

(c) To calculate the number of nitrogen atoms in the sample, we look at the ratio of the elements in the chemical formula. The ratio for nitrogen in ammonia is 1 atom N/1 molecule NH_3. Therefore, the number of N atoms is the same as the number of NH_3 molecules, 1.06×10^{24} N atoms.

(d) From the chemical formula, we know that there are 3 moles of H for each mole of NH_3. This is the formula ratio needed to calculate the moles of H in the sample of NH_3:

Map: Moles $NH_3 \xrightarrow{\text{formula ratio}}$ Moles H

Moles H $= 1.7612 \ \text{mol } NH_3 \times \dfrac{3 \ \text{mol H}}{1 \ \text{mol } NH_3} = 5.28$ mol H

4.49   The number of atoms per mole of substance depends on the number of atoms in the chemical formula. H_2SO_4 has the most atoms per mole because it has more atoms per molecule. Na has the least atoms per mole.

4.51   To calculate the number of molecules in 0.050 g of water, we need to find the appropriate conversion factors. This can be done using the following conversion map:

Map: Mass in grams $\xrightarrow{MM}$ Moles $\xrightarrow{N_A}$ Number of molecules

Note that the units of molar mass (g/mol) serve as a connection between mass and moles, and the units of Avogadro's number (molecules/mole) serve as a connection between molecules and moles. The molar mass of water is 18.02 g/mol.

Molecules $H_2O \quad = 0.050 \ \text{g } H_2O \times \dfrac{\text{mol } H_2O}{18.02 \ \text{g } H_2O} \times \dfrac{6.022 \times 10^{23} \ \text{molecules } H_2O}{\text{mol } H_2O}$

$= 1.7 \times 10^{21}$ molecules H_2O

4.53   To calculate the number of formula units we use the following conversion map:

Map: Mass in grams $\xrightarrow{MM}$ Moles $\xrightarrow{N_A}$ Number of formula units

(a) $Br_2 = 159.80$ g/mol

Formula units $Br_2 \quad = 250.0 \ \text{g } Br_2 \times \dfrac{\text{mol } Br_2}{159.80 \ \text{g } Br_2} \times \dfrac{6.022 \times 10^{23} \ \text{formula units } Br_2}{\text{mol } Br_2}$

$= 9.421 \times 10^{23}$ formula units Br_2

(b) $MgCl_2$ = 95.21 g/mol

Formula units $MgCl_2$ = 250.0 g $MgCl_2$ $\times \dfrac{mol\ MgCl_2}{159.80\ g\ MgCl_2} \times \dfrac{6.022 \times 10^{23}\ formula\ units\ MgCl_2}{mol\ MgCl_2}$

$= 1.581 \times 10^{24}$ formula units $MgCl_2$

(c) H_2O = 18.02 g/mol

Formula units H_2O = 250.0 g H_2O $\times \dfrac{mol\ H_2O}{18.02\ g\ H_2O} \times \dfrac{6.022 \times 10^{23}\ formula\ units\ H_2O}{mol\ H_2O}$

$= 8.355 \times 10^{24}$ formula units H_2O

(d) Fe = 55.85 g/mol

Formula units Fe = 250.0 g Fe $\times \dfrac{mol\ Fe}{58.85\ g\ Fe} \times \dfrac{6.022 \times 10^{23}\ formula\ units\ Fe}{mol\ Fe}$

$= 2.696 \times 10^{24}$ formula units Fe

4.55   To calculate the number of atoms or ions of each element in 140.0 g of each substance, the best strategy is to first calculate the number of formula units. Once you have determined this, use the formula ratios to calculate the number of atoms or ions. In part (b), for example, you will calculate the formula units of $Ca(NO_3)_2$ and then use formula ratios to calculate the number of Ca^{2+} and NO_3^- ions. Two problem solving maps are applied:
    Number of formula units:

Mass in grams $\xrightarrow{\ MM\ }$ Moles $\xrightarrow{\ N_A\ }$ Number of formula units

Number of ions or atoms:

Number of formula units $\xrightarrow{\ formula\ ratio\ }$ ions or atoms

(a) Since there we are only looking for atoms of one element, we can combine the two problem solving maps. The appropriate conversions are: H_2 = 2.016 g/mol; 2 H atom = 1 H_2 molecule

Atoms = 140.0 g H_2 $\times \dfrac{mol\ H_2}{2.016\ g\ H_2} \times \dfrac{6.022 \times 10^{23}\ H_2}{mol\ H_2} \times \dfrac{2\ H\ atom}{H_2} = 8.364 \times 10^{25}$ H atoms

(b) $Ca(NO_3)_2$ = 164.10 g/mol

Formula units $Ca(NO_3)_2$ =

140.0 g $Ca(NO_3)_2$ $\times \dfrac{mol\ Ca(NO_3)_2}{164.10\ Ca(NO_3)_2} \times \dfrac{6.022 \times 10^{23}\ Ca(NO_3)_2}{mol\ Ca(NO_3)_2} = 5.138 \times 10^{23}\ Ca(NO_3)_2$

Ca^{2+} ions: 1 Ca^{2+} ion = 1 $Ca(NO_3)_2$

Ca^{2+} ions = 5.138×10^{23} $Ca(NO_3)_2$ $\times \dfrac{1\ Ca^{2+}}{1\ Ca(NO_3)_2} = 5.138 \times 10^{23}\ Ca^{2+}$ ions

NO_3^- ions: 2 NO_3^- ions = 1 $Ca(NO_3)_2$

NO_3^- ions = 5.138×10^{23} $Ca(NO_3)_2$ $\times \dfrac{2\ NO_3^-}{1\ Ca(NO_3)_2} = 1.028 \times 10^{24}\ NO_3^-$ ions

(c) $N_2O_2 = 60.02$ g/mol

Formula units N_2O_2 = 140.0 g N_2O_2 $\times \dfrac{\text{mol } N_2O_2}{60.02 \text{ g } N_2O_2} \times \dfrac{6.022 \times 10^{23} \text{ } N_2O_2}{\text{mol } N_2O_2}$ = 1.405×10^{24} N_2O_2

N atoms = 1.405×10^{24} N_2O_2 $\times \dfrac{2 \text{ N}}{1 \text{ } N_2O_2}$ = 2.809×10^{24} N atoms

O atoms = 1.405×10^{24} N_2O_2 $\times \dfrac{2 \text{ O}}{1 \text{ } N_2O_2}$ = 2.809×10^{24} O atoms

(d) $K_2SO_4 = 174.26$ g/mol

Formula units = 140.0 g K_2SO_4 $\times \dfrac{\text{mol } K_2SO_4}{174.26 \text{ g } K_2SO_4} \times \dfrac{6.022 \times 10^{23} \text{ } K_2SO_4}{\text{mol } K_2SO_4}$ = 4.838×10^{23} K_2SO_4

K^+ ions: 2 K^+ ions = 1 K_2SO_4

K^+ ions = 4.838×10^{23} K_2SO_4 $\times \dfrac{2 \text{ } K^+}{1 \text{ } K_2SO_4}$ = 9.676×10^{23} K^+ ions

SO_4^{2-} ions: 1 SO_4^{2-} ion = 1 K_2SO_4

SO_4^{2-} ions = 4.838×10^{23} K_2SO_4 $\times \dfrac{1 \text{ } SO_4^{2-}}{1 \text{ } K_2SO_4}$ = 4.838×10^{23} SO_4^{2-} ions

4.57   To calculate the mass of 6.4×10^{22} molecules of SO_2 we use the following conversion map:

Map: Number of molecules $\xrightarrow{\quad N_A \quad}$ Moles $\xrightarrow{\quad MM \quad}$ Mass in grams

The molar mass of SO_2 is 64.06 g/mol.

Mass SO_2 = 6.4×10^{22} molecules SO_2 $\times \dfrac{1 \text{ mol } SO_2}{6.022 \times 10^{23} \text{ molecules } SO_2} \times \dfrac{64.06 \text{ g } SO_2}{1 \text{ mol } SO_2}$ = 6.8 g SO_2

4.59   The substance that has the most nitrogen atoms per 25.0-g sample will also have the highest number of moles of nitrogen atoms. We use the following conversion map:

Map: Mass in grams $\xrightarrow{\quad MM \quad}$ Moles $\xrightarrow{\quad \text{formula ratio} \quad}$ Moles of N atoms

The molar mass of NH_3 is 17.03 g/mol and there is 1 mol N atoms per mol NH_3.

Moles N atoms = 25.0 g NH_3 $\times \dfrac{\text{mol } NH_3}{17.03 \text{ g } NH_3} \times \dfrac{1 \text{ mol N atoms}}{\text{mol } NH_3}$ = 1.47 mol N atoms

The molar mass of NH_4Cl is 53.49 g/mol and there is 1 mol N atoms per mole NH_4Cl.

Moles N atoms = 25.0 g NH_4Cl $\times \dfrac{\text{mol } NH_4Cl}{53.49 \text{ g } NH_4Cl} \times \dfrac{1 \text{ mol N atoms}}{\text{mol } NH_4Cl}$ = 0.467 mol N atoms

The molar mass of NO_2 is 46.01 g/mol and there is 1 mol N atoms per mole NO_2.

$$\text{Moles N atoms} = 25.0 \text{ g } NO_2 \times \frac{\text{mol } NO_2}{46.01 \text{ g } NO_2} \times \frac{1 \text{ mol N atoms}}{\text{mol } NO_2} = 0.543 \text{ mol N atoms}$$

The molar mass of N_2O_3 is 76.02 g/mol and there are 2 mol N atoms per mole N_2O_3.

$$\text{Moles N atoms} = 25.0 \text{ g } N_2O_3 \times \frac{\text{mol } N_2O_3}{76.02 \text{ g } N_2O_3} \times \frac{2 \text{ mol N atoms}}{\text{mol } N_2O_3} = 0.658 \text{ mol N atoms}$$

Because NH_3 has the highest number of moles of nitrogen atoms, we know that it has the most nitrogen atoms per 25.0-g sample.

4.61   No. Molecules of the same substance have the same percent compositions.

4.63   The empirical formula shows the relative amounts of each atom in a compound, expressed as small whole numbers. The molecular formula shows the exact numbers of each atom present in one molecule of that compound.

4.65   The empirical and molecular formulas are different if the subscripts in the molecular formula are all divisible by a common factor other than 1. For example, in the formula H_2O_2, both subscripts are divisible by 2, so the empirical formula (HO) is different than the molecular formula. Of the substances listed, those with different empirical and molecular formulas are:

Molecular	Empirical
H_2O_2	HO
N_2O_4	NO_2

4.67   The empirical and molecular formulas are the same if the subscripts in the molecular formula are not divisible by a common factor other than 1. Often, it is desirable to simplify the molecular formula in order to determine if the molecular and empirical formulas are different. For part (d) $HO_2CC_4H_8CO_2H$ can be simplified to $C_6H_{10}O_4$.

Molecular Formula	Common Factor	Empirical Formula
(a) P_4O_{10}	2	P_2O_5
(b) Cl_2O_5	none	same as molecular
(c) $PbCl_4$	none	same as molecular
(d) $C_6H_{10}O_4$	2	$C_3H_5O_2$

4.69   The empirical and molecular formulas are the same if the subscripts in the molecular formula are not divisible by a common factor other than 1.

Molecular Formula	Common Factor	Empirical Formula
(a) $C_6H_4Cl_2$	2	C_3H_2Cl
(b) C_6H_5Cl	none	same as molecular
(c) N_2O_5	none	same as molecular

4.71 NO_2 and N_2O_4 have the identical empirical formulas (NO_2). The empirical formulas of the other compounds are different than NO_2.

Molecular Formula	Common Factor	Empirical Formula
N_2O	none	N_2O
NO	none	NO
NO_2	none	**NO_2**
N_2O_3	none	N_2O_3
N_2O_4	2	**NO_2**
N_2O_5	none	N_2O_5

4.73   The empirical formula shows the whole-number ratio of moles of each element in a compound. To determine the empirical formula of a compound, we calculate the number of moles of each element in the sample, and then determine the mole ratios. When presented with percent composition data, it is easiest to assume that we have exactly 100 grams of the substance.

(a)   The percent composition of our sample is 72.36% Fe and 27.64%O. If we have a 100-gram sample, it will contain 72.36 g Fe and 27.64 g O. We can convert these masses to the equivalent number of moles as follows:

$$\text{Moles Fe} = 72.36 \ \cancel{g \ Fe} \times \frac{1 \text{ mol Fe}}{55.85 \ \cancel{g \ Fe}} = 1.296 \text{ mol Fe}$$

$$\text{Moles O} = 27.64 \ \cancel{g \ O} \times \frac{1 \text{ mol O}}{16.00 \ \cancel{g \ O}} = 1.728 \text{ mol O}$$

Next divide the number of moles of each element by the smallest of the molar amounts (i.e. 1.296 mol Fe).

$$\frac{\text{moles Fe}}{\text{moles Fe}} = \frac{1.296 \text{ mol Fe}}{1.296 \text{ mol Fe}} = \frac{1 \text{ mol Fe}}{1 \text{ mol Fe}}$$

$$\frac{\text{moles O}}{\text{moles Fe}} = \frac{1.728 \text{ mol O}}{1.296 \text{ mol Fe}} = \frac{1.333 \text{ mol O}}{1 \text{ mol Fe}}$$

Notice that one of the ratios is not a whole number. Since the subscripts of chemical formulas must be whole numbers (i.e. you can't have fractions of atoms), we multiply each ratio by 3 to convert the ratio into a whole number. We find that the empirical formula has 4 moles of oxygen for every 3 moles of iron. The empirical formula is Fe_3O_4.

(b)   The percent composition of our sample is 58.53% C, 4.09% H, 11.38% N, and 25.99% O. If we have a 100 gram sample, it will contain 58.53 g C, 4.09 g H, 11.38 g N, and 25.99 g O. We can convert these masses to the equivalent number of moles as follows:

$$\text{Moles C} = 58.53 \ \cancel{g \ C} \times \frac{1 \text{ mol C}}{12.01 \ \cancel{g \ C}} = 4.873 \text{ mol C}$$

$$\text{Moles H} = 4.09 \ \cancel{g \ H} \times \frac{1 \text{ mol H}}{1.008 \ \cancel{g \ H}} = 4.06 \text{ mol H}$$

$$\text{Moles N} = 11.38 \text{ g N} \times \frac{1 \text{ mol N}}{14.01 \text{ g N}} = 0.8123 \text{ mol N}$$

$$\text{Moles O} = 25.99 \text{ g O} \times \frac{1 \text{ mol O}}{16.00 \text{ g O}} = 1.624 \text{ mol O}$$

Next divide the number of moles of each element by the smallest of the molar amounts (i.e. 0.8123 mol N).

$$\frac{\text{moles C}}{\text{moles N}} = \frac{4.873 \text{ mol C}}{0.8123 \text{ mol N}} = \frac{6.000 \text{ mol C}}{1 \text{ mol N}}$$

$$\frac{\text{moles H}}{\text{moles N}} = \frac{4.06 \text{ mol H}}{0.8123 \text{ mol N}} = \frac{5.00 \text{ mol H}}{1 \text{ mol N}}$$

$$\frac{\text{moles N}}{\text{moles N}} = \frac{0.8123 \text{ mol N}}{0.8123 \text{ mol N}} = \frac{1 \text{ mol N}}{1 \text{ mol N}}$$

$$\frac{\text{moles O}}{\text{moles N}} = \frac{1.624 \text{ mol O}}{0.8123 \text{ mol N}} = \frac{2.000 \text{ mol O}}{1 \text{ mol N}}$$

The empirical formula has 6 moles of carbon, 5 moles of hydrogen, and 2 moles of oxygen for every 1 mole of nitrogen. The empirical formula is $C_6H_5NO_2$.

4.75   The percent composition of our sample is 73.19% C, 19.49% O, and 7.37% H. If we have a 100-gram sample, it will contain 73.19 g C, 19.49 g O, and 7.37 g H. To determine the empirical formula, we convert the masses of each element to the equivalent number of moles, and then determine the relative number of moles of each element in the substance.

$$\text{Moles C} = 73.19 \text{ g C} \times \frac{1 \text{ mol C}}{12.01 \text{ g C}} = 6.094 \text{ mol C}$$

$$\text{Moles O} = 19.49 \text{ g O} \times \frac{1 \text{ mol O}}{16.00 \text{ g O}} = 1.218 \text{ mol O}$$

$$\text{Moles H} = 7.37 \text{ g H} \times \frac{1 \text{ mol H}}{1.008 \text{ g H}} = 7.31 \text{ mol H}$$

Next divide the number of moles of each element by the smallest of the molar amounts (i.e. 1.218 mol O).

$$\frac{\text{moles C}}{\text{moles O}} = \frac{6.094 \text{ mol C}}{1.218 \text{ mol O}} = \frac{5.003 \text{ mol C}}{1 \text{ mol O}}$$

$$\frac{\text{moles O}}{\text{moles O}} = \frac{1.218 \text{ mol O}}{1.218 \text{ mol O}} = \frac{1 \text{ mol O}}{1 \text{ mol O}}$$

$$\frac{\text{moles H}}{\text{moles O}} = \frac{7.31\text{ mol H}}{1.218\text{ mol O}} = \frac{6.00\text{ mol H}}{1\text{ mol O}}$$

Rounding the mole ratios to the nearest whole numbers, we find that chemical formula has 5 moles of carbon, 6 moles of hydrogen, and 1 mole of oxygen. The empirical formula for eugenol is C_5H_6O.

4.77   The percent composition of our sample is 37.01% C, 2.22% H, 18.50% N, and 42.27% O. Assuming a 100 gram sample, it will contain 37.01 g C, 2.22 g H, 18.50 g N, and 42.27 g O. To determine the empirical formula, we convert the masses of each element to the equivalent number of moles, and then determine the relative number of moles of each element in the substance.

$$\text{Moles C} = 37.01\text{ g C} \times \frac{1\text{ mol C}}{12.01\text{ g C}} = 3.082\text{ mol C}$$

$$\text{Moles H} = 2.22\text{ g H} \times \frac{1\text{ mol H}}{1.008\text{ g H}} = 2.202\text{ mol H}$$

$$\text{Moles N} = 18.50\text{ g N} \times \frac{1\text{ mol N}}{14.01\text{ g N}} = 1.320\text{ mol N}$$

$$\text{Moles O} = 42.27\text{ g O} \times \frac{1\text{ mol O}}{16.00\text{ g O}} = 2.642\text{ mol O}$$

Next divide the number of moles of each element by the smallest of the molar amounts (i.e. 1.320 mol N).

$$\frac{\text{moles C}}{\text{moles N}} = \frac{3.082\text{ mol C}}{1.320\text{ mol N}} = \frac{2.335\text{ mol C}}{1\text{ mol N}}$$

$$\frac{\text{moles H}}{\text{moles N}} = \frac{2.202\text{ mol H}}{1.320\text{ mol N}} = \frac{1.668\text{ mol H}}{1\text{ mol N}}$$

$$\frac{\text{moles N}}{\text{moles N}} = \frac{1.320\text{ mol N}}{1.320\text{ mol N}} = \frac{1\text{ mol N}}{1\text{ mol N}}$$

$$\frac{\text{moles O}}{\text{moles N}} = \frac{2.642\text{ mol O}}{1.320\text{ mol N}} = \frac{2.002\text{ mol O}}{1\text{ mol N}}$$

Notice that two of the ratios are not whole numbers. The fractional portions (0.333 and 0.668) represent one-third (1/3) and two-thirds (2/3). We can convert these values to whole numbers by multiplying each ratio by 3. The empirical formula has 7 moles of carbon, 5 moles of hydrogen, and 6 moles of oxygen for every 3 moles of nitrogen. The empirical formula is $C_7H_5N_3O_6$.

4.79   The percent composition by mass and the molar mass are needed to determine the molecular formula.

4.81   The ratio of the molar mass to the mass calculated from the empirical formula gives the information necessary to determine the molecular formula. The mass of the empirical formula, CH_2O is:

Mass of 1 mol of C	= 1 mol × 12.01 g/mol	=	12.01 g
Mass of 2 mol of H	= 2 mol × 1.008 g/mol	=	2.016 g
Mass of 1 mol of O	= 1 mol × 16.00 g/mol	=	16.00 g
Mass of 1 mol of CH_2O			30.03 g

The ratio of the molar mass to empirical formula mass is:

$$\text{Molar mass ratio} = \frac{90 \text{ g/mol}}{30.03 \text{ g/mol}} = 3$$

Multiplying each of the subscripts in CH_2O by three, we obtain the molecular formula $C_3H_6O_3$.

4.83   The strategy for determining the molecular formula from percent composition and molar mass involves two key steps. First, determine the empirical formula and the molar mass of the empirical formula from the percent composition data. Next, use the ratio of the molar mass of the compound to the empirical formula mass to determine the molecular formula.

The percent composition of our sample is 40.00% C, 6.72% H, and 53.29% O. A 100-gram sample of the compound will contain 40.00 g C, 6.72 g H, and 53.29 g O. To determine the empirical formula, we convert the masses of each element to the equivalent number of moles, and then determine the relative number of moles of each element in the compound.

$$\text{Moles C} = 40.00 \text{ gC} \times \frac{1 \text{ mol C}}{12.01 \text{ gC}} = 3.331 \text{ mol C}$$

$$\text{Moles H} = 6.72 \text{ gH} \times \frac{1 \text{ mol H}}{1.008 \text{ gH}} = 6.67 \text{ mol H}$$

$$\text{Moles O} = 53.29 \text{ gO} \times \frac{1 \text{ mol O}}{16.00 \text{ gO}} = 3.331 \text{ mol O}$$

Next divide the number of moles of each element by the smallest of the molar amounts (i.e. 3.331 mol C or 3.331 mol O).

$$\frac{\text{moles C}}{\text{moles O}} = \frac{3.331 \text{ mol C}}{3.331 \text{ mol O}} = \frac{1 \text{ mol C}}{1 \text{ mol O}}$$

$$\frac{\text{moles H}}{\text{moles O}} = \frac{6.67 \text{ mol H}}{3.331 \text{ mol O}} = \frac{2.00 \text{ mol H}}{1 \text{ mol O}}$$

$$\frac{\text{moles O}}{\text{moles O}} = \frac{3.331 \text{ mol O}}{3.331 \text{ mol O}} = \frac{1 \text{ mol O}}{1 \text{ mol O}}$$

The empirical formula has 2 moles of hydrogen for each mole of oxygen and carbon. The empirical formula is CH_2O. The mass of the empirical formula is calculated as:

Mass of 1 mol of C	= 1 mol × 12.01 g/mol =	12.01 g
Mass of 2 mol of H	= 2 mol × 1.008 g/mol =	2.016 g
Mass of 1 mol of O	= 1 mol × 16.00 g/mol =	16.00 g
Mass of 1 mol of CH_2O		30.03 g

The ratio of the molar mass to the empirical formula molar mass is:

$$\text{Molar mass ratio} = \frac{180 \text{ g/mol}}{30.03 \text{ g/mol}} = 6.0$$

Multiplying the subscripts in CH_2O by 6, we obtain the molecular formula $C_6H_{12}O_6$.

4.85   To calculate the percent composition of a compound from the chemical formula, assume a sample size of 1 mole. Calculate the mass of each element in 1 mole of the compound, divide by the molar mass of the compound and multiply by 100.

(a) Percent composition of SO_2

Mass of 1 mol of S	= 1 mol × 32.06 g/mol =	32.06 g
Mass of 2 mol of O	= 2 mol × 16.00 g/mol =	32.00 g
Mass of 1 mol of SO_2	=	64.06 g

$$\% \text{ S} = \frac{32.06 \text{ g}}{64.06 \text{ g}} \times 100\% = 50.05\% \text{ S}$$

$$\% \text{ O} = \frac{32.00 \text{ g}}{64.06 \text{ g}} \times 100\% = 49.95\% \text{ O}$$

(b) Percent composition of $CuCl_2$

Mass of 1 mol of Cu	= 1 mol × 63.55 g/mol =	63.55 g
Mass of 2 mol of Cl	= 2 mol × 35.45 g/mol =	70.90 g
Mass of 1 mol of $CuCl_2$		134.45 g

$$\% \text{ Cu} = \frac{63.55 \text{ g}}{134.45 \text{ g}} \times 100\% = 47.27\% \text{ Cu}$$

$$\% \text{ Cl} = \frac{70.90 \text{ g}}{134.45 \text{ g}} \times 100\% = 52.73\% \text{ Cl}$$

(c) Percent composition of Na_3PO_4

Mass of 3 mol of Na = 3 mol × 22.99 g/mol = 68.97 g
Mass of 1 mol of P   = 1 mol × 30.97 g/mol = 30.97 g
Mass of 4 mol of O   = 4 mol × 16.00 g/mol = 64.00 g
Mass of 1 mol of Na_3PO_4 163.94 g

$$\% \text{ Na} = \frac{68.97 \text{ g}}{163.94 \text{ g}} \times 100\% = 42.07\% \text{ Na}$$

$$\% \text{ P} = \frac{30.97 \text{ g}}{163.94 \text{ g}} \times 100\% = 18.89\% \text{ P}$$

$$\% \text{ O} = \frac{64.00 \text{ g}}{163.94 \text{ g}} \times 100\% = 39.04\% \text{ O}$$

(d) Percent composition of $Mg(NO_3)_2$

Mass of 1 mol of Mg = 1 mol × 24.31 g/mol = 24.31 g
Mass of 2 mol of N   = 2 mol × 14.01 g/mol = 28.02 g
Mass of 6 mol of O   = 6 mol × 16.00 g/mol = 96.00 g
Mass of 1 mol of $Mg(NO_3)_2$ 148.33 g

$$\% \text{ Mg} = \frac{24.31 \text{ g}}{148.33 \text{ g}} \times 100\% = 16.39 \% \text{ Mg}$$

$$\% \text{ N} = \frac{28.02 \text{ g}}{148.33 \text{ g}} \times 100\% = 18.89 \% \text{ N}$$

$$\% \text{ O} = \frac{96.00 \text{ g}}{148.33 \text{ g}} \times 100\% = 64.72 \% \text{ O}$$

4.87   A comparison of the percent composition of each mineral shows that cuprite, CuO, has the highest percentage of copper (79.89% Cu) with chalcocite, Cu_2S, an extremely close second at 79.84% Cu.

(a) Cu_2S

Mass of 2 mol of Cu = 2 mol × 63.55 g/mol = 127.1 g
Mass of 1 mol of S   = 1 mol × 32.06 g/mol = 32.06 g
Mass of 1 mol of Cu_2S = 159.2 g

$$\% \text{Cu} = \frac{127.1 \text{ g}}{159.2 \text{ g}} \times 100\% = 79.84\% \text{ Cu}$$

(b) $Cu_2(CO_3)(OH)_2$

Mass of 2 mol of Cu	= 2 mol × 63.55 g/mol	= 127.1 g
Mass of 1 mol of C	= 1 mol × 12.01 g/mol	= 12.01 g
Mass of 5 mol of O	= 5 mol × 16.00 g/mol	= 80.00 g
Mass of 2 mol of H	= 2 mol × 1.008 g/mol	= 2.016 g
Mass of 1 mol of $Cu_2(CO_3)(OH)_2$		221.1 g

$$\%Cu = \frac{127.1\,g}{221.1\,g} \times 100\% = 57.49\%\ Cu$$

(c) CuO

Mass of 1 mol of Cu	= 1 mol × 63.55 g/mol	= 63.55 g
Mass of 1 mol of O	= 1 mol × 16.00 g/mol	= 16.00 g
Mass of 1 mol of CuO		79.55 g

$$\%Cu = \frac{63.55\,g}{79.55\,g} \times 100\% = 79.89\%\ Cu$$

(d) $Cu_3(CO_3)_2(OH)_2$

Mass of 3 mol of Cu	= 3 mol × 63.55 g/mol	= 190.65 g
Mass of 2 mol of C	= 2 mol × 12.01 g/mol	= 24.02 g
Mass of 8 mol of O	= 8 mol × 16.00 g/mol	= 128.0 g
Mass of 2 mol of H	= 2 mol × 1.008 g/mol	= 2.016 g
Mass of 1 mol of $Cu_3(CO_3)_2(OH)_2$		344.7 g

$$\%Cu = \frac{190.7\,g}{344.7\,g} \times 100\% = 55.32\%\ Cu$$

4.89   A solution is a homogenous mixture of two or more substances. Some common solutions: clear drinks (coffee and tea), window cleaner, soapy water, tap water, air, brass (a homogenous mixture of copper and zinc).

4.91   Water is the solvent because it is present in the larger amount. Calcium chloride, $CaCl_2$, is the solute because it is present in the smaller amount. Pure calcium chloride is a solid, but when it is added to water, it dissociates into its ions as it is dissolving (as shown in the figure).

4.93   Dilute and concentrated are relative terms. A concentrated solution has a relatively high solute concentration while a dilute solution has a lower solute concentration.

4.95   Concentration describes the relationship between the quantities of solute and solvent in a solution. Molarity (number of moles of solute per liter of solution) and percent concentration (number of grams of solute per gram of solution) are quantitative ways of expressing concentration.

4.97   Solution A is more concentrated. The volume shown in both figures is the same, but solution A shows more solute particles.

4.99   The molarity of a solution is calculated from the number of moles of solute and volume of solution (measured in liters). The volumes are given, but the solute quantities are given in grams. In each case, we must convert the mass of solute into moles using the molar mass. The overall process can be described by the following map:

Map: Grams of solute $\xrightarrow{\text{MM}}$ Moles of solute $\xrightarrow{\text{volume}}$ Molarity

(a) CH_3CO_2H (60.05 g/mol)

$$\text{Moles of } CH_3CO_2H = 122 \text{ g } CH_3CO_2H \times \frac{1 \text{ mol } CH_3CO_2H}{60.05 \text{ g } CH_3CO_2H} = 2.03 \text{ mol } CH_3CO_2H$$

$$\text{Molarity} = \frac{2.03 \text{ mol } CH_3CO_2H}{1.00 \text{ L solution}} = 2.03 \text{ } M\,CH_3CO_2H$$

(b) $C_{12}H_{22}O_{11}$ (342.3 g/mol)

$$\text{Moles of } C_{12}H_{22}O_{11} = 185 \text{ g } C_{12}H_{22}O_{11} \times \frac{1 \text{ mol } C_{12}H_{22}O_{11}}{342.3 \text{ g } C_{12}H_{22}O_{11}} = 0.540 \text{ mol } C_{12}H_{22}O_{11}$$

$$\text{Molarity} = \frac{0.540 \text{ mol } C_{12}H_{22}O_{11}}{1.00 \text{ L solution}} = 0.540 \text{ } M\,C_{12}H_{22}O_{11}$$

(c) HCl (36.46 g/mol)

$$\text{Moles of } HCl = 70.0 \text{ g } HCl \times \frac{1 \text{ mol } HCl}{36.46 \text{ g } HCl} = 1.92 \text{ mol } HCl$$

$$\text{Molarity} = \frac{1.92 \text{ mol } HCl}{0.600 \text{ L solution}} = 3.20 \text{ } M\,HCl$$

(d) This problem is slightly different than the others because the volume is not given in liters. KOH (56.11 g/mol)

$$\text{Moles of } KOH = 45.0 \text{ g } KOH \times \frac{1 \text{ mol } KOH}{56.11 \text{ g } KOH} = 0.802 \text{ mol } KOH$$

Convert the volume to liters:

$$\text{Volume (L)} = 250.0 \text{ mL} \times \frac{1 \text{ L}}{1000 \text{ mL}} = 0.2500 \text{ L}$$

$$\text{Molarity} = \frac{0.802 \text{ mol } KOH}{0.2500 \text{ L solution}} = 3.21 \text{ } M\,KOH$$

4.101   To calculate moles of substance, it helps to remember that molarity is the conversion between moles and volume (L). Since you are given molarity and volume, you can calculate moles of a substance. The moles of ions is calculated using the formula ratios of ions per formula unit.

Volume $\xrightarrow{M}$ Moles of solute

$$\text{Moles } Na_2SO_4 = 150.0 \text{ mL } Na_2SO_4 \times \frac{10^{-3} \text{ L } Na_2SO_4}{1 \text{ mL } Na_2SO_4} \times \frac{0.124 \text{ mol } Na_2SO_4}{\text{L } Na_2SO_4} = 0.0186 \text{ mol } Na_2SO_4$$

Moles of Na_2SO_4 $\xrightarrow{2 \text{ mol } Na^+ = 1 \text{ mol } Na_2SO_4}$ mol Na^+

Mol Na^+ = 0.0186 mol Na_2SO_4 $\times \dfrac{2 \text{ mol } Na^+}{1 \text{ mol } Na_2SO_4}$ = 0.0372 mol Na^+

Moles of Na_2SO_4 $\xrightarrow{1 \text{ mol } SO_4^{2-} = 1 \text{ mol } Na_2SO_4}$ mol SO_4^{2-}

Mol SO_4^{2-} = 0.0186 mol Na_2SO_4 $\times \dfrac{1 \text{ mol } SO_4^{2-}}{1 \text{ mol } Na_2SO_4^{2-}}$ = 0.0186 mol SO_4^{2-}

4.103   Multiplying the volume of solution (in liters) by the molarity gives you the moles of solute:

Map: Volume of solution $\xrightarrow{\text{molarity}}$ Moles of solute

Make sure to convert the volume to liters. Then use the molar mass of the substance to convert the moles of solute to mass:

Map: Moles of solute $\xrightarrow{MM}$ Grams of solute

(a) 250.0 mL of 1.50 M KCl (MM = 74.55 g/mol)

Volume of solution = 250.0 mL $\times \dfrac{1 \text{ L}}{1000 \text{ mL}}$ = 0.2500 L

Moles KCl = 0.2500 L $\times \dfrac{1.50 \text{ mol KCl}}{1 \text{ L}}$ = 0.375 mol KCl

Grams KCl = 0.375 mol KCl $\times \dfrac{74.55 \text{ g KCl}}{1 \text{ mol KCl}}$ = 28.0 g KCl

(b) 250.0 mL of 2.05 M Na_2SO_4 (MM = 142.04 g/mol)

Volume of solution = 250.0 mL $\times \dfrac{1 \text{ L}}{1000 \text{ mL}}$ = 0.2500 L

Moles Na_2SO_4 = 0.2500 L $\times \dfrac{2.05 \text{ mol } Na_2SO_4}{1 \text{ L}}$ = 0.512 mol Na_2SO_4

Grams Na_2SO_4 = 0.5125 mol Na_2SO_4 $\times \dfrac{142.04 \text{ g } Na_2SO_4}{1 \text{ mol } Na_2SO_4}$ = 72.8 g Na_2SO_4

4.105   We convert moles of solute to volume of solution using molarity:

Map: Moles of solute $\xrightarrow{\text{molarity}}$ Volume of solution

Make sure you check to determine that your answer makes sense. If the number of moles of solute you need is greater than the molarity of the solution, you will need more than one liter of solution. Conversely, if the number of moles of solute you need is smaller than the molarity of the solution, you will need less than one liter of solution.

(a) Volume of solution = 0.250 mol $AlCl_3$ $\times \dfrac{1 \text{ L}}{0.250 \text{ mol } AlCl_3}$ = 1.00 L

(b)   Volume of solution = $0.250 \; \cancel{mol \; HCl} \times \dfrac{1 \; L}{3.00 \; \cancel{mol \; HCl}} = 0.0833 \; L$

4.107   Count the copper ions in the image before and after the addition of water. Before the addition of water, there are 10 copper ions. After the addition of water, there are two copper ions shown in the same volume. When diluting solutions, the concentration is inversely proportional to volume. Since the concentration decreased by a factor of five (i.e. 2/10 = 1/5) the volume must have increased by five (i.e.10/2 = 5). Starting with a volume of 10.0 mL, the final volume is 50.0 mL. 40.0 mL of water was added to the original solution.

4.109   The relationship between the concentration and molarity of the dilute and concentrated solutions is given as:

$$M_{con}V_{con} = M_{dil}V_{dil}$$

To calculate the volume added, you need to first determine the volume of the dilute solution, and then calculate the increase in volume.

$$V_{dil} = \frac{M_{con}V_{con}}{M_{dil}}$$

To help you solve this problem it is important to identify the appropriate variables:

$M_{con} = 0.1074 \; M$
$V_{con} = 935.0 \; mL$
$M_{dil} = 0.1000 \; M$

Since the volume of the concentrated solution is given in milliliters, we first convert the volume to units of liters:

$$V_{con} = 935.0 \; \cancel{mL} \times \dfrac{1 \; L}{1000 \; \cancel{mL}} = 0.9350 \; L$$

$$V_{dil} = \frac{\left(0.1074 \; \cancel{M}\right)\left(0.9350 \; L\right)}{0.1000 \; \cancel{M}} = 1.004 \; L$$

The final volume required is 1.004 L. A volume of 0.0692 L should be added. You should note that this answer is reasonable since the volume of the dilute solution should always be larger than the more concentrated solution.

4.111   The relationship between the concentration and molarity of the dilute and concentrated solutions is given as:

$$M_{con}V_{con} = M_{dil}V_{dil}$$

To calculate the molarity of the dilute solution, we solve the equation for M_{dil}.

$$M_{dil} = \frac{M_{con}V_{con}}{V_{dil}}$$

(a) To help you solve this problem correctly it is important to identify the appropriate variables in the problem:

$M_{con} = 0.1832\ M$

$V_{con} = 24.75\ mL$

$V_{dil} = 250.0\ mL$

At this point, there are two ways to calculate the molarity of the dilute solution. Since the volumes are given in milliliters, it is not necessary to convert the units to liters. Notice how the mL units cancel in the following calculation:

$$M_{dil} = \frac{(0.1832\ M)(24.75\ \cancel{mL})}{250.0\ \cancel{mL}} = 0.01814\ M$$

Alternatively, you could convert the volumes to units of liters and carry out the same calculation:

$$M_{dil} = \frac{(0.1832\ M)(0.02475\ \cancel{L})}{0.2500\ \cancel{L}} = 0.01814\ M$$

(b) $M_{con} = 1.187\ M$
$V_{con} = 125\ mL = 0.125\ L$ (The volumes need to be converted to the same units.)
$V_{dil} = 0.500\ L$

$$M_{dil} = \frac{(1.187\ M)(0.125\ \cancel{L})}{0.500\ \cancel{L}} = 0.297\ M$$

(c) $M_{con} = 0.2010\ M$
$V_{con} = 10.00\ mL$
$V_{dil} = 50.00\ mL$

$$M_{dil} = \frac{(0.2010\ M)(10.00\ \cancel{mL})}{50.00\ \cancel{mL}} = 0.04020\ M$$

4.113   To calculate the mass of 0.100 mol of $Cu(OH)_2$ we use the molar mass:

Map: Moles $\xrightarrow{\ MM\ }$ Mass in grams

$Cu(OH)_2$

Mass of 1 mol of Cu	=	1 mol × 63.55 g/mol =	63.55 g
Mass of 2 mol of O	=	2 mol × 16.00 g/mol =	32.00 g
Mass of 2 mol of H	=	2 mol × 1.008 g/mol =	2.016 g
Mass of 1 mol of $Cu(OH)_2$			97.57 g

$$\text{Mass } Cu(OH)_2 = 0.100\ \cancel{mol\ Cu(OH)_2} \times \frac{97.57\ g}{1\ \cancel{mol\ Cu(OH)_2}} = 9.76\ g\ Cu(OH)_2$$

4.115 The average mass of one argon atom, in amu, is the value given on the periodic table, 39.95 amu. This is also the mass of one mole of average argon atoms (i.e. 39.95 g/mol). To calculate the average mass of a single Ar atom we look for units of grams per atom. We can derive this result using Avogadro's number to convert g/mol to g/atom:

$$\text{Average Ar atom mass (g)} = \frac{39.95 \text{ g Ar}}{1 \text{ mol Ar}} \times \frac{1 \text{ mol Ar}}{6.022 \times 10^{23} \text{ atom}} = 6.634 \times 10^{-23} \text{ g Ar/atom}$$

4.117 (a) To calculate the number of moles of argon, use the molar mass:

$$\text{Moles Ar} = 36.1 \text{ g} \times \frac{1 \text{ mol Ar}}{39.95 \text{ g}} = 0.904 \text{ mol Ar}$$

Calculate the number of atoms from the number of moles, using Avogadro's number:

$$\text{Atoms Ar} = 0.904 \text{ mol Ar} \times \frac{6.022 \times 10^{23} \text{ atom}}{1 \text{ mol Ar}} = 5.44 \times 10^{23} \text{ atoms}$$

(b) Use the conversion 1 carat = 0.200 g to calculate the mass of carbon in the Hope diamond. Then convert this mass to moles and to number of atoms, using the following conversion map:

$$\text{Map: Mass in carats} \xrightarrow{\;0.200 \text{ g} = 1 \text{ carat}\;} \text{Mass in grams} \xrightarrow{\;MM\;} \text{Moles} \xrightarrow{\;N_A\;} \text{Number of atoms}$$

$$\text{Moles C} = 44.5 \text{ carat} \times \frac{0.200 \text{ g C}}{1 \text{ carat}} \times \frac{1 \text{ mol C}}{12.01 \text{ g C}} = 0.741 \text{ mol C}$$

$$\text{Atoms C} = 0.741 \text{ mol C} \times \frac{6.022 \times 10^{23} \text{ atom}}{1 \text{ mol C}} = 4.46 \times 10^{23} \text{ atoms C}$$

(c) We can use the density of mercury to convert from volume to mass. Then we can convert from mass to numbers of moles and atoms using the conversion pathway:

$$\text{Map: Mass in grams} \xrightarrow{\;MM\;} \text{Moles} \xrightarrow{\;N_A\;} \text{Number of atoms}$$

$$\text{Mass Hg} = 2.50 \text{ mL} \times \frac{13.6 \text{ g Hg}}{\text{mL}} = 34.0 \text{ g}$$

$$\text{Moles Hg} = 34.0 \text{ g Hg} \times \frac{1 \text{ mol Hg}}{200.6 \text{ g Hg}} = 0.169 \text{ mol Hg}$$

$$\text{Atoms Hg} = 0.169 \text{ mol Hg} \times \frac{6.022 \times 10^{23} \text{ atom Hg}}{1 \text{ mol Hg}} = 1.02 \times 10^{23} \text{ atoms}$$

4.119 To calculate moles, first convert the mass to grams and then use the molar mass of calcium carbonate (100.09 g/mol) to determine the number of moles

Map: $\text{Mass mg CaCO}_3 \xrightarrow{\ 1\,\text{mg} = 10^{-3}\,\text{g}\ } \text{Mass g CaCO}_3 \xrightarrow{\ 1\,\text{mol} = 100.09\,\text{g}\ } \text{Moles CaCO}_3$

$\text{Moles CaCO}_3 = 750.0 \text{ mg CaCO}_3 \times \dfrac{10^{-3}\ \text{g}}{1\ \text{mg}} \times \dfrac{1\ \text{mol}}{100.09\ \text{g}} = 7.493 \times 10^{-3} \text{ mol CaCO}_3$

4.121 The percent composition of tear gas is 40.25% C, 6.19% H, 8.94% O, and 44.62% Br. A 100-gram sample contains 40.25 g C, 6.19 g H, 8.94 g O, and 44.62 g Br. To determine the empirical formula, we convert the masses of each element to the equivalent number of moles and determine the relative number of moles of each element in the substance.

$\text{Moles C} = 40.25 \text{ g C} \times \dfrac{1\ \text{mol C}}{12.01\ \text{g C}} = 3.351 \text{ mol C}$

$\text{Moles H} = 6.19 \text{ g H} \times \dfrac{1\ \text{mol H}}{1.008\ \text{g H}} = 6.14 \text{ mol H}$

$\text{Moles O} = 8.94 \text{ g O} \times \dfrac{1\ \text{mol O}}{16.00\ \text{g O}} = 0.559 \text{ mol O}$

$\text{Moles Br} = 44.62 \text{ g Br} \times \dfrac{1\ \text{mol Br}}{79.90\ \text{g Br}} = 0.5584 \text{ mol Br}$

Next divide the number of moles of each element by the smallest of the molar amounts (i.e. 0.5584 mol Br).

$\dfrac{\text{Moles C}}{\text{Moles Br}} = \dfrac{3.351 \text{ mol C}}{0.5584 \text{ mol Br}} = \dfrac{6.001 \text{ mol C}}{1 \text{ mol Br}}$

$\dfrac{\text{Moles H}}{\text{Moles Br}} = \dfrac{6.14 \text{ mol H}}{0.5584 \text{ mol Br}} = \dfrac{11.0 \text{ mol H}}{1 \text{ mol Br}}$

$\dfrac{\text{Moles I}}{\text{Moles Br}} = \dfrac{0.559 \text{ mol O}}{0.5584 \text{ mol Br}} = \dfrac{1.00 \text{ mol O}}{1 \text{ mol Br}}$

The empirical formula has 6 moles of carbon and 11 moles of hydrogen for each mole of O and Br. The empirical formula is $C_6H_{11}OBr$.

4.123 To calculate the number of oxygen atoms, first we calculate the number of moles of H_3PO_4. Then we use the mole ratio and Avogadro's number to calculate the number of oxygen atoms. The molar mass of H_3PO_4 is 97.99 g/mol.

Map: $\text{Mass in grams} \xrightarrow{\ MM\ } \text{Moles} \xrightarrow{\ \text{mole ratio}\ } \text{Moles O} \xrightarrow{\ N_A\ } \text{Number of O atoms}$

$\text{Oxygen Atoms} = 10.00 \text{ g H}_3\text{PO}_4 \times \dfrac{1\ \text{mol H}_3\text{PO}_4}{97.99\ \text{g H}_3\text{PO}_4} \times \dfrac{4\ \text{mol O}}{1\ \text{mol H}_3\text{PO}_4} \times \dfrac{6.022 \times 10^{23}\ \text{atoms O}}{1\ \text{mol O}}$

$= 2.458 \times 10^{23} \text{ atoms O}$

4.125   The molar mass of CO_2 is 44.01 g/mol.

$$\text{Moles } CO_2 = 960 \text{ g } CO_2 \times \frac{1 \text{ mol } CO_2}{44.01 \text{ g } CO_2} \times \frac{6.022 \times 10^{23} \text{ molecules } CO_2}{1 \text{ mol } CO_2} = 1.3 \times 10^{25} \text{ molecules } CO_2$$

4.127   (a) The percent composition of vanillin is 63.15% C, 5.30% H, and 31.55% O. A 100-gram sample will contain 63.15 g C, 5.30 g H, and 31.55 g O. To determine the empirical formula, we convert the mass of each element to the equivalent number of moles, and determine the relative number of moles of each element in the substance.

$$\text{Moles C} = 63.15 \text{ g C} \times \frac{1 \text{ mol C}}{12.01 \text{ g C}} = 5.258 \text{ mol C}$$

$$\text{Moles H} = 5.30 \text{ g H} \times \frac{1 \text{ mol H}}{1.008 \text{ g H}} = 5.258 \text{ mol H}$$

$$\text{Moles O} = 31.55 \text{ g O} \times \frac{1 \text{ mol O}}{16.00 \text{ g O}} = 1.972 \text{ mol O}$$

Next divide the number of moles of each element by the smallest of the molar amounts (i.e., 1.972 mol O).

$$\frac{\text{Moles C}}{\text{Moles O}} = \frac{5.258 \text{ mol C}}{1.972 \text{ mol O}} = \frac{2.667 \text{ mol C}}{1 \text{ mol O}}$$

$$\frac{\text{Moles H}}{\text{Moles O}} = \frac{5.258 \text{ mol H}}{1.972 \text{ mol O}} = \frac{2.667 \text{ mol H}}{1 \text{ mol O}}$$

$$\frac{\text{Moles O}}{\text{Moles O}} = \frac{1.972 \text{ mol O}}{1.972 \text{ mol O}} = \frac{1 \text{ mol O}}{1 \text{ mol O}}$$

The fractional portion of the ratios (0.667) indicates two thirds (i.e., 2/3) and can be converted to whole numbers by multiplying by 3. The empirical formula has 8 moles of carbon, 8 moles of hydrogen for every 3 moles of oxygen. The empirical formula for vanillin is $C_8H_8O_3$. The molar mass of the empirical formula is:

Mass of 8 mol of C   = 8 mol × 12.01 g/mol =   96.08 g
Mass of 8 mol of H   = 8 mol × 1.008 g/mol =   8.064 g
Mass of 3 mol of O   = 3 mol × 16.00 g/mol =   48.00 g
Mass of 1 mol of Na_2CO_3 152.14 g

(b) Since the molar mass of the empirical formula and the vanillin molecule are the same, the molecular formula is also $C_8H_8O_3$.

4.129   To calculate the empirical formula, we need to know the mass of each element in the compound. Since 5.00 g of aluminum produced 9.45 g of aluminum oxide, the principle of conservation of mass tells us that 4.45 g of oxygen has been incorporated into the aluminum oxide (i.e. 9.45 g − 5.00 g = 4.45 g). Starting with those masses, we convert them to moles, and then determine the mole ratio of the elements:

$$\text{Moles Al} = 5.00 \text{ g Al} \times \frac{1 \text{ mol Al}}{26.98 \text{ g Al}} = 0.185 \text{ mol Al}$$

$$\text{Moles O} = 4.45 \text{ g O} \times \frac{1 \text{ mol O}}{16.00 \text{ g O}} = 0.278 \text{ mol O}$$

The mole ratio of aluminum to oxygen (the substance present in smaller number of moles) is:

$$\frac{\text{moles O}}{\text{moles Al}} = \frac{0.278 \text{ mol O}}{0.185 \text{ mol Al}} = \frac{1.50 \text{ mol O}}{1 \text{ mol Al}}$$

The fractional portion of the number indicates one half (1/2). Multiplying both parts of the ratio by 2 will produce a whole number. The empirical formula has 2 moles of aluminum for every 3 moles of oxygen. The empirical formula is Al_2O_3.

4.131   First we must calculate the molar mass of calcium nitrate so that we can use it to convert between moles and grams:

$$MM_{Ca(NO_3)_2} = 40.08 \text{ g/mol} + 2(14.01 \text{ g/mol}) + 6(16.00 \text{ g/mol}) = 164.10 \text{ g/mol}$$

Then we multiply molar mass by moles to get grams:

$$\text{Mass}_{Ca(NO_3)_2} = 0.742 \text{ mol} \times \frac{164.10 \text{ g } Ca(NO_3)_2}{1 \text{ mol } Ca(NO_3)_2} = 122 \text{ g}$$

4.133   Both compounds have the same ratio of calcium to anion, so we can compare the molar masses of the anions to determine relative mass percent calcium.

$$MM_{PO_4^{3-}} = 30.97 \text{ g/mol} + 4(16.00 \text{ g/mol}) = 94.97 \text{ g/mol}$$

$$MM_{C_6H_5O_7^{3-}} = 6(12.01 \text{ g/mol}) + 5(1.008 \text{ g/mol}) + 7(16.00 \text{ g/mol}) = 189.1 \text{ g/mol}$$

The phosphate ion contributes a lower mass to its calcium compound compared to the citrate ion, so $Ca_3(PO_4)_2$ has the greater mass percent calcium.

4.135   This is similar to the question: Which box contains more balls, a 10-pound box of baseballs or a 10-pound box of ping-pong balls? Many more of the ping-pong balls are needed to make a total mass of 10 pounds than the heavier baseballs. Similarly more of the lighter-weight F_2 molecules are needed to make a 1.0 gram sample than the heavier-weight SF_2 and CF_4 molecules. We can also reason this out mathematically: Moles are proportional to number of molecules, so if we determine which sample has the greater number of moles, that is also the sample with the greater number of molecules. To calculate moles of each substance, we divide 1.0 g by the molar mass of each substance. The larger the molar mass the smaller the calculated number of moles. The molar mass of F_2 is the smallest, so a 1.0 gram sample of F_2 has the greatest number of moles and molecules.

4.137 (a) 98% of 20.0 g is: $0.98 \times 20.0 \text{ g} = 19.6 \text{ g}$

(b) To determine the mass of Na in 19.6 g NaCl, we must first determine the mass percent Na in the sample:

$$\% \text{ Na} = \frac{\text{Mass of 1 mol Na}}{\text{Mass of 1 mol NaCl}} \times 100 = \frac{22.99 \text{ g Na}}{58.44 \text{ g NaCl}} \times 100 = 39.34\% \text{ Na}$$

Mass Na in 19.6 g NaCl $= 0.3934 \times 19.6 \text{ g} = 7.71 \text{ g Na}$

(c) Since sodium and chlorine are the only elements in the compound, we can calculate the mass of chlorine by subtracting the mass of sodium from the total mass of the compound:

Mass Cl in 19.6 g NaCl $= 19.6 \text{ g NaCl} - 7.71 \text{ g Na} = 11.9 \text{ g Cl}$

4.139 In each case we will calculate the number of formula units using Avogadro's number, and then we will multiply by the number of Cl^- ions per formula unit to determine the number of Cl^- ions in each quantity.

(a) Number of Cl^- ions =

$$1.0 \text{ mol AlCl}_3 \times \frac{6.022 \times 10^{23} \text{ AlCl}_3 \text{ formula units}}{1 \text{ mol AlCl}_3} \times \frac{3 \text{ Cl}^- \text{ ions}}{1 \text{ AlCl}_3 \text{ formula unit}} = 1.8 \times 10^{24} \text{ Cl}^- \text{ ions}$$

(b) Number of Cl^- ions =

$$0.25 \text{ mol AlCl}_3 \times \frac{6.022 \times 10^{23} \text{ AlCl}_3 \text{ formula units}}{1 \text{ mol}} \times \frac{3 \text{ Cl}^- \text{ ions}}{1 \text{ AlCl}_3 \text{ formula unit}} = 4.5 \times 10^{23} \text{ Cl}^- \text{ ions}$$

(c) Number of Cl^- ions =

$$0.25 \text{ mol MgCl}_2 \times \frac{6.022 \times 10^{23} \text{ MgCl}_2 \text{ formula units}}{1 \text{ mol MgCl}_2} \times \frac{2 \text{ Cl}^- \text{ ions}}{1 \text{ MgCl}_2 \text{ formula unit}} = 3.0 \times 10^{23} \text{ Cl}^- \text{ ions}$$

Chapter 5 – Chemical Reactions and Equations

5.1   (a) single-displacement reaction; (b) anhydrous; (c) molecular equation; (d) decomposition reaction;
        (e) balanced equation; (f) reactant; (g) spectator ion; (h) combustion reaction; (i) precipitate

5.3   Substances that are formed as a result of the reaction are the products. This reaction produces solid aluminum
        oxide, $Al_2O_3(s)$. You should pay attention to the substance formed **and** the state of the material if it is given.
        The reactants are the substances used to form the products. The reactants are aluminum metal, $Al(s)$, and
        oxygen gas, $O_2(g)$ (i.e. "Aluminum metal reacts with oxygen gas").

5.5   Image A represents the reactants because it contains both fluorine and xenon. Images B-D show product
        molecules, but only image C represents the products. In a chemical reaction, mass must be conserved. In
        image A there are two xenon atoms and eight fluorine *atoms*. In order for mass to be conserved, the same
        number of each type of atom must be present in the products.

5.7   Mass must be conserved in a chemical reaction. In order for mass to be conserved, the same number of each
        type of atom must be present in both the products and reactants. The image is not accurate because mass is not
        conserved. To fix the image, you first need to count the number of each type of atom. Here's a summary of the
        atoms present in the reactants and products:

Atom	Reactants	Products
H	10	12
N	4	4

The numbers of hydrogen atoms do not match in the reactants and products. If you add one hydrogen molecule
to the reactant image, the numbers will balance and the molecular-level diagram will be accurate.

The numbers of oxygen atoms do not match in the reactants and products. If you add one hydrogen molecule to
the reactant image and one molecule of water (H_2O) to the product image, the numbers will balance and the
molecular-level diagram will be accurate.

Atom	Reactants	Products
H	12	12
O	6	6

5.9   The law of conservation of mass is obeyed when the same number of each type of atom is present in both the
        products and reactants. In the product image you could have 5 molecules and one unreacted atom.

5.11   One oxygen molecule (two oxygen atoms) and two sulfur dioxide molecules must react for every two sulfur trioxide molecules (SO_3) produced. Since there are four sulfur dioxide molecules, both oxygen molecules will react. This means that none of the reactants remain after the reaction, and we should show four sulfur trioxide molecules as the reaction products.

5.13   Three signs that a chemical reaction is occurring are evident: (1) a brown gas is forming, (2) bubbles are being produced, and (3) the solution is changing color (assuming that the liquid was not green to start with).

5.15   We are starting and ending with CO_2, so no new substances are formed. A chemical reaction has not taken place. What we are observing is a change in the physical state of CO_2.

5.17   There is a rearrangement of the atoms into new substances. Since new substances are formed, a chemical reaction has taken place.

5.19   The solution shown on the left has the spheres placed closer together, while in the solution shown on the right the spheres have moved farther apart. Since no new chemical substances have formed, a chemical reaction has not taken place. This is what you would expect to see in dilution.

5.21   A chemical equation is a chemist's shorthand way of showing what happens during a chemical reaction. It identifies the formulas of the reactants and products and demonstrates how mass is conserved during the reaction. You can think of a chemical equation as a chemist's recipe for making new substances. The reactants are the ingredients (i.e. melted chocolate or solid chocolate chips) and the products are what are produced by the reaction (i.e. fudge or cookies). Like a recipe, the chemical equation indicates the states of the materials and the quantities of substances that are used (the reactants) and produced (the products).

5.23   To decide if a reaction has taken place, we have to determine whether a new substance has been formed.
       (a) This equation represents a chemical reaction. Carbonic acid, $H_2CO_3(aq)$, has been formed from the reaction of carbon dioxide (CO_2) and water (H_2O).
       (b) This equation represents a physical change. The chemical compositions of solid water (ice) and liquid water are the same, so a chemical reaction did not occur.
       (c) This equation represents a chemical reaction. Even though the same atoms are present in the reactant and product, the order in which they are connected is different so the substances are different compounds.

5.25   Balancing a chemical equation demonstrates how mass is conserved during a reaction. A balanced equation is a quantitative tool for determining the amounts of reactant(s) used and of product(s) produced.

5.27   In balancing the equation, we add coefficients so that the number of each type of atom is the same on the reactant and product sides of the equation.

(a)   To start with, we write the skeletal equation for the reaction:

$$NaH(s) + H_2O(l) \rightarrow H_2(g) + NaOH(aq)$$

When we count the number of each type of atom in the reactants and products, we find that this equation is already balanced.

Atoms in Reactants			Atoms in Products		
1 Na	3 H	1 O	1 Na	3 H	1 O

(b)   We start by writing the skeletal equation:

$$Al(s) + Cl_2(g) \rightarrow AlCl_3(s) \qquad Unbalanced$$

We count the number of each type of atom on each side of the unbalanced equation to determine where to start balancing the equation.

Atoms in Reactants		Atoms in Products	
1 Al	2 Cl	1 Al	3 Cl

The chlorines are not balanced. To balance, we need to find a common factor. The smallest factor would be 6. We use coefficients of 3 and 2 on $Cl_2(g)$ and $AlCl_3(s)$, respectively

$$Al(s) + \underline{3Cl_2(g)} \rightarrow \underline{2AlCl_3(s)}$$

Atoms in Reactants		Atoms in Products	
1 Al	6 Cl	2 Al	6 Cl

Now we balance the Al on the reactant side, using a coefficient of 2.

$$\underline{2Al(s)} + 3Cl_2(g) \rightarrow 2AlCl_3(s)$$

Atoms in Reactants		Atoms in Products	
2 Al	6 Cl	2 Al	6 Cl

Since the number of each type of atom on the reactant and product sides of the equation is the same, the equation is balanced.

$$\underline{2Al(s)} + 3Cl_2(g) \rightarrow 2AlCl_3(s) \quad Balanced$$

5.29   There are two nitrogen molecules (N_2) and six chlorine molecules (Cl_2) in the reactant image, and 4 NCl_3 molecules in the product image. The large spacing between the molecules allows us to assume that the reactants and products are in the gas state. The balanced chemical equation is:

$$2N_2(g) + 6Cl_2(g) \rightarrow 4NCl_3(g)$$

The coefficients of a balanced equation should be written as the smallest possible whole numbers. Since each of the coefficients is divisible by 2, we reduce them by dividing each by 2.

$$N_2(g) + 3Cl_2(g) \rightarrow 2NCl_3(g) \qquad\qquad Balanced$$

5.31   Magnesium is a solid and it burns in the presence of oxygen, $O_2(g)$, from the air. The product is also a solid (indicated by the statement "ash-like substance"). Image B matches the reactants most closely because it shows the correct states of materials in the reactants. Also, if we assume that magnesium oxide is formed, the chemical formula for the product would be $MgO(s)$. This also matches image B. The skeletal equation for this reaction is:

$Mg(s) + O_2(g) \rightarrow MgO(s)$ *Unbalanced*

Atoms in Reactants		Atoms in Products	
1 Mg	2 O	1 Mg	1 O

We add a coefficient of 2 to MgO to balance the oxygen:

$Mg(s) + O_2(g) \rightarrow \underline{2MgO(s)}$

Atoms in Reactants		Atoms in Products	
1 Mg	2 O	2 Mg	2 O

Then we add a coefficient of 2 to Mg to complete the balancing process:

$\underline{2Mg(s)} + O_2(g) \rightarrow 2MgO(s)$

Atoms in Reactants		Atoms in Products	
2 Mg	2 O	2 Mg	2 O

$2Mg(s) + O_2(g) \rightarrow 2MgO(s)$ *Balanced*

5.33   According to the balanced chemical equation for this reaction, for each molecule of H_2 and I_2 that are consumed in the reaction, two HI molecules are produced. Since three H_2 and three I_2 molecules are shown, we must show six molecules of HI in the product image of the molecular-level diagram.

5.35   Our molecular-level diagram should include two molecules of NO_2 as the reactants and one molecule of N_2O_4 as the products. The law of conservation of mass must be satisfied for the diagram to be accurate. In each image there are two nitrogen atoms and four oxygen atoms.

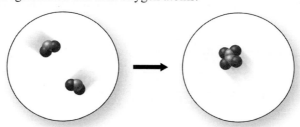

5.37 (a) We start by writing the skeletal equation:

$Al(s) + Cl_2(g) \rightarrow AlCl_3(s)$ *Unbalanced*

Start by taking an inventory of atoms present on each side to help determine where to start balancing the equation.

Atoms in Reactants		Atoms in Products	
1 Al	2 Cl	1 Al	3 Cl

Only the chlorines are not balanced. To balance, we need to find a common factor for 2 and 3. The smallest factor would be 6. We use coefficients of 3 and 2 on $Cl_2(g)$ and $AlCl_3(s)$, respectively

$Al(s) + \underline{3Cl_2(g)} \rightarrow \underline{2AlCl_3(s)}$

Atoms in Reactants		Atoms in Products	
1 Al	6 Cl	2 Al	6 Cl

We recheck our inventory and find that we need to balance the Al on the reactant side using a coefficient of 2.

$\underline{2Al(s)} + 3Cl_2(g) \rightarrow 2AlCl_3(s)$

Atoms in Reactants		Atoms in Products	
2 Al	6 Cl	2 Al	6 Cl

Since the number of each type of atom on the reactant and product sides of the equation is the same, the equation is balanced.

$2Al(s) + 3Cl_2(g) \rightarrow 2AlCl_3(s)$ *Balanced*

(b) We start by writing the skeletal equation:
$Pb(NO_3)_2(aq) + K_2CrO_4(aq) \rightarrow PbCrO_4(s) + KNO_3(aq)$ *Unbalanced*

In the inventory it is preferable to keep polyatomic ions together as groups. This works when the polyatomic ions are not changed in the reaction:

Reactants				Products			
1 Pb	2 NO$_3$	2 K	1 CrO$_4$	1 Pb	1 NO$_3$	1 K	1 CrO$_4$

Start by balancing nitrates or potassium ions. Using a two in front of KNO_3 balances both potassiums and nitrates!

$Pb(NO_3)_2(aq) + K_2CrO_4(aq) \rightarrow PbCrO_4(s) + 2KNO_3(aq)$

Reactants				Products			
1 Pb	2 NO$_3$	2 K	1 CrO$_4$	1 Pb	1 NO$_3$	2 K	1 CrO$_4$

(c) We start by writing the skeletal equation:

Li(s) + H$_2$O(l) → LiOH(aq) + H$_2$(g) *Unbalanced*

Atoms in Reactants			Atoms in Products		
1 Li	2 H	1 O	1 Li	3 H	1 O

Notice that the hydrogens are the only atoms not balanced, but on the product side, the hydrogens appear in two different substances. If we change the number of H$_2$ and H$_2$O molecules, the reaction will continue to be unbalanced. The solution is to use a coefficient of 2 on the H$_2$O and see where it might lead you.

Li(s) + 2H$_2$O(l) → LiOH(aq) + H$_2$(g) *Unbalanced*

Atoms in Reactants			Atoms in Products		
1 Li	4 H	2 O	1 Li	3 H	1 O

Next balance the oxygens:

Li(s) + 2H$_2$O(l) → 2LiOH(aq) + H$_2$(g) *Unbalanced*

Atoms in Reactants			Atoms in Products		
1 Li	4 H	2 O	2 Li	4 H	2 O

The last step is to balance the lithium atoms and check.

2Li(s) + 2H$_2$O(l) → 2LiOH(aq) + H$_2$(g) *Balanced*

Atoms in Reactants			Atoms in Products			
2 Li	4 H	2 O	2 Li	4 H	2	O

(d) We start by writing the skeletal equation:

C$_6$H$_{14}$(g) + O$_2$(g) → CO$_2$(g) + H$_2$O(g)   *Unbalanced*

Atoms in Reactants			Atoms in Products		
6C	14H	2O	1C	2H	3O

In the combustion reactions of hydrocarbons it is usually easiest to begin by balancing the carbon and leave the balancing of oxygen for the last step. This follows the general concepts of balancing the atoms that appear in the smallest number of substances first (you could have also started with hydrogen) and balancing the substances appearing as pure elements last.

C$_6$H$_{14}$(g) + O$_2$(g) → 6CO$_2$(g) + H$_2$O(g) *Unbalanced*

Atoms in Reactants			Atoms in Products		
6C	14H	2O	6C	2H	13O

C$_6$H$_{14}$(g) + O$_2$(g) → 6CO$_2$(g) + 7H$_2$O(g)   *Unbalanced*

Atoms in Reactants			Atoms in Products		
6C	14H	2O	6C	14H	19O

To balance the oxygen, we use a coefficient of 19/2 for O_2:

$$C_6H_{14}(g) + 19/2\ O_2(g) \rightarrow \underline{6CO_2(g)} + 7H_2O(g)$$

While this equation is balanced, we need to remove the fraction by multiplying the entire equation by a factor of two.

$$2C_6H_{14}(g) + 19\ O_2(g) \rightarrow 12CO_2(g) + 14H_2O(g)\quad Balanced$$

Atoms in Reactants			Atoms in Products		
12C	28H	38O	12C	28H	38O

5.39 (a) We start by writing the skeletal equation:

$$CuCl_2(aq) + AgNO_3(aq) \rightarrow Cu(NO_3)_2(aq) + AgCl(s)\qquad Unbalanced$$

When balancing chemical equations that contain polyatomic ions which do not undergo chemical change during the reaction, treat the ions as whole units in the same way that you would treat individual atoms.

Reactants				Products			
1 Cu	2 Cl	1 Ag	1 NO_3^-	1 Cu	1 Cl	1 Ag	1 NO_3^-

You can balance this equation by starting with either Cl or NO_3^-.

$$CuCl_2(aq) + AgNO_3(aq) \rightarrow Cu(NO_3)_2(aq) + \underline{2AgCl(s)}$$

Reactants				Products			
1 Cu	2 Cl	1 Ag	1 NO_3^-	1 Cu	2 Cl	2 Ag	2 NO_3^-

$$CuCl_2(aq) + \underline{2AgNO_3(aq)} \rightarrow Cu(NO_3)_2(aq) + 2AgCl(s)\quad Balanced$$

Reactants				Products			
1 Cu	2 Cl	2 Ag	2 NO_3^-	1 Cu	2 Cl	2 Ag	2 NO_3^-

Since the number of each type of atom on both sides of the equation is the same, the equation is balanced.

(b) We start by writing the skeletal equation:

$$S_8(s) + O_2(g) \rightarrow SO_2(g)\qquad Unbalanced$$

Reactants		Products	
8 S	2 O	1 S	2 O

Begin by balancing S, because O already appears to be balanced.

$$S_8(s) + O_2(g) \rightarrow \underline{8SO_2(g)}$$

Reactants		Products	
8 S	2 O	8 S	16 O

Add a coefficient of 8 for O_2 to balance the equation.

$S_8(s) + \underline{8O_2(g)} \rightarrow 8SO_2(g)$ *Balanced*

Reactants		Products	
8 S	16 O	8 S	16 O

Since the number of each type of atom on both sides of the equation is the same, the equation is balanced.

(c) We start by writing the skeletal equation:

$C_3H_8(g) + O_2(g) \rightarrow CO_2(g) + H_2O(g)$ *Unbalanced*

Reactants			Products		
3 C	8 H	2 O	1 C	2 H	3 O

In combustion reactions of hydrocarbons (hydrocarbon + molecular oxygen → carbon dioxide + water), we begin by balancing the carbon. This follows the general concept of balancing the atoms that appear in the smallest number of substances first (you could have also started with hydrogen). We add a coefficient of 3 for CO_2.

$C_3H_8(g) + O_2(g) \rightarrow \underline{3CO_2(g)} + H_2O(g)$

Reactants			Products		
3 C	8 H	2 O	3 C	2 H	7 O

Continue balancing all the other types of atoms in C_3H_8 before proceeding to balance the oxygen. We add a coefficient of 4 for H_2O.

$C_3H_8(g) + O_2(g) \rightarrow 3CO_2(g) + \underline{4H_2O(g)}$

Reactants			Products		
3 C	8 H	2 O	3 C	8 H	10 O

Finally, add a coefficient of 5 for O_2.

$C_3H_8(l) + \underline{5O_2(g)} \rightarrow 3CO_2(g) + 4H_2O(g)$ *Balanced*

Reactants			Products		
3 C	8 H	10 O	3 C	8 H	10 O

Since the number of each type of atom on both sides of the equation is the same, the equation is balanced.

5.41   Copper and silver, Cu and Ag, are metals, so they, like virtually all the metals, occur in the solid state. Silver nitrate is an ionic compound. The formulas for silver ion and nitrate ion are Ag^+ and NO_3^-, respectively (see Figures 3.12 and 3.17 if you need help with the ion formulas). The formula for an aqueous solution of silver nitrate is $AgNO_3(aq)$. From the ion name, copper(II), we know that the formula for copper ion is Cu^{2+}. The formula for an aqueous solution of copper(II) nitrate is $Cu(NO_3)_2(aq)$. The unbalanced chemical equation is:

$Cu\ (s) + AgNO_3(aq) \rightarrow Cu(NO_3)_2(aq) + Ag(s)$ *Unbalanced*

When balancing chemical equations that contain polyatomic ions which do not undergo chemical change during the reaction, treat the ions as whole units in the same way that you would treat individual atoms.

Reactants			Products		
1 Cu	1 Ag	1 NO_3^-	1 Cu	1 Ag	2 NO_3^-

Balance NO_3^- by placing a coefficient of 2 in front of $AgNO_3$.

Cu (s) + $\underline{2AgNO_3}$(aq) → $Cu(NO_3)_2$(aq) + Ag(s)

Reactants			Products		
1 Cu	2 Ag	2 NO_3^-	1 Cu	1 Ag	2 NO_3^-

The equation is balanced when we add a coefficient of 2 for Ag(s).

Cu (s) + $2AgNO_3$(aq) → $Cu(NO_3)_2$(aq) + $\underline{2Ag}$(s) *Balanced*

5.43   Table 5.1 lists the characteristics of the reactants and products for the different classifications of reactions.
Decomposition: Reactants: 1 compound; Products: 2 elements or smaller compounds
Combination: Reactants: 2 elements or compounds; Products: 1 compound
Single-displacement: Reactants: 1 element and 1 compound; Products: 1 element and 1 compound
Double-displacement: Reactants: 2 compounds; Products: 2 compounds

5.45   For help classifying reactions, refer to Table 5.1. We can classify two of the types based on the number of products and reactants in the chemical equation. In combination reactions, 2 substances combine to form 1 new substance; in decomposition reactions, 1 compound forms several new substances. Equations indicating 2 reactants and 2 products represent either single- or double-displacement reactions. If one of the reactants and one of the products is an element, the reaction is single-displacement.
(a) combination
(b) single-displacement
(c) decomposition

5.47   From the description we learn that there are two reactants: sodium chloride and lead(II) nitrate. These are both compounds. There are also two products: lead(II) chloride and sodium nitrate. Since these are also compounds, the reaction is double-displacement (2 compounds as reactants and 2 compounds as products).

5.49   (a) There are two different substances in the reactant image and only one substance in the product image. This is a combination reaction.
(b) The dark red and gray spheres of solid represent two different elements. The gray spheres are displacing the dark red spheres. One compound and 1 element are producing another compound and element. This is a single-displacement reaction.

5.51   For help classifying reactions, refer to Table 5.1. We can classify two of the types based on the number of products and reactants in the chemical equation. In combination reactions, 2 substances combine to form 1 new substance. In decomposition reactions, 1 compound forms several new substances. Equations indicating 2 reactants and 2 products represent either single- or double-displacement reactions. If 1 of the reactants and 1 of the products is an element, the reaction is single-displacement.
(a) double-displacement: $CaCl_2$(aq) + Na_2SO_4(aq) → $CaSO_4$(s) + 2NaCl(aq)
(b) single-displacement: Ba(s) + 2HCl(aq) → $BaCl_2$(aq) + H_2(g)
(c) combination: N_2(g) + $3H_2$(g) → $2NH_3$(g)
(d) This reaction does not easily fit the classification scheme given in Table 5.1; however, it is most like a single-displacement reaction. CO is displacing iron from FeO: FeO(s) + CO(g) $\xrightarrow{\text{heat}}$ Fe(s) + CO_2(g)
(e) combination: CaO(s) + H_2O(l) → $Ca(OH)_2$(aq)
(f) double-displacement: Na_2CrO_4(aq) + $Pb(NO_3)_2$(aq) → $PbCrO_4$(s) + $2NaNO_3$(aq)
(g) single-displacement: 2KI(aq) + Cl_2(g) → 2KCl(aq) + I_2(aq)
(h) decomposition: $2NaHCO_3$(s) $\xrightarrow{\text{heat}}$ Na_2CO_3(s) + CO_2(g) + H_2O(g)

5.53 Carbonate-containing compounds (except Group IA (1)), when heated, produce the metal oxide and carbon dioxide gas (Table 5.2). In this case, we know that nickel has a 2+ charge to balance the charge of the 2– carbonate ion. The oxide will be NiO. The balanced chemical equation is $NiCO_3(s) \rightarrow NiO(s) + CO_2(g)$.

5.55 See Table 5.2.

(a) $CaCO_3(s) \xrightarrow{heat} CaO(s) + CO_2(g)$

Metal carbonates (except Group IA (1)) decompose to produce carbon dioxide and the metal oxide. We determine the formula of the metal oxide from the charge on the metal and the 2– charge of the oxide ion (i.e. Ca^{2+} and O^{2-} produce CaO).

(b) $CuSO_4 \cdot 5H_2O(s) \xrightarrow{heat} CuSO_4(s) + 5H_2O(g)$

Hydrates decompose by separation of the water from the ionic compound.

5.57 See Table 5.3

$2Mg(s) + O_2(g) \rightarrow 2MgO(s)$

Combination reactions result in the formation of 1 product from 2 reactants. If the reactants are a metal and a nonmetal, we can reliably predict that the metal will form a cation and the nonmetal will form an anion. When oxygen reacts with elements, the oxides of those elements are formed. With metals, the products are the metal oxides. Magnesium is a Group IIA (2) metal. This means that it will form an ion with a 2+ charge. The formula for oxide ion is O^{2-}. The formula for the metal oxide product in this reaction is MgO.

5.59 Combination reactions result in the formation of 1 product from 2 reactants (see Table 5.3). If the reactants are a metal and a nonmetal, you can reliably predict that the metal will form a cation and the nonmetal will form an anion. In other situations, simply try to combine the two reactants to form a product you recognize.

(a) $3Ca(s) + N_2(g) \rightarrow Ca_3N_2(s)$

Calcium and nitrogen (metal and nonmetal) form the Ca^{2+} and N^{3-} ions based on their positions on the periodic table. We predict that the product they form is Ca_3N_2.

(b) $2K(s) + Br_2(l) \rightarrow 2KBr(s)$

Metals and nonmetals often react to form metal salts.

(c) $4Al(s) + 3O_2(g) \rightarrow 2Al_2O_3(s)$

Aluminum and oxygen (metal and nonmetal) form Al^{3+} and O^{2-} ions, based on their positions on the periodic table. We predict that the product they form is Al_2O_3.

5.61 In a single-displacement reaction, 1 element displaces its ionic counterpart from a compound. That is, a metal displaces the metal ion, or a nonmetal displaces the nonmetal ion. To determine whether a reaction occurs, we compare the activities of the metals involved (Figure 5.21). If the metal is more active than the metal whose ion appears in the compound, a reaction occurs. In these cases, the formula for the ionic product is determined by the metal ion that is produced. If you can't predict the charge (i.e. it is not listed in Figure 3.12), assume that it has a charge of 2+. For example if Fe displaces Cu, Fe will form a 2+ ion. In this type of single-displacement reaction, the formula for the anion remains unchanged.

(a) Zn is higher on the activity series than Ag so a reaction occurs. Zn forms a 2+ ion, and Ag^+ is displaced, forming silver metal, Ag(s).

$Zn(s) + 2AgNO_3(aq) \rightarrow 2Ag(s) + Zn(NO_3)_2(aq)$

(b) Na is higher on the activity series than Fe so a reaction occurs. Na forms a 1+ ion, and Fe^{2+} is displaced, forming iron metal, Fe(s).

$2Na(s) + FeCl_2(s) \rightarrow 2NaCl(s) + Fe(s)$

5.63   In a single-displacement reaction, 1 element displaces its ionic counterpart from a compound. That is, a metal displaces the metal ion or a nonmetal displaces the nonmetal ion. The formulas for the products are determined by the element or ions that are produced (i.e. metal or diatomic molecules, etc.). When Zn displaces Sn, Zn will form a 2+ ion (Figure 3.12). In this single-displacement reaction, the formula for the anion remains unchanged. Since the formula for chloride ion is Cl^-, the product is $ZnCl_2$. The balanced equation is:

$$Zn(s) + SnCl_2(aq) \rightarrow ZnCl_2(aq) + Sn(s)$$

5.65   In reactions of metals with hydrochloric acid solution or water, one product will always be hydrogen gas (H_2). To determine whether a reaction takes place, compare the activity of the metal (not the metal ion found in the product) with the activity of hydrogen gas (Figure 5.21). If the metal is higher in the activity series than hydrogen gas, a reaction will take place. Some metals are more active than others. Those that are more active react under less extreme conditions.

(a) Ca is more active than H_2. Because it is so high on the activity series, it will displace hydrogen from water and form an acidic solution. As noted on the activity series shown in Figure 5.21, if a substance reacts with cold water, it also reacts with steam and acid.
Ca reacts with water: $Ca(s) + 2H_2O(l) \rightarrow Ca(OH)_2(aq) + H_2(g)$
Ca reacts with HCl: $Ca(s) + 2HCl(aq) \rightarrow CaCl_2(aq) + H_2(g)$

(b) Fe is more active than H_2. This means that Fe can displace H_2 under the proper conditions. As is noted in Figure 5.21, Fe is not active enough to displace hydrogen in cold water, but can do so from either hydrochloric acid solution or from steam.
Fe reacts with steam: $Fe(s) + 2 H_2O(g) \rightarrow Fe(OH)_2(s) + H_2(g)$
Fe reacts with HCl: $Fe(s) + 2HCl(aq) \rightarrow FeCl_2(aq) + H_2(g)$

(c) No reaction. Copper is below H_2 on the activity series. There are no reaction conditions which favor a reaction of copper with water or HCl solutions.

5.67   In a single-displacement reaction, 1 element displaces its ionic counterpart from a compound. That is, a metal displaces the metal ion, or a nonmetal displaces the nonmetal ion. To determine whether a reaction occurs, we compare the activities of the metals involved (Figure 5.21). If the metal is more active than the metal whose ion appears in the compound, a reaction occurs.
(a) Magnesium is more active than aluminum. The reaction occurs.
(b) Zinc is less active than magnesium. No reaction occurs.
(c) Copper is less active than lead. No reaction occurs.
(d) Nickel is more active than silver. The reaction occurs.

5.69   To determine solubility, refer to the solubility rules given in Table 5.4. Except for compounds containing NH_4^+ and/or Group IA (1) cations (including Na^+ and K^+), the solubility rules focus on the anions in ionic compounds. It is worth memorizing that Group IA (1), nitrates, acetates, and ammonium ion compounds are soluble, because we encounter them so frequently when we study chemistry.
(a) $CuCl_2$: The anion is Cl^-; the compound is soluble.
(b) $AgNO_3$: The anion is NO_3^-; the compound is soluble.
(c) $PbCl_2$: The anion is Cl^-; most chlorides are soluble, but $PbCl_2$ is an exception. It is insoluble.
(d) $Cu(OH)_2$: The anion is OH^-; the compound is insoluble.

5.71   For each reaction we: 1) determine the ion formulas, 2) predict the products based on ionic charges, and 3) determine the states of the products. If one or both products is insoluble, is a gas, or is a molecule (for example, water), a reaction takes place. Finally, we balance the resulting equation.

         (a) Reactants: $K_2CO_3(aq)$ and $BaCl_2(aq)$
            Ions: K^+, CO_3^{2-}, Ba^{2+}, Cl^-
            Products: $BaCO_3$ and KCl
            Solubility (from Table 5.4): $BaCO_3(s)$ and $KCl(aq)$

            Balanced equation: $K_2CO_3(aq) + BaCl_2(aq) \rightarrow BaCO_3(s) + 2KCl(aq)$

         (b) Reactants: $CaS(aq)$ and $Hg(NO_3)_2(aq)$
            Ions: Ca^{2+}, S^{2-}, Hg^{2+}, NO_3^-
            Products: $Ca(NO_3)_2$ and HgS
            Solubility (from Table 5.4): $Ca(NO_3)_2(aq)$ and $HgS(s)$

            Balanced equation: $CaS(aq) + Hg(NO_3)_2(aq) \rightarrow Ca(NO_3)_2(aq) + HgS(s)$

         (c) Reactants: $Pb(NO_3)_2(aq)$ and $K_2SO_4(aq)$
            Ions: Pb^{2+}, NO_3^-, K^+, SO_4^{2-}
            Products: $PbSO_4$ and KNO_3
            Solubility (from Table 5.4): $PbSO_4(s)$ and $KNO_3(aq)$

            Balanced equation: $Pb(NO_3)_2(aq) + K_2SO_4(aq) \rightarrow PbSO_4(s) + 2KNO_3(aq)$

5.73   For each reaction, 1) determine the ion formulas, 2) predict the products based on ion charges, and 3) determine the states of the products. If one or both products is insoluble, is a gas, or is a molecule (for example, water), a reaction takes place. Balance the resulting equations.

     (a) Reactants: $CaCO_3(s)$ and $H_2SO_4(aq)$
         Ions: Ca^{2+}, CO_3^{2-}, H^+, SO_4^{2-}
         Products: $CaSO_4$ and H_2CO_3
         According to the solubility rules (Table 5.4), calcium sulfate is insoluble in water. Carbonic acid decomposes to carbon dioxide and water (as described in Table 5.2). Since a precipitate is formed and carbonic acid decomposes to form water and carbon dioxide, we predict that the following reaction takes place:
         $CaCO_3(s) + H_2SO_4(aq) \rightarrow CaSO_4(s) +$ **$H_2CO_3(aq)$**   *carbonic acid decomposes*
         $CaCO_3(s) + H_2SO_4(aq) \rightarrow CaSO_4(s) +$ **$H_2O(l)$ $+ CO_2(g)$**   *Balanced*

     (b) Reactants: $SnCl_2(aq)$ and $AgNO_3(aq)$
         Ions: Sn^{2+} (based on charge of chloride), Cl^-, Ag^+, NO_3^-
         Products: $Sn(NO_3)_2$ and $AgCl$
         According to the solubility rules (Table 5.4), nitrates are soluble in water. While most chlorides are soluble in water, $AgCl$ is an exception. Since a precipitate is formed, we can predict that the following reaction takes place:
         $SnCl_2(aq) + 2AgNO_3(aq) \rightarrow Sn(NO_3)_2(aq) + 2AgCl(s)$ *Balanced*

5.75   To identify the precipitate, we first determine the products of the reaction, and then check the solubility of the products using the solubility rules (Table 5.4). The precipitate will be the compound that is insoluble in water. We deduce the reactants from the description given in the problem.
     Reactants: Na_2SO_4 and $Pb(NO_3)_2$
     Ions: Na^+, SO_4^{2-}, Pb^{2+}, NO_3^-
     Products: $PbSO_4$ and $NaNO_3$
     Solubility (from Table 5.4): $PbSO_4(s)$, $NaNO_3(aq)$
     Lead(II) sulfate is the white solid that forms when the two solutions are mixed.

5.77   From the problem description we can write:
$$CaCl_2(aq) + K_2CO_3(aq) \rightarrow$$
In double-displacement reactions, the cations switch places to form new products (i.e. the metal ions displace one another). To avoid common errors, and to be able to spot mistakes easily, it is helpful if we write the complete formulas for each ion before we determine the products.
   Reactants: $CaCl_2(aq)$ and $K_2CO_3(aq)$
   Ions: Ca^{2+}, Cl^-, K^+, CO_3^{2-}
Considering the charges of the ions, we can write the chemical formulas for the products. The products will be calcium carbonate, $CaCO_3$, and potassium chloride, KCl. Using the solubility rules (Table 5.4) we can determine whether these products are soluble or insoluble in water (i.e. whether we should write (aq), or (s), respectively). From the solubility rules, we determine that carbonates are insoluble (i.e. $CaCO_3(s)$) and chlorides are generally soluble (i.e. $KCl(aq)$).

$$CaCl_2(aq) + K_2CO_3(aq) \rightarrow CaCO_3(s) + 2KCl(aq) \qquad \textit{Balanced}$$

5.79   Chemical reactions occur for a "reason." For many reactions, the reason is the nature of the products that form. This "driving force" of a reaction can often be identified by formation of a precipitate, formation of a gas (water-insoluble gas), or formation of molecular compounds, such as water. Water-insoluble gases will leave a solution, providing an additional driving force for the reaction. The driving force for each reaction is:
(a) precipitation of $HgS(s)$
(b) formation of the insoluble gas $H_2S(g)$
(c) formation of the molecular compound $H_2O(l)$, and precipitation of $BaSO_4(s)$

5.81   Neutralization reactions involve the reaction of an acid with a base. The products are a salt (i.e. ionic compound) and water. Just as with precipitation reactions we: 1) determine the ionic formulas, 2) predict the products based on ion charges, and 3) determine the states of the products. Acid-base reactions are usually driven by the formation of water, a molecular compound. We must use the solubility rules to determine whether or not the salt will dissolve in water.
(a) Reactants: $H_2S(aq)$ and $Cu(OH)_2(s)$
   Ions: H^+, S^{2-}, Cu^{2+}, OH^-
   Products: $H_2O(l)$ and $CuS(s)$ (determined from solubility rules)
   Balanced equation: $H_2S(aq) + Cu(OH)_2(s) \rightarrow CuS(s) + 2H_2O(l)$
(b) CH_4 is not an acid, so no reaction is expected.

(c) Reactants: $KHSO_4(aq)$ and $KOH(aq)$
   Ions: K^+, H^+, SO_4^{2-}, OH^-
   Products: $H_2O(l)$ and K_2SO_4. Note that the H^+ reacts with OH^-, leaving K^+ and SO_4^{2-} to form the salt, which is water-soluble.
   Balanced equation: $KHSO_4(aq) + KOH(aq) \rightarrow K_2SO_4(aq) + H_2O(l)$

5.83   Oxygen, $O_2(g)$ is always one of the reactants of a combustion reaction.

5.85   The combustion reaction products are the oxides of those elements. For metals, we can predict the oxide formula based on the common charges of the metal ions and the charge of oxide ion, O^{2-}. Metals that form several different cations can form several different oxides.
(a) Cesium only forms a 1+ ion, so the oxide that forms is $Cs_2O(s)$.
(b) Lead forms either a 2+ or 4+ ion (Figure 3.24), so the oxides that lead can form are $PbO(s)$ and $PbO_2(s)$.
(c) Aluminum only forms a 3+ ion, so the formula of aluminum oxide is $Al_2O_3(s)$.
(d) Combustion of hydrogen produces $H_2O(g)$.
(e) Carbon produces two different oxides, $CO(g)$ and $CO_2(g)$.

5.87   The combustion reaction products of compounds are the oxides of the elements which make up the compound. In many cases, there are multiple products.
        (a) CH_4 (and all hydrocarbons) combusts to form oxides of carbon and hydrogen. The products are $CO(g)$ or $CO_2(g)$ and $H_2O(g)$.
        (b) The combustion of CO results in formation of $CO_2(g)$.
        (c) The combustion of aluminum metal results in formation of solid aluminum oxide solid, $Al_2O_3(s)$.
        (d) CH_3OH (and all compounds containing C, H, and O) combusts to form $CO(g)$ or $CO_2(g)$ and $H_2O(g)$.

5.89   An electrolyte produces ions when it dissolves in water. A nonelectrolyte does not produce ions when it dissolves in water.

5.91   Aqueous solutions of soluble ionic compounds, acids, or bases are electrolytes because the substances produce ions when they dissolve.
        (a) $NaOH(aq)$: base; electrolyte
        (b) $HCl(aq)$: acid; electrolyte
        (c) $C_{12}H_{22}O_{11}$: not a soluble ionic compound, acid, or base; nonelectrolyte

5.93   The substance is a nonelectrolyte because it does not appear to have dissociated in the solution. In order to be an electrolyte, ions must form in the solution.

5.95   Soluble ionic compounds, acids, and bases produce ions when they dissolve in water.
        (a) $C_6H_{12}O_6$: not a soluble ionic compound, acid, or base; does not form ions in solution
        (b) CH_4: not a soluble ionic compound, acid, or base; does not form ions in solution
        (c) NaCl: soluble ionic compound; forms Na^+ and Cl^- ions in solution; $Na^+(aq)$ and $Cl^-(aq)$

5.97   Molecular equations: We write the chemical formulas for all substances in their complete form (for example, $O_2(g)$, $NaCl(aq)$, $Ag(s)$).
        Ionic Equations: We write the formulas of all soluble salts, strong acids and bases as individual, solvated ions (for example, $NaCl(aq)$ becomes $Na^+(aq)$ and $Cl^-(aq)$).
        Net ionic equations: We remove all spectator ions from the ionic equation.

5.99   Spectator ions are those ions that do not actually participate in the chemical change that occurs during the reaction. These are the ions that appear both as reactants and as products in an ionic equation.

5.101   (a) From the solubility rules (Table 5.4) we find that all the substances are soluble ionic compounds except Ag_2SO_4 and AgCl, which are insoluble in water.

             $2NaCl(aq) + Ag_2SO_4(s) \rightarrow Na_2SO_4(aq) + 2AgCl(s)$ *Molecular equation*
             Then we write the formulas of all soluble ionic compounds as ions, and eliminate the spectator ions.

             $2Na^+(aq) + 2Cl^-(aq) + Ag_2SO_4(s) \rightarrow 2Na^+(aq) + SO_4^{2+}(aq) + 2AgCl(s)$ *Ionic equation*

             $\cancel{2Na^+}(aq) + 2Cl^-(aq) + Ag_2SO_4(s) \rightarrow \cancel{2Na^+}(aq) + SO_4^{2+}(aq) + 2AgCl(s)$

             This leaves the balanced net ionic equation.
             $2Cl^-(aq) + Ag_2SO_4(s) \rightarrow 2AgCl(s) + SO_4^{2-}(aq)$ *Net ionic equation*

(b) From the solubility rules we know that $Cu(OH)_2$ is insoluble, while the other ionic compounds are soluble. Water is a liquid.

$$Cu(OH)_2(s) + 2HCl(aq) \rightarrow CuCl_2(aq) + 2H_2O(l) \qquad \textit{Molecular equation}$$

Next we write the ionic equation by writing all the aqueous ionic compounds and strong acids (Table 3.10) in ionic form.

$$Cu(OH)_2(s) + 2H^+(aq) + 2Cl^-(aq) \rightarrow Cu^{2+} + 2Cl^-(aq) + 2H_2O(l)$$

Finally, we write the net ionic equation by identifying and eliminating the spectator ions.

$$Cu(OH)_2(s) + 2H^+(aq) + \cancel{2Cl^-}(aq) \rightarrow Cu^{2+} + \cancel{2Cl^-}(aq) + 2H_2O(l)$$

$$Cu(OH)_2(s) + 2H^+(aq) \rightarrow Cu^{2+}(aq) + 2H_2O(l) \qquad \textit{Net ionic equation}$$

(c) From the solubility rules we know that only $BaCl_2$ is soluble, while the other ionic compounds are insoluble.

$$BaCl_2(aq) + Ag_2SO_4(s) \rightarrow BaSO_4(s) + 2AgCl(s) \qquad \textit{Molecular equation}$$

Only $BaCl_2$ forms ions in solution. Since none of the products form ions, there are no spectator ions.

$$Ba^{2+}(aq) + 2Cl^-(aq) + Ag_2SO_4(s) \rightarrow BaSO_4(s) + 2AgCl(s) \qquad \textit{Net ionic equation}$$

5.103 From the solution descriptions, we know that there are four ions present in the solution before any precipitation takes place: Na^+, CO_3^{2-}, Ca^{2+}, and Br^-.
(a) The only possible precipitate is $CaCO_3$ (see Table 5.4). The other combinations represent soluble compounds.
(b) $Na_2CO_3(aq) + CaBr_2(aq) \rightarrow CaCO_3(s) + 2NaBr(aq)$
(c) $2Na^+(aq) + CO_3^{2-}(aq) + Ca^{2+}(aq) + 2Br^-(aq) \rightarrow CaCO_3(s) + 2Na^+(aq) + 2Br^-(aq)$
(d) $Na^+(aq)$ and $Br^-(aq)$
(e) $CO_3^{2-}(aq) + Ca^{2+}(aq) \rightarrow CaCO_3(s)$

5.105 To write the net ionic equation for this reaction, we start by writing the molecular equation. From the description in the question we have:

$$CaCl_2(aq) + 2AgNO_3(aq) \rightarrow Ca(NO_3)_2(aq) + 2AgCl(s) \qquad \textit{Molecular equation}$$

Next we write the ionic equation by separating the formulas of all the water-soluble ionic compounds into their respective ions. All the ions written in the equation below are solvated by water.

$$Ca^{2+}(aq) + 2Cl^-(aq) + 2Ag^+(aq) + 2NO_3^-(aq) \rightarrow Ca^{2+}(aq) + 2NO_3^-(aq) + 2AgCl(s)$$

Next we eliminate the spectator ions.

$$\cancel{Ca^{2+}}(aq) + 2Cl^-(aq) + 2Ag^+(aq) + \cancel{2NO_3^-}(aq) \rightarrow \cancel{Ca^{2+}}(aq) + \cancel{2NO_3^-}(aq) + 2AgCl(s)$$
$$2Cl^-(aq) + 2Ag^+(aq) \rightarrow 2AgCl(s)$$

Because each of the coefficients is divisible by two, we simplify the net ionic equation to:

$$Ag^+(aq) + Cl^-(aq) \rightarrow AgCl(s) \qquad \textit{Net ionic equation}$$

A quick way to check the answer is to see if the total charge on the reactant and product sides of the equation is the same. In this case, the total charge is zero for both the reactants (Ag^+ and Cl^-) and the products.

5.107   The reactants are sodium metal (you can tell from its well ordered structure) and $H_2O(l)$. The products are Na^+, OH^-, and H_2. Na^+ and OH^- are ions dissolved in water. Molecular hydrogen is a gas. The balanced ionic equation is:

$$2Na(s) + 2H_2O(l) \rightarrow 2Na^+(aq) + 2OH^-(aq) + H_2(g)$$

There are no spectator ions, so the balanced net ionic equation and the balanced ionic equations are identical.

$$2Na(s) + 2H_2O(l) \rightarrow 2Na^+(aq) + 2OH^-(aq) + H_2(g) \quad \textit{Net ionic equation}$$

5.109   (a)   First we write the molecular equation and determine the solubility of ionic reactants and products (Table 5.4).

$$Cu(s) + 2AgNO_3(aq) \rightarrow Cu(NO_3)_2(aq) + 2Ag(s) \qquad \textit{Molecular equation}$$

Next we write the ionic equation by writing all the aqueous ionic compounds in ionic form.

$$Cu(s) + 2Ag^+(aq) + 2NO_3^-(aq) \rightarrow Cu^{2+}(aq) + 2NO_3^-(aq) + 2Ag(s)$$

Finally, we eliminate the spectator ions and write the net ionic equation.

$$Cu(s) + 2Ag^+(aq) + \cancel{2NO_3^-}(aq) \rightarrow Cu^{2+}(aq) + \cancel{2NO_3^-}(aq) + 2Ag(s)$$

$$Cu(s) + 2Ag^+(aq) \rightarrow Cu^{2+}(aq) + 2Ag(s) \qquad \textit{Net ionic equation}$$

(b)   First we write the molecular equation and determine the solubility of ionic reactants and products (Table 5.4).

$$FeO(aq) + 2HCl(aq) \rightarrow FeCl_2(aq) + H_2O(l) \qquad \textit{Molecular equation}$$

Next we write the ionic equation by writing all the aqueous ionic compounds and strong acids (Table 3.10) in ionic form.

$$Fe^{2+}(aq) + O^{2-}(aq) + 2H^+(aq) + Cl^-(aq) \rightarrow Fe^{2+}(aq) + 2Cl^-(aq) + H_2O(l)$$

Finally, we eliminate the spectator ions and write the net ionic equation.

$$\cancel{Fe^{2+}}(aq) + O^{2-}(aq) + 2H^+(aq) + \cancel{2Cl^-}(aq) \rightarrow \cancel{Fe^{2+}}(aq) + \cancel{2Cl^-}(aq) + H_2O(l)$$

$$O^{2-}(aq) + 2H^+(aq) \rightarrow H_2O(l) \qquad \textit{Net ionic equation}$$

5.111   For each reaction we: 1) determine the ion formulas, 2) predict the products based on ionic charges, and 3) determine the states for the products. If one or both products is insoluble, is a gas, or is a molecule (for example, water), a reaction takes place.

(a)   Reactants: $Sr(NO_3)_2(aq)$ and $H_2SO_4(aq)$

Ions: Sr^{2+}, NO_3^-, H^+, SO_4^{2-}

Products: $SrSO_4$ and HNO_3

Solubility (Table 5.4, Table 3.10): $Sr(NO_3)_2(aq)$, $H_2SO_4(aq)$, $SrSO_4(s)$, $HNO_3(aq)$

$$Sr(NO_3)_2(aq) + H_2SO_4(aq) \rightarrow SrSO_4(s) + 2HNO_3(aq) \qquad \textit{Molecular equation}$$

Next we write the ionic equation by separating all the aqueous ionic compounds and strong acids into their respective ions.

$$Sr^{2+} + 2NO_3^-(aq) + 2H^+(aq) + SO_4^{2-}(aq) \rightarrow SrSO_4(s) + 2H^+(aq) + 2NO_3^-(aq)$$

Finally, we eliminate the spectator ions:

$$Sr^{2+} + \cancel{2NO_3^-}(aq) + \cancel{2H^+}(aq) + SO_4^{2-}(aq) \rightarrow SrSO_4(s) + \cancel{2H^+}(aq) + \cancel{2NO_3^-}(aq)$$

$$Sr^{2+}(aq) + SO_4^{2-}(aq) \rightarrow SrSO_4(s) \qquad \textit{Net ionic equation}$$

(b) Reactants: $Zn(NO_3)_2$ and Na_2SO_4
    Ions: Zn^{2+}, NO_3^-, Na^+, SO_4^{2-}
    Products: $ZnSO_4$ and $NaNO_3$
    Solubility: Both products are aqueous ionic compounds.
    The products do not precipitate. In addition, no molecular products are formed. No reaction takes place.

(c) Reactants: $CuSO_4(aq)$ and $BaS(s)$
    Ions: Cu^{2+}, SO_4^{2-}, Ba^{2+}, S^{2-}
    Products: CuS and $BaSO_4$
    Solubility: $CuSO_4(aq)$, $BaS(s)$, $CuS(s)$, $BaSO_4(s)$
    $CuSO_4(aq) + BaS(s) \rightarrow CuS(s) + BaSO_4(s)$ *Molecular equation*

Next we write the ionic equation by separating all the aqueous ionic compounds into their respective ions.

$Cu^{2+}(aq) + SO_4^{2-}(aq) + BaS(s) \rightarrow CuS(s) + BaSO_4(s)$ *Ionic & net ionic equation*

Since only one reactant is soluble, there are no spectator ions, so the ionic and net ionic equations are the same.

(d) Reactants: $NaHCO_3$ and HCH_3CO_2
    Ions: Na^+, HCO_3^-, H^+, and $CH_3CO_2^-$ (see **note** below)
    Products: $NaCH_3CO_2$ and H_2CO_3 which decomposes to $H_2O(l)$ and $CO_2(g)$
    Solubility: $NaHCO_3(aq)$, $HCH_3CO_2(aq)$, $NaCH_3CO_2(aq)$: Acids are generally soluble. Other solubilities are obtained from Table 5.4.
    A reaction takes place because of the formation of water and carbon dioxide.
    $NaHCO_3(aq) + CH_3CO_2H (aq) \rightarrow NaCH_3CO_2(aq) + CO_2(g) + H_2O(l)$ *Molecular equation*

Next we write the ionic equation by separating all the aqueous ionic compounds into their respective ions. **Note** that HCH_3CO_2 is a weak acid and does not completely ionize, however when it is ionized, H^+ and $CH_3CO_2^-$ are formed and we use this information to predict what products are formed.

$Na^+(aq) + HCO_3^-(aq) + HCH_3CO_2(aq) \rightarrow Na^+(aq) + CH_3CO_2^-(aq) + CO_2(g) + H_2O(l)$

Finally, we eliminate the spectator ions.

$\cancel{Na^+(aq)} + HCO_3^-(aq) + HCH_3CO_2(aq) \rightarrow \cancel{Na^+(aq)} + CH_3CO_2^-(aq) + CO_2(g) + H_2O(l)$

$HCO_3^-(aq) + CH_3CO_2H (aq) \rightarrow CH_3CO_2^-(aq) + CO_2(g) + H_2O(l)$ *Net Ionic Equation*

5.113 Because we are starting with 2 reactants and forming 1 product, this is a combination reaction (Table 5.1).

5.115 For each reaction we: 1) determine the ion formulas, 2) predict the products based on ionic charges, and 3) determine the states for the products. If one or both products is insoluble, is a gas, or is a molecule (for example, water), a reaction takes place.
    (a) Reactants: $ZnSO_4(aq)$ and $Ba(NO_3)_2(aq)$
        Ions: Zn^{2+}, SO_4^{2-}, Ba^{2+}, NO_3^-
        Products: $BaSO_4$ and $Zn(NO_3)_2$
        Solubility (Table 5.4): $BaSO_4(s)$ and $Zn(NO_3)_2(aq)$
        $ZnSO_4(aq) + Ba(NO_3)_2(aq) \rightarrow BaSO_4(s) + Zn(NO_3)_2(aq)$ *Balanced*

    (b) Reactants: $Ca(NO_3)_2(aq)$ and $K_3PO_4(aq)$
        Ions: Ca^{2+}, NO_3^-, K^+, PO_4^{3-}
        Products: $Ca_3(PO_4)_2$ and KNO_3
        Solubility (Table 5.4): $Ca_3(PO_4)_2(s)$ and $KNO_3(aq)$

        $3Ca(NO_3)_2(aq) + 2K_3PO_4(aq) \rightarrow Ca_3(PO_4)_2(s) + 6KNO_3(aq)$ *Balanced*

(c) Reactants: $ZnSO_4(aq)$ and $BaCl_2(aq)$
Ions: Zn^{2+}, SO_4^{2-}, Ba^{2+}, Cl^-
Products: $ZnCl_2$ and $BaSO_4$
Solubility (Table 5.4): $ZnCl_2(aq)$ and $BaSO_4(s)$

$ZnSO_4(aq) + BaCl_2(aq) \rightarrow BaSO_4(s) + ZnCl_2(aq)$ *Balanced*

(d) Reactants: $KOH(aq)$ and $MgCl_2(aq)$
Ions: K^+, OH^-, Mg^{2+}, Cl^-
Products: KCl and $Mg(OH)_2$
Solubility (Table 5.4): $KCl(aq)$ and $Mg(OH)_2(s)$

$2KOH(aq) + MgCl_2(aq) \rightarrow 2KCl(aq) + Mg(OH)_2(s)$ *Balanced*

(e) Reactants: $CuSO_4(aq)$ and $BaS(aq)$
Ions: Cu^{2+}, SO_4^{2-}, Ba^{2+}, S^{2-}
Products: CuS and $BaSO_4$
Solubility (Table 5.4): $CuS(s)$ and $BaSO_4(s)$

$CuSO_4(aq) + BaS(aq) \rightarrow CuS(s) + BaSO_4(s)$ *Balanced*

5.117   Potassium and sodium are in the same family (Group IA (1)), so we expect that they will react similarly with water. Water reacts with potassium to form potassium hydroxide and hydrogen gas.

$2K(s) + 2H_2O(l) \rightarrow 2KOH(aq) + H_2(g)$

5.119   A precipitate of silver chloride will form if a compound which forms chloride ions is added to the solution.
(a) Molecular chlorine does not form chloride ions when placed in water, so we don't expect to see a precipitate.
(b) When sodium chloride is added to water, it dissolves, forming sodium and chloride ions. A precipitate of $AgCl(s)$ should form.
(c) Lead chloride is not soluble in water, so we cannot expect there to be enough Cl^- ions added to the solution to cause AgCl to precipitate.
(d) Carbon tetrachloride is a molecular compound; it does not form chloride ions when added to water. The only molecular compounds that produce ions are acids and bases.

5.121   Certain combinations of strong acid and strong base produce the net ionic equation:

$H^+(aq) + OH^-(aq) \rightarrow H_2O(l)$

Some examples are solutions of HCl and NaOH, HNO_3 and KOH, HCl and KOH. This reaction occurs because strong acids and bases completely ionize in aqueous solution. For example:

$HCl(aq) + NaOH(aq) \rightarrow NaCl(aq) + H_2O(l)$ *Molecular equation*
$H^+(aq) + \cancel{Cl^-}(aq) + \cancel{Na^+}(aq) + OH^-(aq) \rightarrow \cancel{Na^+}(aq) + \cancel{Cl^-}(aq) + H_2O(l)$ *Ionic equation*
$H^+(aq) + OH^-(aq) \rightarrow H_2O(l)$ *Net ionic equation*

5.123   No reaction takes place because copper is less active than iron (Figure 5.21).

5.125 To show that KI is a strong electrolyte, we must show that it is an ionic solid that separates into its ions in solution. To show that CO is a nonelectrolyte, we must show that the CO molecules do not form ions in solution. Some representative images are shown below.

Solution of KI

Solution of CO

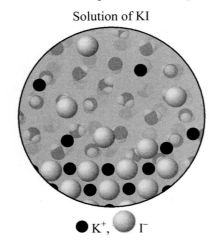
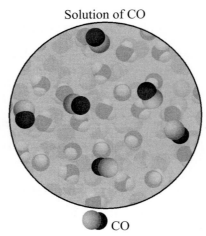

● K⁺, ◐ I⁻   ◐ CO

5.127 We can select any metal higher in the activity series (Figure 5.21). (a) Al; (b) Zn; (c) Mg; (d) Sn

5.129 We can apply the solubility rules (Table 5.4) to separate ions.
(a) Adding a potassium chromate solution will cause iron(III) chromate to precipitate, leaving Al^{3+} in solution.
(b) Adding a sodium sulfate solution will cause barium sulfate to precipitate, leaving Mg^{2+} in solution.
(c) Adding a silver perchlorate solution will cause silver chloride to precipitate, leaving NO_3^- in solution.
(d) Adding water will cause $MgSO_4$ to dissolve, separating it from solid barium sulfate.

5.131 We can look at Table 5.4, the solubility rules, to find an anion that will make an insoluble compound with Ag^+ but be soluble with Ba^{2+}. Most chlorides are soluble but AgCl is insoluble. If we add sodium chloride (NaCl), the Ag^+ ions will precipitate as AgCl leaving the Ba^{2+} ions in solution. Next we look for an insoluble compound of Ba^{2+}. There are several, including $BaSO_4$. Adding sodium sulfate (Na_2SO_4) to the solution will precipitate the Ba^{2+} ions as $BaSO_4$.

5.133 In a single-displacement reaction, one free element displaces another element from a compound. For example, a metal displaces an ion of another metal, or a nonmetal displaces an ion of another nonmetal. To determine whether a reaction occurs, we compare the activities of the metals involved (Figure 5.21). If the metal is more active than the metal whose ion appears in the compound, a reaction occurs. Nickel is more active than copper, so we expect a reaction to occur. In the lab we would observe the formation of copper on the surface of the nickel metal. The Cu^{2+} ions in solution give it a blue color while nickel ions give no color to the solution. As the Cu^{2+} ions react to form solid copper, the blue color of the solution would become lighter. As the nickel atoms form nickel ions they are removed from the surface of the metal. Over time we might observe that the nickel metal will appear to disintegrate.

5.135 Hydrochloric acid and sodium hydroxide solutions are both clear and colorless. They react in a double-displacement, neutralization reaction in which aqueous sodium chloride and water are formed. Aqueous sodium chloride is also clear and colorless so there wouldn't be an observable change unless we inserted a thermometer. We would notice a change in temperature due to the formation of water. The formation of more water molecules in an environment that already has a lot of water would not produce a noticeable change in volume.

5.137   We start by writing the skeletal equation:

$C_3H_8(g) + O_2(g) \rightarrow CO_2(g) + H_2O(g)$ *Unbalanced*

Atoms in Reactants			Atoms in Products		
3 C	8 H	2 O	1 C	2 H	3 O

In the combustion reactions of hydrocarbons it is usually easiest to begin by balancing the carbon and leave the balancing of oxygen for the last step. This follows the general concepts of balancing the atoms that appear in the smallest number of substances first (you could have also started with hydrogen) and balancing the substances appearing as pure elements last. Start with a coefficient of 3 in front of CO_2.

$C_3H_8(g) + O_2(g) \rightarrow \underline{3CO_2}(g) + H_2O(g)$ *Unbalanced*

Atoms in Reactants			Atoms in Products		
3 C	8 H	2 O	3 C	2 H	5 O

To balance the hydrogen, we use a coefficient of 4 for H_2O

$C_3H_8(g) + O_2(g) \rightarrow 3CO_2(g) + \underline{4H_2O}(g)$ *Unbalanced*

Atoms in Reactants			Atoms in Products		
3 C	8 H	2 O	3 C	8 H	10 O

To balance the oxygen, we use a coefficient of 5 for O_2

$C_3H_8(g) + \underline{5O_2}(g) \rightarrow 3CO_2(g) + 4H_2O(g)$ *Balanced*

Atoms in Reactants			Atoms in Products		
3 C	8 H	10 O	3 C	8 H	10 O

5.139   During this reaction atoms of elemental copper with a charge of zero are converted to Cu^{2+} ions. Each copper atom that reacts loses two electrons. As a copper atom loses electrons, Ag^+ ions are converted to neutral silver atoms. Each silver ion that reacts gains one electron.

Chapter 6 – Quantities in Chemical Reactions

6.1   (a) stoichiometry; (b) heat; (c) endothermic reaction; (d) specific heat; (e) actual yield

6.3   (a) The balanced chemical equation for the metabolism of glucose is:

$$C_6H_{12}O_6 + 6O_2 \rightarrow 6CO_2 + 6H_2O$$

        (b) The coefficients of the balanced chemical equation give the proportions of the reactants and products in a reaction. From the balanced chemical equation, we see that the proportion of carbon dioxide to glucose is 6 molecules of CO_2 to 1 molecule of $C_6H_{12}O_6$. When 12 molecules of $C_6H_{12}O_6$ react, 72 (12 × 6) molecules of CO_2 form:

$$12 \text{ molecules } C_6H_{12}O_6 \times \frac{6 \text{ molecules } CO_2}{1 \text{ molecule } C_6H_{12}O_6} = 72 \text{ molecules } CO_2$$

        (c) From the balanced equation we see that glucose reacts with oxygen in a ratio of 1 $C_6H_{12}O_6$ molecule to 6 O_2 molecules. If 30 molecules of O_2 react, then 5 $C_6H_{12}O_6$ molecules will react:

$$30 \text{ molecules } O_2 \times \frac{1 \text{ molecule } C_6H_{12}O_6}{6 \text{ molecules } C_6H_{12}O_6} = 5 \text{ molecules } C_6H_{12}O_6$$

6.5   The coefficients in a balanced chemical equation represent the relative ratios, or proportions, of reactants consumed and products produced. They can represent the actual numbers of molecules, or the number of moles of each substance that reacts.

6.7   The coefficients of the balanced chemical equation represent the amounts of each substance that react and are produced. We can interpret these coefficients on a molecule-to-molecule basis, on a mole-to-mole basis, or even on a dozen-to-dozen basis.
        (a) For each formula unit of $Mg(NO_3)_2(s)$ that dissolves, one $Mg^{2+}(aq)$ ion and two $NO_3^-(aq)$ ions form.
        (b) For each mole of $Mg(NO_3)_2(s)$ that dissolves, one mole of $Mg^{2+}(aq)$ ions and two moles of $NO_3^-(aq)$ ions form.

6.9   The best representation is (d); however, we could also have chosen (c) because it has the reactants and products in the correct proportions, although not in the smallest whole-number ratio. In reaction (a) the reactants are not diatomic and the product only has one atom of B. In reaction (b), mass is not conserved (the equation is not balanced).

6.11   (a) We take the mole ratio of O_2 to C_6H_6 directly from the balanced chemical equation.

$$\text{Mole ratio} = \frac{15 \text{ mol } O_2}{2 \text{ mol } C_6H_6}$$

        (b) We calculate the number of moles of O_2 that react with 1 mole of C_6H_6 using the following problem solving map:

$$\text{Moles } C_6H_6 \xrightarrow{\text{mole ratio}} \text{Moles } O_2$$

        When cancelling units, make sure that the units you want to finish with (mol O_2) are in the numerator (top) of the conversion factor, and that the units you want to cancel (mol C_6H_6) are in the denominator (bottom):

$$\text{Moles of } O_2 = 1 \text{ mol } C_6H_6 \times \frac{15 \text{ mol } O_2}{2 \text{ mol } C_6H_6} = 7.5 \text{ mol } O_2$$

(c)   We can use the same problem solving map here that we used in (b):

$$\text{Moles of } O_2 = 0.38 \text{ mol } C_6H_6 \times \frac{15 \text{ mol } O_2}{2 \text{ mol } C_6H_6} = 2.8 \text{ mol } O_2$$

6.13   (a)   There are two possible ways of writing the mole ratios for HNO_3 and Al: $\dfrac{6 \text{ mol } HNO_3}{2 \text{ mol Al}}$ and

$\dfrac{2 \text{ mol Al}}{6 \text{ mol } HNO_3}$. The correct mole ratio for the conversion of mol Al to mol HNO_3 is $\dfrac{6 \text{ mol } HNO_3}{2 \text{ mol Al}}$

because it allows for the proper cancellation of units:

$$\text{Moles of } HNO_3 = \text{mol Al} \times \frac{6 \text{ mol } HNO_3}{2 \text{ mol Al}}$$

(b)   As in part (a) there are two different forms of the mole ratio that we can write for H_2 and Al: $\dfrac{3 \text{ mol } H_2}{2 \text{ mol Al}}$ and

$\dfrac{2 \text{ mol Al}}{3 \text{ mol } H_2}$. We determine the correct mole ratio for the conversion of mol Al to mol H_2 by the cancellation

of units:

$$\text{Moles of } H_2 = \text{mol Al} \times \frac{3 \text{ mol } H_2}{2 \text{ mol Al}}$$

(c)   The correct mole ratio is $\dfrac{2 \text{ mol Al}}{3 \text{ mol } H_2}$.

$$\text{Moles of Al} = \text{mol } H_2 \times \frac{2 \text{ mol Al}}{3 \text{ mol } H_2}$$

6.15   Because atoms in the reactant molecules rearrange to form different product molecules, the number of moles of molecules (a) is generally not conserved. In the combustion reaction of propane, there are six reactant molecules that produce seven product molecules. For all reactions, (b), (c), and (d) are conserved. A balanced chemical equation is a statement of the law of conservation of mass for a chemical reaction. Mass is conserved because atoms are not created or destroyed in the reaction.

6.17   (a)   There are two ways of writing each of the mole ratios for each product (a total of 8 different mole ratios):

$$\frac{7 \text{ mol C}}{2 \text{ mol } C_7H_5(NO_2)_3} \text{ and } \frac{2 \text{ mol } C_7H_5(NO_2)_3}{7 \text{ mol C}}$$

$$\frac{7 \text{ mol CO}}{2 \text{ mol } C_7H_5(NO_2)_3} \text{ and } \frac{2 \text{ mol } C_7H_5(NO_2)_3}{7 \text{ mol CO}}$$

$$\frac{3 \text{ mol } N_2}{2 \text{ mol } C_7H_5(NO_2)_3} \text{ and } \frac{2 \text{ mol } C_7H_5(NO_2)_3}{3 \text{ mol } N_2}$$

$$\frac{5 \text{ mol } H_2O}{2 \text{ mol } C_7H_5(NO_2)_3} \text{ and } \frac{2 \text{ mol } C_7H_5(NO_2)_3}{5 \text{ mol } H_2O}$$

(b) To calculate the number of moles of product, we choose the mole ratio that allows us to cancel moles of $C_7H_5(NO_2)_3$ and produces the desired units (i.e. mol C, mol CO, mol N_2, and mol H_2O).

$$\text{Moles of C} = 1.00 \ \cancel{\text{mol } C_7H_5(NO_2)_3} \times \frac{7 \ \text{mol C}}{2 \ \cancel{\text{mol } C_7H_5(NO_2)_3}} = 3.50 \ \text{mol C}$$

For mole-to-mole conversions, you can usually tell when you have the correct answer. For example, in the balanced equation, the coefficient for C is 7 and the coefficient for $C_7H_5(NO_2)_3$ is 1. Therefore, there should be more moles of C than moles of $C_7H_5(NO_2)_3$. In this case, we started with 1.00 mol of $C_7H_5(NO_2)_3$ and produced 3.50 mol of C.

$$\text{Moles of CO} = 1.00 \ \cancel{\text{mol } C_7H_5(NO_2)_3} \times \frac{7 \ \text{mol CO}}{2 \ \cancel{\text{mol } C_7H_5(NO_2)_3}} = 3.50 \ \text{mol CO}$$

$$\text{Moles of } N_2 = 1.00 \ \cancel{\text{mol } C_7H_5(NO_2)_3} \times \frac{3 \ \text{mol } N_2}{2 \ \cancel{\text{mol } C_7H_5(NO_2)_3}} = 1.50 \ \text{mol } N_2$$

$$\text{Moles of } H_2O = 1.00 \ \cancel{\text{mol } C_7H_5(NO_2)_3} \times \frac{5 \ \text{mol } H_2O}{2 \ \cancel{\text{mol } C_7H_5(NO_2)_3}} = 2.50 \ \text{mol } H_2O$$

(c) To calculate the number of moles of product, we choose the mole ratio which allows for the proper cancellation of units.

$$\text{Moles of C} = 6.25 \ \cancel{\text{mol } C_7H_5(NO_2)_3} \times \frac{7 \ \text{mol C}}{2 \ \cancel{\text{mol } C_7H_5(NO_2)_3}} = 21.9 \ \text{mol C}$$

$$\text{Moles of CO} = 6.25 \ \cancel{\text{mol } C_7H_5(NO_2)_3} \times \frac{7 \ \text{mol CO}}{2 \ \cancel{\text{mol } C_7H_5(NO_2)_3}} = 21.9 \ \text{mol CO}$$

$$\text{Moles of } N_2 = 6.25 \ \cancel{\text{mol } C_7H_5(NO_2)_3} \times \frac{3 \ \text{mol } N_2}{2 \ \cancel{\text{mol } C_7H_5(NO_2)_3}} = 9.38 \ \text{mol } N_2$$

$$\text{Moles of } H_2O = 6.25 \ \cancel{\text{mol } C_7H_5(NO_2)_3} \times \frac{5 \ \text{mol } H_2O}{2 \ \cancel{\text{mol } C_7H_5(NO_2)_3}} = 15.6 \ \text{mol } H_2O$$

6.19 (a) The molar mass calculation for $CuSO_4 \cdot 5H_2O$ is shown below (see section 4.2 of your text for a review). $CuSO_4 \cdot 5H_2O$ (Note that the mass of 1 water molecule was calculated separately as 18.016 g/mol.)

Mass of 1 mol of Cu $= 1 \ \text{mol} \times 63.55 \ \text{g/mol} = 63.55 \ \text{g}$
Mass of 1 mol of S $= 1 \ \text{mol} \times 32.06 \ \text{g/mol} = 32.06 \ \text{g}$
Mass of 4 mol of O $= 4 \ \text{mol} \times 16.00 \ \text{g/mol} = 64.00 \ \text{g}$
Mass of 5 mol of H_2O $= 5 \ \text{mol} \times 18.016 \ \text{g/mol} = 90.08 \ \text{g}$
Mass of 1 mol of $CuSO_4 \cdot 5H_2O$ $= 249.69 \ \text{g}$

Molar mass = 249.69 g/mol

(b) $CuSO_4$

Mass of 1 mol of Cu $= 1 \ \text{mol} \times 63.55 \ \text{g/mol} = 63.55 \ \text{g}$
Mass of 1 mol of S $= 1 \ \text{mol} \times 32.06 \ \text{g/mol} = 32.06 \ \text{g}$
Mass of 4 mol of O $= 4 \ \text{mol} \times 16.00 \ \text{g/mol} = 64.00 \ \text{g}$
Mass of 1 mol of $CuSO_4$ $= 159.61 \ \text{g}$

Molar mass = 159.61 g/mol

(c) This mass-mass conversion follows the general problem solving map:

$$\text{Grams A} \xrightarrow{\ MM\ A\ } \text{Moles A} \xrightarrow{\ \text{mole ratio}\ } \text{Moles B} \xrightarrow{\ MM\ B\ } \text{Grams B}$$

$$\text{Grams } CuSO_4 \cdot 5H_2O \xrightarrow{\ MM\ CuSO_4 \cdot 5H_2O\ } \text{Moles } CuSO_4 \cdot 5H_2O$$

$$\xrightarrow{\ \text{mole ratio}\ } \text{Moles } CuSO_4 \xrightarrow{\ MM\ CuSO_4\ } \text{Grams } CuSO_4$$

Note that there are three steps in the problem solving map and three conversion factors (two molar masses and one mole ratio). We start with the mass of $CuSO_4 \cdot 5H_2O$ and use the molar mass of $CuSO_4 \cdot 5H_2O$ so that the grams of $CuSO_4 \cdot 5H_2O$ cancel, leaving us with moles of $CuSO_4 \cdot 5H_2O$. After we complete the first step of the problem solving map, our calculation should look like:

$$1.00 \text{ g } CuSO_4 \cdot 5H_2O \times \frac{1 \text{ mol } CuSO_4 \cdot 5H_2O}{249.69 \text{ g } CuSO_4 \cdot 5H_2O}$$

In the second step, we apply the mole conversion factor so that moles of $CuSO_4 \cdot 5H_2O$ cancel, leaving moles of $CuSO_4$. At this point our calculation looks like:

$$1.00 \text{ g } CuSO_4 \cdot 5H_2O \times \frac{1 \text{ mol } CuSO_4 \cdot 5H_2O}{249.69 \text{ g } CuSO_4 \cdot 5H_2O} \times \frac{1 \text{ mol } CuSO_4}{1 \text{ mol } CuSO_4 \cdot 5H_2O}$$

Notice that the units of the answer at each step are the units indicated in the problem solving map. That is, after using the mole ratio we should be left with moles of $CuSO_4$. In the final step we convert moles of $CuSO_4$ to grams of $CuSO_4$. The complete calculation looks like this:

$$\text{Mass of } CuSO_4 = 1.00 \text{ g } CuSO_4 \cdot 5H_2O \times \frac{1 \text{ mol } CuSO_4 \cdot 5H_2O}{249.69 \text{ g } CuSO_4 \cdot 5H_2O} \times \frac{1 \text{ mol } CuSO_4}{1 \text{ mol } CuSO_4 \cdot 5H_2O} \times \frac{159.61 \text{ g } CuSO_4}{1 \text{ mol } CuSO_4}$$

$$= 0.639 \text{ g } CuSO_4$$

6.21 (a) The mass increase is directly related to the mass of oxygen incorporated into the product. Since the sodium increased in mass by 2.05 g, this must be the mass of oxygen that reacted with the sodium.

(b) The mass-mass conversion involves three steps.

$$\text{Grams } O_2 \xrightarrow{\ MM\ O_2\ } \text{Moles } O_2 \xrightarrow{\ \text{mole ratio}\ } \text{Moles Na} \xrightarrow{\ MM\ Na\ } \text{Grams Na}$$

We calculate the molar masses of O_2 (32.00 g/mol) and Na (22.99 g/mol) using information on the periodic table. The mole ratio comes from the balanced chemical equation given in the problem (4 mol Na:1 mol O_2).

$$\text{Mass of Na} = 2.05 \text{ g } O_2 \times \frac{1 \text{ mol } O_2}{32.00 \text{ g } O_2} \times \frac{4 \text{ mol Na}}{1 \text{ mol } O_2} \times \frac{22.99 \text{ g Na}}{1 \text{ mol Na}} = 5.89 \text{ g Na}$$

(c) There are several different methods for solving this problem. One is to use the law of conservation of mass. Since we know the masses of sodium [part (b)] and oxygen [part (a)] that reacted, the mass of the product must be the sum of these masses of reactants:

Total mass = 5.89 g + 2.05 g = 7.94 g

We can also use mass-mass conversions to get this same result. Starting with either the mass of oxygen or sodium, we can calculate the mass of product using the following problem solving map:

$$\text{Grams O}_2 \xrightarrow{\; MM\ O_2 \;} \text{Moles O}_2 \xrightarrow{\; \text{mole ratio} \;} \text{Moles Na}_2\text{O} \xrightarrow{\; MM\ Na_2O \;} \text{Grams Na}_2\text{O}$$

We calculate the molar masses of O_2 (32.00 g/mol) and Na_2O (61.98 g/mol) from information on the periodic table. The mole ratio comes from the balanced chemical equation given in the problem (2 mol Na_2O:1 mol O_2).

$$\text{Mass of Na}_2\text{O} = 2.05 \text{ g O}_2 \times \frac{1 \text{ mol O}_2}{32.00 \text{ g O}_2} \times \frac{2 \text{ mol Na}_2\text{O}}{1 \text{ mol O}_2} \times \frac{61.98 \text{ g Na}_2\text{O}}{1 \text{ mol Na}_2\text{O}} = 7.94 \text{ g Na}_2\text{O}$$

6.23   (a) We are looking for the mass of CO that will react with 50.0 g of I_2O_5 according to the equation:

$$I_2O_5(s) + 5CO(g) \rightarrow I_2(s) + 5CO_2(g)$$

The problem solving map begins with the mass of I_2O_5 (50.0 g) and ends with the mass of CO.

$$\text{Grams I}_2\text{O}_5 \xrightarrow{\; MM\ I_2O_5 \;} \text{Moles I}_2\text{O}_5 \xrightarrow{\; \text{mole ratio} \;} \text{Moles CO} \xrightarrow{\; MM\ CO \;} \text{Grams CO}$$

The molar masses of I_2O_5 and CO are 333.8 g/mol and 28.01 g/mol, respectively. We find the mole ratio from the balanced chemical equation.

$$\text{Mass of CO} = 50.0 \text{ g I}_2\text{O}_5 \times \frac{1 \text{ mol I}_2\text{O}_5}{333.8 \text{ g I}_2\text{O}_5} \times \frac{5 \text{ mol CO}}{1 \text{ mol I}_2\text{O}_5} \times \frac{28.01 \text{ g CO}}{1 \text{ mol CO}} = 21.0 \text{ g CO}$$

(b) To calculate the mass of I_2 in the respirator we need the mole ratio of I_2 to I_2O_5 in this reaction, which we get from the balanced chemical equation, and the molar mass of I_2 (253.8 g/mol).

$$\text{Grams I}_2\text{O}_5 \xrightarrow{\; MM\ I_2O_5 \;} \text{Moles I}_2\text{O}_5 \xrightarrow{\; \text{mole ratio} \;} \text{Moles I}_2 \xrightarrow{\; MM\ I_2 \;} \text{Grams I}_2$$

$$\text{Mass of I}_2 = 50.0 \text{ g I}_2\text{O}_5 \times \frac{1 \text{ mol I}_2\text{O}_5}{333.8 \text{ g I}_2\text{O}_5} \times \frac{1 \text{ mol I}_2}{1 \text{ mol I}_2\text{O}_5} \times \frac{253.8 \text{ g I}_2}{1 \text{ mol I}_2} = 38.0 \text{ g I}_2$$

6.25   Before we can use any chemical equation, we must balance it. Then, for this problem we must determine the molar masses of the reactants in question. Because we know the mass of the first reactant (0.600 g), the problem solving map starts with the mass of the first reactant and ends with the mass of the second reactant. Letting R1 be the first reactant and R2 be the second reactant, the problem solving map will look like this:

$$\text{Grams R1} \xrightarrow{\; MM\ R1 \;} \text{Moles R1} \xrightarrow{\; \text{mole ratio} \;} \text{Moles R2} \xrightarrow{\; MM\ R2 \;} \text{Grams R2}$$

(a) Balanced equation: $2Cr(s) + 3Cl_2(g) \rightarrow 2CrCl_3(s)$

Molar masses: Cr (52.00 g/mol), Cl_2 (70.90 g/mol)

$$\text{Mass of Cl}_2 = 0.600 \text{ g Cr} \times \frac{\text{mol Cr}}{52.00 \text{ g Cr}} \times \frac{3 \text{ mol Cl}_2}{2 \text{ mol Cr}} \times \frac{70.90 \text{ g Cl}_2}{1 \text{ mol Cl}_2} = 1.23 \text{ g Cl}_2$$

(b) Balanced equation: $4RbO_2(s) + 2H_2O(l) \rightarrow 3O_2(g) + 4RbOH(s)$

Molar masses: RbO_2 (117.47 g/mol), H_2O (18.02 g/mol)

$$\text{Mass of H}_2\text{O} = 0.600 \text{ g RbO}_2 \times \frac{\text{mol RbO}_2}{117.47 \text{ g RbO}_2} \times \frac{2 \text{ mol H}_2\text{O}}{4 \text{ mol RbO}_2} \times \frac{18.02 \text{ g H}_2\text{O}}{1 \text{ mol H}_2\text{O}} = 0.0460 \text{ g H}_2\text{O}$$

(c) Balanced equation: $C_5H_{12}(g) + 8O_2(g) \rightarrow 5CO_2(g) + 6H_2O(g)$

Molar masses: C_5H_{12} (72.15 g/mol), O_2 (32.00 g/mol)

$$\text{Mass of } O_2 = 0.600 \text{ g } C_5H_{12} \times \frac{1 \text{ mol } C_5H_{12}}{72.15 \text{ g } C_5H_{12}} \times \frac{8 \text{ mol } O_2}{1 \text{ mol } C_5H_{12}} \times \frac{32.00 \text{ g } O_2}{1 \text{ mol } O_2} = 2.13 \text{ g } O_2$$

(d) Balanced equation: $2Li(s) + Cl_2(g) \rightarrow 2LiCl(s)$

Molar masses: Li (6.94 g/mol), Cl_2 (70.90 g/mol)

$$\text{Mass of } Cl_2 = 0.600 \text{ g Li} \times \frac{1 \text{ mol Li}}{6.94 \text{ g Li}} \times \frac{1 \text{ mol } Cl_2}{2 \text{ mol Li}} \times \frac{70.90 \text{ g } Cl_2}{1 \text{ mol } Cl_2} = 3.06 \text{ g } Cl_2$$

6.27   We want to calculate mass of $CaCO_3$ that reacts with (neutralizes) 0.020 mol of HCl. The problem solving map should start with what we know, the number of moles of HCl, and end with what we're looking for, the mass of $CaCO_3$.

$$\text{Moles HCl} \xrightarrow{\text{mole ratio}} \text{Moles } CaCO_3 \xrightarrow{MM \ CaCO_3} \text{Grams } CaCO_3$$

The molar mass of $CaCO_3$ is 100.09 g/mol and we find the mole ratio from the balanced chemical equation (given in the problem):

$$CaCO_3(s) + 2HCl(aq) \rightarrow CaCl_2(aq) + CO_2(g) + H_2O(l)$$

$$\text{Mass of } CaCO_3 = 0.020 \text{ mol HCl} \times \frac{1 \text{ mol } CaCO_3}{2 \text{ mol HCl}} \times \frac{100.09 \text{ g } CaCO_3}{1 \text{ mol } CaCO_3} = 1.0 \text{ g } CaCO_3$$

6.29   The fuel is the limiting reactant. Typically, air contains 20% O_2 and there is usually an inexhaustible supply. However, there is a limited supply of fuel.

6.31   The equation for making a turkey sandwich can be summarized as:

2 pieces of bread + 1 piece of turkey = 1 sandwich

There is a 2:1 ratio of pieces of bread to pieces of turkey required for making sandwiches, and we are starting with 24 pieces of bread and 15 pieces of turkey. Since there are not twice as many pieces of bread as there are pieces of turkey (that would be 30 pieces of bread), the bread is the limiting reactant. With 24 pieces of bread we can make 12 sandwiches (2 pieces of bread for each sandwich). We also know that making 12 sandwiches requires 12 pieces of turkey. Since there are 15 pieces of turkey, there is an excess of turkey (3 pieces). (a) 12 sandwiches; (b) bread is limiting; (c) 3 pieces of turkey

6.33   The balanced chemical equation tells us that 3 molecules of F_2 react with 1 molecule of N_2. To determine the limiting reactant, we can calculate the number of molecules of F_2 necessary to react with the given number of molecules of N_2. By comparing the amount of N_2 needed to react with the amount given in the problem, we can determine the limiting reactant.

(a) First, we calculate the number of F_2 molecules required to react with the 9 molecules of N_2:

$$\text{Molecules of } F_2 = 9 \text{ molecules } N_2 \times \frac{3 \text{ molecules } F_2}{1 \text{ molecule } N_2} = 27 \text{ molecules of } F_2$$

Because the number of F_2 molecules available (9) is less than the number needed to react completely with 27 N_2 molecules, F_2 is the limiting reactant.

(b) Molecules of F_2 = 5 ~~molecules N_2~~ $\times \dfrac{3 \text{ molecules } F_2}{1 \text{ ~~molecule N_2~~}}$ = 15 molecules of F_2

Since there are more F_2 molecules (20) than are needed to react completely with the N_2, N_2 is the limiting reactant. That is, when 5 molecules of N_2 react, 15 molecules of F_2 will also react. There will be 5 molecules of F_2 remaining.

(c) Molecules of F_2 = 6 ~~molecules N_2~~ $\times \dfrac{3 \text{ molecules } F_2}{1 \text{ ~~molecule N_2~~}}$ = 18 molecules of F_2

Since the number of F_2 molecules available (18) is the same as the number needed to react with 6 N_2 molecules, both are limiting (i.e. the reaction stops when both run out). Both F_2 and N_2 react completely.

6.35   (a) There are 3 O_2 molecules and 8 H_2 molecules shown in the diagram. For each molecule of O_2 that reacts, 2 H_2 molecules also react, and 2 H_2O molecules are produced. We can calculate the number of H_2 molecules that will be consumed when 3 O_2 molecules react:

Molecules of H_2 = 3 ~~molecules O_2~~ $\times \dfrac{2 \text{ molecules } H_2}{1 \text{ ~~molecule O_2~~}}$ = 6 molecules of H_2

Because we have 8 H_2 molecules available, which is 2 more than we need, O_2 is the limiting reactant. The number of water molecules that are produced by the reaction is determined by the limiting reactant:

Molecules of H_2O = 3 ~~molecules O_2~~ $\times \dfrac{2 \text{ molecules } H_2O}{1 \text{ ~~molecule O_2~~}}$ = 6 molecules of H_2O

In the "after" diagram there should be 6 molecules of H_2O, 2 hydrogen molecules, and no oxygen molecules.

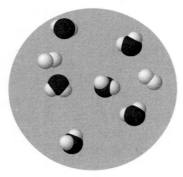

(b) Oxygen is the limiting reactant.

(c) Hydrogen is left over following the reaction.

6.37   Starting with 6 molecules of C_2H_6 we can calculate the number of O_2 molecules that would react:

$$\text{Molecules of } O_2 = 6 \text{ molecules } C_2H_6 \times \frac{7 \text{ molecules } O_2}{2 \text{ molecules } C_2H_6} = 21 \text{ molecules of } O_2$$

Since there are only 18 O_2 molecules, O_2 is the limiting reactant. When 14 O_2 molecules react, 4 C_2H_6 molecules will also react. According to the equation, oxygen reacts in groups of 7 molecules at a time so four of the O_2 molecules will remain unreacted (see table below). For every 7 O_2 molecules that react, 2 C_2H_6 molecules will react (based on the balanced equation). This means 4 molecules of C_2H_6 were used. The products are formed in proportion to the number of molecules of limiting reactant available:

$$\text{Molecules of } CO_2 = 14 \text{ molecules } O_2 \times \frac{4 \text{ molecules } CO_2}{7 \text{ molecules } O_2} = 8 \text{ molecules of } CO_2$$

$$\text{Molecules of } H_2O = 14 \text{ molecules } O_2 \times \frac{6 \text{ molecules } H_2O}{7 \text{ molecules } O_2} = 12 \text{ molecules of } H_2O$$

	$2C_2H_6(g)$ +	$7O_2(g)$ →	$4CO_2(g)$ +	$6H_2O(g)$
Initially mixed	6 molecules	18 molecules	0 molecules	0 molecules
How much reacts	4 molecules	14 molecules	---	---
Composition of final mixture	2 molecules	4 molecules	8 molecules	12 molecules

6.39   We can calculate the number of moles of O_2 required to react with 3.10 mol of C_4H_{10}:

$$\text{Moles of } O_2 = 3.10 \text{ mol } C_4H_{10} \times \frac{13 \text{ mol } O_2}{2 \text{ mol } C_4H_{10}} = 20.2 \text{ mol of } O_2$$

Since the initial reaction mixture contains fewer than 20.2 mol O_2, we know that O_2 is the limiting reactant. We calculate the number of moles of C_4H_{10} that will react with the O_2 (limiting reactant) to determine how much C_4H_{10} remains:

$$\text{Moles of } C_4H_{10} = 13.0 \text{ mol } O_2 \times \frac{2 \text{ mol } C_4H_{10}}{13 \text{ mol } O_2} = 2.00 \text{ mol of } C_4H_{10}$$

We started with 3.10 mol C_4H_{10} and when the reaction ends, 1.10 mol will remain. We calculate the number of moles of product from the number of moles of the limiting reactant:

$$\text{Moles of } CO_2 = 13.0 \text{ mol } O_2 \times \frac{8 \text{ mol } CO_2}{13 \text{ mol } O_2} = 8.00 \text{ mol of } CO_2$$

$$\text{Moles of } H_2O = 13.0 \text{ mol } O_2 \times \frac{10 \text{ mol } H_2O}{13 \text{ mol } O_2} = 10.0 \text{ mol of } H_2O$$

	$2C_4H_{10}(g)$ +	$13O_2(g)$ →	$8CO_2(g)$ +	$10H_2O(g)$
Initially mixed	3.10 mol	13.0 mol	0.00 mol	0.00 mol
How much reacts	2.00 mol	13.0 mol	---	---
Composition of final mixture	1.10 mol	0.0 mol	8.00 mol	10.0 mol

6.41　To determine the limiting reactant, we can calculate the number of moles of O_2 required to react with the specified number of moles of P_4. If the number of moles of O_2 available is less than the calculated number, then O_2 is the limiting reactant. If the number of moles of O_2 available is greater than the calculated number, then P_4 is the limiting reactant. If the number of moles of O_2 available is equal to the calculated number, then both reactants will run out at the same time and both limit the amount of product produced. (You will come to the same conclusions if you calculate the number of moles of P_4 needed to react exactly with the number of moles of O_2 given in the problem.)

Balanced equation (from question): $P_4(g) + 5O_2(g) \rightarrow P_4O_{10}(s)$

(a)　Moles of O_2 = 0.50 $\text{mol } P_4 \times \dfrac{5 \text{ mol } O_2}{1 \text{ mol } P_4}$ = 2.5 mol O_2

Since 5.0 moles of O_2 are available, P_4 is the limiting reactant.

(b)　Moles of O_2 = 0.20 $\text{mol } P_4 \times \dfrac{5 \text{ mol } O_2}{1 \text{ mol } P_4}$ = 1.0 mol O_2

Since 1.0 mol of O_2 is available, both reactants will be completely consumed at the same time.

(c)　Moles of O_2 = 0.25 $\text{mol } P_4 \times \dfrac{5 \text{ mol } O_2}{1 \text{ mol } P_4}$ = 1.3 mol O_2

Since only 0.75 mol of O_2 are available, O_2 is the limiting reactant.

6.43　(a)　To determine the limiting reactant, we can calculate the mass of F_2 required to react with the specified mass of N_2. If the calculated mass of F_2 is greater than the available mass, then F_2 is the limiting reactant. If the calculated mass is smaller than the available mass, then N_2 is the limiting reactant. If the calculated and available masses are the same, then both reactants will be completely consumed at the same time and both limit the amount of product produced.

Balanced equation (given): $N_2(g) + 3 F_2(g) \rightarrow 2NF_3(g)$

Molar Masses: N_2 (28.02 g/mol), F_2 (38.00 g/mol)

Mass of F_2 = 10.0 $\text{g } N_2 \times \dfrac{1 \text{ mol } N_2}{28.02 \text{ g } N_2} \times \dfrac{3 \text{ mol } F_2}{1 \text{ mol } N_2} \times \dfrac{38.00 \text{ g } F_2}{1 \text{ mol } F_2}$ = 40.69 g F_2

Since only 10.0 g of F_2 are available, F_2 is the limiting reactant.

(b)　We determine the mass of NF_3 produced from the mass of the limiting reactant and the molar mass of NF_3 (71.01 g/mol).

Mass of NH_3 = 10.0 $\text{g } F_2 \times \dfrac{1 \text{ mol } F_2}{38.00 \text{ g } F_2} \times \dfrac{2 \text{ mol } NH_3}{3 \text{ mol } F_2} \times \dfrac{71.01 \text{ g } NF_3}{1 \text{ mol } NF_3}$ = 12.5 g NF_3

6.45　We can determine the limiting reactant by calculating the mass of HCl required to react with the available mass of $NaHCO_3$.

Molar masses: HCl (36.46 g/mol), $NaHCO_3$ (84.01 g/mol)

Balanced equation (given): $NaHCO_3(s) + HCl(aq) \rightarrow NaCl(aq) + H_2O(l) + CO_2(g)$

Mass of HCl = 8.0 $\text{g } NaHCO_3 \times \dfrac{1 \text{ mol } NaHCO_3}{84.01 \text{ g } NaHCO_3} \times \dfrac{1 \text{ mol } HCl}{1 \text{ mol } NaHCO_3} \times \dfrac{36.46 \text{ g } HCl}{1 \text{ mol } HCl}$ = 3.5 g HCl

Since the mass of HCl required to react with 8.0 g of $NaHCO_3$ is less than is available, $NaHCO_3$ is the limiting reactant. We calculate the mass of CO_2 produced using the mass of the limiting reactant.

Molar mass: CO_2 (44.01 g/mol)

$$\text{Mass of } CO_2 = 8.0 \text{ g } NaHCO_3 \times \frac{1 \text{ mol } NaHCO_3}{84.01 \text{ g } NaHCO_3} \times \frac{1 \text{ mol } CO_2}{1 \text{ mol } NaHCO_3} \times \frac{44.01 \text{ g } CO_2}{1 \text{ mol } CO_2} = 4.2 \text{ g } CO_2$$

6.47   First, we must determine whether O_2 or C_2H_6 is the limiting reactant. To do this we can calculate the mass of O_2 required to react with 0.260 g of C_2H_6.

Molar masses: C_2H_6 (30.07 g/mol), O_2 (32.00 g/mol)

$$\text{Mass of } O_2 = 0.260 \text{ g } C_2H_6 \times \frac{1 \text{ mol } C_2H_6}{30.07 \text{ g } C_2H_6} \times \frac{7 \text{ mol } O_2}{2 \text{ mol } C_2H_6} \times \frac{32.00 \text{ g } O_2}{1 \text{ mol } O_2} = 0.968 \text{ g } O_2$$

Since we have more than 0.968 g of O_2 available, C_2H_6 is the limiting reactant and the reaction will consume 0.968 g of O_2.

We determine the mass of each product based on the reaction of 0.260 g of C_2H_6.

Molar masses: CO_2 (44.01 g/mol), H_2O (18.02 g/mol)

$$\text{Mass of } CO_2 = 0.260 \text{ g } C_2H_6 \times \frac{1 \text{ mol } C_2H_6}{30.07 \text{ g } C_2H_6} \times \frac{4 \text{ mol } CO_2}{2 \text{ mol } C_2H_6} \times \frac{44.01 \text{ g } CO_2}{1 \text{ mol } CO_2} = 0.761 \text{ g } CO_2$$

$$\text{Mass of } H_2O = 0.260 \text{ g } C_2H_6 \times \frac{1 \text{ mol } C_2H_6}{30.07 \text{ g } C_2H_6} \times \frac{6 \text{ mol } H_2O}{2 \text{ mol } C_2H_6} \times \frac{18.02 \text{ g } H_2O}{1 \text{ mol } H_2O} = 0.467 \text{ g } H_2O$$

	$2C_2H_6(g)$ +	$7O_2(g)$ →	$4CO_2(g)$ +	$6H_2O(g)$
Initially mixed	0.260 g	1.00 g	0.00 g	0.00 g
How much reacts	0.260 g	0.968 g	---	---
Composition of final mixture	0.000 g	0.03 g	0.761 g	0.467 g

6.49   (a) We write the balanced chemical equation based on the ions present in the solution:
Ions: Na^+, OH^-, H^+, NO_3^-
Products: H_2O, $NaNO_3$

$NaOH(aq) + HNO_3(aq) \rightarrow H_2O(l) + NaNO_3(aq)$

(b) To determine the limiting reactant, we calculate the mass of NaOH required to react with 10.0 g of HNO_3.

Molar masses: NaOH (40.00 g/mol), HNO_3 (63.02 g/mol)

$$\text{Mass of NaOH} = 10.0 \text{ g } HNO_3 \times \frac{1 \text{ mol } HNO_3}{63.02 \text{ g } HNO_3} \times \frac{1 \text{ mol NaOH}}{1 \text{ mol } HNO_3} \times \frac{40.00 \text{ g NaOH}}{1 \text{ mol NaOH}} = 6.35 \text{ g NaOH}$$

Since less NaOH is required than is available, HNO_3 is the limiting reactant.

(c) Because the acid is the limiting reactant there will be an excess of base in the solution. This means the solution will be basic.

6.51   The amount of copper recovered from the ore is the actual yield.

6.53   The actual yield is almost always lower than the theoretical yield, so the actual percent yield is likely to be less than 100%. The student weighed the solid immediately after filtering it and before it was completely dry, causing its apparent mass to be high.

6.55   (a) The actual yield in this synthesis is the amount of product weighed, 7.44 g.

(b) The theoretical yield is the amount of product that could have been prepared based on the mass of the limiting reagent, 8.95 g.

(c) The percent yield is 83.1%.

$$\text{Percent yield} = \frac{\text{actual yield}}{\text{theoretical yield}} \times 100\%$$

$$= \frac{7.44 \text{ g}}{8.95 \text{ g}} \times 100\% = 83.1\%$$

6.57   To calculate the percent yield we need to know both the actual and theoretical yields of the reaction:

$$\text{Percent yield} = \frac{\text{actual yield}}{\text{theoretical yield}} \times 100\%$$

The 11.5 g of NaCl that was obtained from the experiment is the actual yield. To calculate the theoretical yield we use the mass of the limiting reagent, 5.00 g of Na, and calculate the mass of product (NaCl) that can be produced from the reaction.
Balanced equation (given): $2Na(s) + Cl_2(g) \rightarrow 2NaCl(s)$

Molar masses: Na (22.99 g/mol), NaCl (58.44 g/mol)

$$\text{Mass NaCl} = 5.00 \text{ g Na} \times \frac{1 \text{ mol Na}}{22.99 \text{ g Na}} \times \frac{2 \text{ mol NaCl}}{2 \text{ mol Na}} \times \frac{58.44 \text{ g NaCl}}{1 \text{ mol NaCl}} = 12.7 \text{ g NaCl}$$

Finally, we calculate the percent yield:

$$\text{Percent yield} = \frac{11.5 \text{ g}}{12.7 \text{ g}} \times 100\% = 90.5\%$$

6.59   To calculate the number of moles of I_2 that reacted we need to calculate the theoretical yield. Since the actual yield was 0.80 mol HI and the percent yield is 85%, we can calculate the theoretical yield for this reaction by rearranging the percent yield equation.

$$\text{Percent yield} = \frac{\text{actual yield}}{\text{theoretical yield}} \times 100\%$$

$$\text{Theoretical yield} = \frac{\text{actual yield}}{\text{percent yield}} \times 100\%$$

$$= \frac{0.80 \text{ mol HI}}{85\%} \times 100\% = 0.94 \text{ mol HI}$$

Using the balanced chemical equation we can calculate the number of moles of I_2 that reacted. The balanced chemical equation is:

$$I_2(s) + H_2(g) \rightarrow 2HI(l)$$

$$\text{Moles of } I_2 = 0.94 \, \cancel{\text{mol HI}} \times \frac{1 \text{ mol } I_2}{2 \, \cancel{\text{mol HI}}} = 0.47 \text{ mol } I_2$$

6.61   The mole ratio of the Mg to Br_2 is 1:1. However, since there are twice as many moles of Br_2 (2.0 mol) as of Mg (1.0 mol) in the reaction mixture, Mg is the limiting reactant. The theoretical yield is 1.0 mol $MgBr_2$:

$$\text{Moles of } MgBr_2 = 1.0 \, \cancel{\text{mol Mg}} \times \frac{1 \text{ mol } MgBr_2}{1 \, \cancel{\text{mol Mg}}} = 1.0 \text{ mol } MgBr_2$$

The percent yield is:

$$\text{Percent yield} = \frac{0.84 \text{ mol}}{1.0 \text{ mol}} \times 100\% = 84\%$$

6.63   As the ball rolls downhill, its potential energy is converted to kinetic energy. Some of the kinetic energy is lost as heat (caused by friction and compression).

6.65   In a fuel cell, energy is released as hydrogen and oxygen react to form water. The potential energy in the water molecules is lower than the potential energy in the hydrogen and oxygen molecules. The released energy is used to create electrical current, a form of electrical energy, and the electrical energy is used to run electric motors and produce kinetic energy (move the car).

6.67   The reaction is endothermic. When the chemical reaction absorbs energy, the surrounding substances lose energy. We observe that the surroundings become colder, because they lose the energy absorbed by the reaction.

6.69   Molecules in the gas state have more energy than those in the liquid state. In order for the liquid to evaporate, the molecules going from the liquid to the gas state must absorb some energy from their surroundings. As a result, the surroundings (water and your skin) lose energy and feel colder.

6.71   When the potential energy of reaction products is lower than the potential energy of the reactants a reaction is exothermic.

6.73 $$\text{Energy in cal} \xrightarrow{\;1\,\text{cal} = 4.184\,\text{J}\;} \text{Energy in J}$$

$$\text{Energy in J} = 526 \, \cancel{\text{cal}} \times \frac{4.184 \text{ J}}{1 \, \cancel{\text{cal}}} = 2.20 \times 10^3 \text{ J}$$

6.75   When a change in prefix occurs (i.e. kilo to no prefix) you will most likely want to change to the base unit (joule) before converting to your new base unit (cal). In order to convert from kJ to calories, we first need to convert from kJ to J. Then we can use the conversion for joules to calories.

$$\text{Energy in kJ} \xrightarrow{\;1\,\text{kJ} = 1000\,\text{J}\;} \text{Energy in J} \xrightarrow{\;1\,\text{cal} = 4.184\,\text{J}\;} \text{Energy in cal}$$

$$\text{Energy in cal} = 145 \, \cancel{\text{kJ}} \times \frac{1000 \, \cancel{\text{J}}}{1 \, \cancel{\text{kJ}}} \times \frac{1 \text{ cal}}{4.184 \, \cancel{\text{J}}} = 3.47 \times 10^4 \text{ cal}$$

6.77 Always look carefully at your units. "Cal" is the dietary calorie: 1 Cal = 1 kcal = 1000 cal. Since the number given is already in the base unit (joule) we first convert to the new base unit (calories). Once we are in cal units we can convert to Cal.

$$\text{Energy in J} \xrightarrow{\ 4.184\,J\,=\,1\,cal\ } \text{Energy in cal} \xrightarrow{\ 1000\,cal\,=\,1\,Cal\ } \text{Energy in Cal}$$

$$\text{Energy in Cal} = 876\ \cancel{J}\ \times\ \frac{1\ \cancel{cal}}{4.184\ \cancel{J}}\ \times\ \frac{1\ \text{Cal}}{1000\ \cancel{cal}}\ =\ 0.209\ \text{Cal}$$

6.79 The energy saving is the *difference* in energy provided by 1.0 g of cyclamate and 30.0 g of sucrose. That is:

$$\text{Energy savings} = \text{sucrose energy} - \text{cyclamate energy}$$

The problem asks for the energy difference in Calories, but all the energy values are given in kJ. Should you convert to Calories first and then calculate the difference or is it better to calculate the energy difference and then convert to Calories? Either method will give you the same answer; however, by calculating the difference and then converting to Calories, you actually do fewer calculations.

First we calculate the energy provided by cyclamate:

$$\text{Cyclamate energy in kJ} = 1.00\ \cancel{\text{g cyclamate}}\ \times\ \frac{16.03\ \text{kJ}}{1\ \cancel{\text{g cyclamate}}}\ =\ 16.0\ \text{kJ}$$

The sucrose energy is:

$$\text{Sucrose energy in kJ} = 30.0\ \cancel{\text{g sucrose}}\ \times\ \frac{16.49\ \text{kJ}}{1\ \cancel{\text{g sucrose}}}\ =\ 495\ \text{kJ}$$

Remember not to round numbers until you are done with all of the calculations:

$$\text{Energy Savings} = 495\ \text{kJ} - 16.0\ \text{kJ} = 479\ \text{kJ}$$

Finally, the energy savings is converted to Calories using the following conversion map:

$$\text{Energy in kJ} \xrightarrow{\ 1\,kJ\,=\,1000\,J\ } \text{J} \xrightarrow{\ 4.184\,J\,=\,1\,cal\ } \text{cal} \xrightarrow{\ 1000\,cal\,=\,1\,Cal\ } \text{Cal}$$

$$\text{Energy Savings in Cal} = 479\ \cancel{kJ}\ \times\ \frac{1000\ \cancel{J}}{1\ \cancel{kJ}}\ \times\ \frac{1\ \cancel{cal}}{4.184\ \cancel{J}}\ \times\ \frac{1\ \text{Cal}}{1000\ \cancel{cal}}\ =\ 114\ \text{Cal}$$

If your instructor told you that 1 Cal = 4.184 kJ you could have also calculated the energy savings through:

$$\text{Energy savings in Cal} = 479\ \cancel{kJ}\ \times\ \frac{1\ \text{Cal}}{4.184\ \cancel{kJ}}\ =\ 114\ \text{Cal}$$

6.81 The units of specific heat are (energy units)/(g °C). Specific heat is a physical property of a substance that tells us how much energy is required to change the temperature of 1 g of the substance by 1°C. The specific heat of lead (0.129 J/g °C) is the lower than those of both copper (0.377 J/g °C) and aluminum (0.895 J/g °C). This means that the temperature of the lead sample will increase more than the temperatures of either the aluminum or copper samples after input of the same amount of heat. The final temperature of the lead sample will be the highest and the final temperature of the copper sample will be higher than that of the aluminum sample.

6.83   The equation relating energy and specific heat is:

$q = m \times C \times \Delta T$ where $\Delta T = T_f - T_i$

$q = ?$ $C = 0.377$ J/(g °C) (Table 6.2)

$m = 528$ g $\Delta T = 49.8°C - 22.3°C = 27.5°C$

$$q = \left(528 \; \cancel{g}\right)\left(0.377 \frac{J}{\cancel{g} \; \cancel{°C}}\right)\left(27.5 \; \cancel{°C}\right) = 5470 \text{ J (or } 5.47 \times 10^3 \text{ J)}$$

We can also carry out the calculation using the specific heat given with units of calories/g °C:

$$q = \left(528 \; \cancel{g}\right)\left(0.0900 \frac{cal}{\cancel{g} \; \cancel{°C}}\right)\left(27.5 \; \cancel{°C}\right) = 1310 \text{ cal (or } 1.31 \times 10^3 \text{ cal)}$$

6.85   The equation relating energy and specific heat is:

$q = m \times C \times \Delta T$ where $\Delta T = T_f - T_i$

$q = ?$ $C = 2.02$ J/(g °C)

$m = 1.25$ g $\Delta T = 102.1°C - 185.3°C = -83.2°C$

$$q = \left(1.25 \; \cancel{g}\right)\left(2.02 \frac{J}{\cancel{g} \; \cancel{°C}}\right)\left(-83.2 \; \cancel{°C}\right) = -210. \text{ J or } -2.10 \times 10^2 \text{ J}$$

6.87   A calorimeter should have good insulation and provide a means of accurately and precisely measuring temperature changes for the reactions that occur in it. The temperature measuring device should respond quickly to temperature changes.

6.89   (a) In order for the water temperature to increase, the water must gain heat energy. This means that the rock must have lost heat.

(b) The law of conservation of energy requires that the total energy change must be zero. That is:

$q_{rock} + q_{water} = 0$

$q_{rock} = -q_{water}$

This means that the energy changes of the rock and the water must be of the same magnitudes, but with the opposite signs. Although we might not be able to accurately measure the energy change of the rock, we can measure the temperature change of the water and use it to calculate the associated energy change.

$q_{water} = m \times C \times \Delta T$ where $\Delta T = T_f - T_i$

$q = ?$ $C = 4.184$ J/(g °C) (Table 6.2)

$m = 60.0$ g $\Delta T = 30.1°C - 25.0°C = 5.1°C$

$$q_{water} = \left(60.0 \; \cancel{g}\right)\left(4.184 \frac{J}{\cancel{g} \; \cancel{°C}}\right)\left(5.1 \; \cancel{°C}\right) = 1280 \text{ J} = 1.3 \times 10^3 \text{ J}$$

The energy change of the rock is -1.3×10^3 J (the energy change has the same magnitude as that of the water, but with the opposite sign).

6.91   (a) Since the temperature of the calorimeter & water decreased, the copper's temperature must have been lower than 45.0 °C.
       (b) The copper gained the heat lost by the water as the water cooled.
       (c) The gain in heat by the copper can be calculated from the energy loss of the water since

$q_{Cu} + q_{water} = 0$ (conservation of energy)

$q_{Cu} = -q_{water}$

$q_{water} = m \times C \times \Delta T$ where $\Delta T = T_f - T_i$

$q = ?$ $C = 4.184$ J/(g °C) (Table 6.2)

$m = 75.0$ g $\Delta T = 36.2°C - 45.0°C = -8.8°C$

$$q_{water} = (75.0 \text{ g})\left(4.184\frac{\text{J}}{\text{g °C}}\right)(-8.8 °C) = -2.8 \times 10^3 \text{ J}$$

$q_{Cu} = -q_{water} = 2.8 \times 10^3 \text{ J}$

6.93   Logically we know the temperature will be between 80.0°C and 25.0°C. Since there are 70.0 g (70% of the total water) of the hot water and 30.0 grams of the cold water, we can expect that the change of temperature is 70.0% of the way between 80.0°C and 25.0°C.
       $\Delta T = 0.700 \times (80.0°C - 25.0°C) = 38.5°C$. The final temperature should be 63.5°C.

       We can also show this mathematically. We can write a conservation of energy statement for the hot and cold waters that are mixed:

$q_{hot} + q_{cold} = 0$

$q_{hot} = m_{hot} \times C_{water} \times \Delta T_{hot}$

$\Delta T_{hot} = T_f - 80.0°C$

$q_{cold} = m_{cold} \times C_{water} \times \Delta T_{cold}$

$\Delta T_{cold} = T_f - 25.0$

$q_{hot} + q_{cold} = m_{hot} \times C_{water} \times \Delta T_{hot} + m_{cold} \times C_{water} \times \Delta T_{cold} = 0$

Substituting we have

$(70.0 \text{ g}) \times C_{water} \times (T_f - 80.0 °C) + (30.0 \text{ g}) \times C_{water} \times (T_f - 25.0 °C) = 0$

Interestingly, the heat capacity of water can be cancelled out of the equation along with the units. This helps to simplify the equation to:

$70.0 \times (T_f - 80.0 °C) + 30.0 \times (T_f - 25.0 °C) = 0$

Now solve for T_f:

$70.0T_f - 5600 + 30.0T_f - 750 = 0$

$100.0T_f = 6350$

$T_f = 63.5 °C$

6.95 Because a chemical reaction is a process, we can only measure the effect it has on the surroundings. The temperature change that occurs during a chemical reaction results from the process of forming new substances.

6.97 The chemical reaction is the system (the object being studied), and the water and calorimeter are the surroundings.

6.99 (a) When a reaction releases energy it is exothermic. (b) Because the reaction is exothermic, the sign of the heat change is negative. The energy change is −525 kJ/mol CO.

6.101 (a) We assume that all of the heat released by the burning coal is absorbed by the water.

Energy change for water:

$q_{water} = m \times C \times \Delta T$ where $\Delta T = T_f - T_i$

$q = ?$ $C = 4.184$ J/(g °C) (Table 6.2)

$m = 2010$ g $\Delta T = 41.5°C - 24.0°C = 17.5°C$

$$q_{water} = (2010 \text{ g}) \left(4.184 \frac{J}{\text{g °C}} \right) (17.5 °C) = 1.47 \times 10^5 \text{ J}$$

Converting from J to kJ

$$\text{Energy in kJ} = 1.47 \times 10^5 \text{ J} \times \frac{1 \text{ kJ}}{1000 \text{ J}} = 147 \text{ kJ}$$

The reaction released 147 kJ of energy.

(b) We assume that the water absorbed the same amount of heat that the reaction released. The sum of the heat change of the reaction plus the heat change of the water must equal zero.

$q_{reaction} + q_{water} = 0$

$q_{reaction} = -q_{water}$

The heat change for the reaction is −147 kJ. Because 6.00 g of coal burned in the reaction, the heat released per gram of coal is −24.5 kJ/g.

$$\text{Heat per gram} = \frac{-147 \text{ kJ}}{6.00 \text{ g}} = -24.5 \text{ kJ/g}$$

6.103 (a) The law of conservation of energy tells us that the heat released by the combustion of the peanut is absorbed by the water in the calorimeter.

Energy change for water:

$q_{water} = m \times C \times \Delta T$ where $\Delta T = T_f - T_i$

$q = ?$ $C = 4.184$ cal/(g °C) (Table 6.2)

$m = 1200$ g $\Delta T = 30.25°C - 25.0°C = 5.25°C$

$$q_{water} = (1200 \text{ g}) \left(4.184 \frac{cal}{\text{g °C}} \right) (5.25 °C) = 26400 \text{ J} = 2.6 \times 10^4 \text{ J}$$

$q_{nut} = -q_{water} = -2.6 \times 10^4 \text{ J}$

The peanut released 2.6×10^4 J of heat when burned.

(b) To convert the energy released from joules to calories, and to Calories, we use the appropriate conversion factors:

$1 \text{ cal} = 4.184 \text{ J}$

$1 \text{ Cal} = 1000 \text{ cal}$

Energy in cal $= 26400 \text{ J} \times \dfrac{1 \text{ cal}}{4.184 \text{ J}} = 6300 \text{ cal} = 6.3 \times 10^3 \text{ cal}$

Energy in Cal $= 6300 \text{ cal} \times \dfrac{1 \text{ Cal}}{1000 \text{ cal}} = 6.3 \text{ Cal}$

(c) The energy released per gram of peanut is 3.2 Cal/g.

Energy per gram $= \dfrac{6.3 \text{ Cal}}{2.00 \text{ g}} = 3.2 \text{ Cal/g}$

6.105 We can use the heat change per mole of a substance (i.e., -393.7 kJ/mol CO_2) as a conversion factor to convert between the number of moles of substance that react and the energy produced by the reaction.

Heat change $= 0.650 \text{ mol} \times \dfrac{-393.7 \text{ kJ}}{1 \text{ mol}} = -256 \text{ kJ}$

6.107 We can write two mole ratios for each product (a total of 8 different mole ratios):

$\dfrac{12 \text{ mol } CO_2}{4 \text{ mol } C_3H_5O_9N_3}$ and $\dfrac{4 \text{ mol } C_3H_5O_9N_3}{12 \text{ mol } CO_2}$

$\dfrac{6 \text{ mol } N_2}{4 \text{ mol } C_3H_5O_9N_3}$ and $\dfrac{4 \text{ mol } C_3H_5O_9N_3}{6 \text{ mol } N_2}$

$\dfrac{1 \text{ mol } O_2}{4 \text{ mol } C_3H_5O_9N_3}$ and $\dfrac{4 \text{ mol } C_3H_5O_9N_3}{1 \text{ mol } O_2}$

$\dfrac{10 \text{ mol } H_2O}{4 \text{ mol } C_3H_5O_9N_3}$ and $\dfrac{4 \text{ mol } C_3H_5O_9N_3}{10 \text{ mol } H_2O}$

(a) To calculate the number of moles of each product formed from the decomposition of 1.00 mol nitroglycerine, we choose the mole ratio that allows us to cancel moles of $C_3H_5O_9N_3$ and leaves us with the desired units (i.e. mol CO, mol N_2, mol O_2, and mol H_2O).

Moles of CO_2 = 1.00 mol $C_3H_5O_9N_3 \times \dfrac{12 \text{ mol } CO_2}{4 \text{ mol } C_3H_5O_9N_3} = 3.00 \text{ mol } CO_2$

Moles of N_2 = 1.00 mol $C_3H_5O_9N_3 \times \dfrac{6 \text{ mol } N_2}{4 \text{ mol } C_3H_5O_9N_3} = 1.50 \text{ mol } N_2$

Moles of O_2 = 1.00 mol $C_3H_5O_9N_3 \times \dfrac{1 \text{ mol } O_2}{4 \text{ mol } C_3H_5O_9N_3} = 0.250 \text{ mol } O_2$

Moles of H_2O = 1.00 mol $C_3H_5O_9N_3 \times \dfrac{10 \text{ mol } H_2O}{4 \text{ mol } C_3H_5O_9N_3} = 2.50 \text{ mol } H_2O$

(b) To calculate the number of moles of each product formed from the decomposition of 2.50 mol of nitroglycerine, we follow the procedure of part (a) but substitute 2.50 mol of $C_3H_5O_9N_3$ for 1 mol of $C_3H_5O_9N_3$.

$$\text{Moles of } CO_2 = 2.50 \ \cancel{\text{mol } C_3H_5O_9N_3} \times \frac{12 \text{ mol } CO_2}{4 \ \cancel{\text{mol } C_3H_5O_9N_3}} = 7.50 \text{ mol } CO_2$$

$$\text{Moles of } N_2 = 2.50 \ \cancel{\text{mol } C_3H_5O_9N_3} \times \frac{6 \text{ mol } N_2}{4 \ \cancel{\text{mol } C_3H_5O_9N_3}} = 3.75 \text{ mol } N_2$$

$$\text{Moles of } O_2 = 2.50 \ \cancel{\text{mol } C_3H_5O_9N_3} \times \frac{1 \text{ mol } O_2}{4 \ \cancel{\text{mol } C_3H_5O_9N_3}} = 0.625 \text{ mol } O_2$$

$$\text{Moles of } H_2O = 2.50 \ \cancel{\text{mol } C_3H_5O_9N_3} \times \frac{10 \text{ mol } H_2O}{4 \ \cancel{\text{mol } C_3H_5O_9N_3}} = 6.25 \text{ mol } H_2O$$

6.109 (a) In most mass-mass conversion problems we start with the mass of one substance (in this case, 10.0 g of $CaCl_2$) and a balanced equation to solve for the mass of a second substance (in this case, $AgNO_3$). Using the following problem solving map:

$$\text{Grams } CaCl_2 \xrightarrow{\ MM \ CaCl_2\ } \text{Moles } CaCl_2 \xrightarrow{\ \text{mole ratio}\ } \text{Moles } AgNO_3 \xrightarrow{\ MM \ AgNO_3\ } \text{Grams } AgNO_3$$

Molar masses: $AgNO_3$ (169.9 g/mol), $CaCl_2$ (110.98 g/mol)

Balanced equation (given): $2AgNO_3(s) + CaCl_2(aq) \rightarrow 2AgCl(s) + Ca(NO_3)_2(aq)$

$$\text{Mass } AgNO_3 = 10.0 \ \cancel{\text{g } CaCl_2} \times \frac{1 \ \cancel{\text{mol } CaCl_2}}{110.98 \ \cancel{\text{g } CaCl_2}} \times \frac{2 \ \cancel{\text{mol } AgNO_3}}{1 \ \cancel{\text{mol } CaCl_2}} \times \frac{169.9 \text{ g } AgNO_3}{1 \ \cancel{\text{mol } AgNO_3}} = 30.6 \text{ g } AgNO_3$$

(b) Following a similar problem solving approach, we determine that when all the $AgNO_3$ has reacted, 25.8 g AgCl should have precipitated.

Molar Mass: AgCl (143.4 g/mol)

$$\text{Mass of } AgCl = 10.0 \ \cancel{\text{g } CaCl_2} \times \frac{1 \ \cancel{\text{mol } CaCl_2}}{110.98 \ \cancel{\text{g } CaCl_2}} \times \frac{2 \ \cancel{\text{mol } AgCl}}{1 \ \cancel{\text{mol } CaCl_2}} \times \frac{143.4 \text{ g } AgCl}{1 \ \cancel{\text{mol } AgCl}} = 25.8 \text{ g } AgCl$$

6.111 To calculate the mass of oxygen required for the reaction, we first need to calculate the mass of octane that reacts. We do this using the density and volume of octane. Because density is reported in units of g/mL, first we must convert the volume of octane from gallons to milliliters.

$$\text{Problem solving map: Volume in gal} \xrightarrow{\ 1 \text{ gal} = 3.79 \text{ L}\ } \text{Volume in L} \xrightarrow{\ 1000 \text{ mL} = 1 \text{L}\ } \text{Volume in mL}$$

$$\text{Volume in mL} = 1.00 \ \cancel{\text{gal}} \times \frac{3.79 \ \cancel{L}}{1 \ \cancel{\text{gal}}} \times \frac{1000 \text{ mL}}{1 \ \cancel{L}} = 3.79 \times 10^3 \text{ mL}$$

Now we can use the density of octane to calculate its mass.

$$\text{Volume octane} \xrightarrow{\ 0.703 \text{ g}/\text{mL}\ } \text{Mass octane}$$

$$\text{Grams of octane} = 3.79 \times 10^3 \ \cancel{\text{mL}} \times \frac{0.703 \text{ g}}{1 \ \cancel{\text{mL}}} = 2.66 \times 10^3 \text{ g}$$

Now we can calculate the mass of O_2 required for the reaction.

Problem solving map:

Grams C_8H_{18} $\xrightarrow{MM\ C_8H_{18}}$ Moles C_8H_{18} $\xrightarrow{\text{mole ratio}}$ Moles O_2 $\xrightarrow{MM\ O_2}$ Grams O_2

Molar masses: C_8H_{18} (114.22 g/mol), O_2 (32.00 g/mol)

Balanced equation: $2C_8H_{18}(l) + 25O_2(g) \rightarrow 16CO_2(g) + 18H_2O(l)$

Mass of O_2 = 2.66×10^3 g $C_8H_{18} \times \dfrac{1\ \text{mol}\ C_8H_{18}}{114.22\ \text{g}\ C_8H_{18}} \times \dfrac{25\ \text{mol}\ O_2}{2\ C_8H_{18}} \times \dfrac{32.00\ \text{g}\ O_2}{1\ \text{mol}\ O_2} = 9.33 \times 10^3$ g O_2

6.113 (a) Chlorine reacts with metals to form ionic compounds. In this case, the metal is potassium. Because potassium forms 1+ ions and chlorine atoms form 1– ions, the formula of the reaction product is KCl. The balanced chemical equation is:

$2K(s) + Cl_2(g) \rightarrow 2KCl(s)$

(b) Because we are given the mass of reactants, we must first convert from grams to moles for each reactant.

Molar Masses: K (39.10 g/mol), Cl_2 (70.90 g/mol)

Moles of K = 10.0 g K $\times \dfrac{1\ \text{mol}\ K}{39.10\ \text{g}\ K}$ = 0.256 mol K

Moles of Cl_2 = 5.0 g $Cl_2 \times \dfrac{1\ \text{mol}\ Cl_2}{70.90\ \text{g}\ Cl_2}$ = 0.071 mol Cl_2

Next we determine the moles of Cl_2 needed to react with 0.256 mol K.

Moles of Cl_2 = 0.256 mol K $\times \dfrac{1\ \text{mol}\ Cl_2}{2\ \text{mol}\ K}$ = 0.128 mol Cl_2

We need 0.128 mol Cl_2 but we only have 0.071 mol Cl_2 so chlorine is the limiting reactant.

(c) There should be some gray solid (K metal) and a white solid (KCl) inside the container. The greenish-yellow Cl_2 gas will have disappeared because it was completely consumed in the reaction.

(d) Based on the mass of the limiting reactant (Cl_2) we determine that 11 g of KCl will form.

Molar mass: KCl (74.55 g/mol)

Grams of KCl = 0.071 mol $Cl_2 \times \dfrac{2\ \text{mol}\ KCl}{1\ \text{mol}\ Cl_2} \times \dfrac{74.55\ \text{g}\ KCl}{1\ \text{mol}\ KCl}$ = 11 g KCl

(e) Since we know how much Cl_2 reacted, we can calculate the corresponding mass of K used by the reaction:

Grams of K = 0.071 mol $Cl_2 \times \dfrac{2\ \text{mol}\ K}{1\ \text{mol}\ Cl_2} \times \dfrac{39.10\ \text{g}\ K}{1\ \text{mol}\ K}$ = 5.5 g K

Subtracting 5.5 g K from the starting amount of 10.0 g K, we find that 4.5 g of K remain in the flask mixed with 5.5 g of KCl.

6.115 (a) The mole ratio is: $\dfrac{1\,mol\,O_2}{2\,mol\,H_2}$.

(b) The reaction should consume 7.9 g O_2 and produce 8.9 g H_2O, as shown below:

Molar masses: H_2 (2.016 g/mol), O_2 (32.00 g/mol), H_2O (18.02 g/mol)

$$\text{Grams of } O_2 = 1.0 \text{ g } H_2 \times \frac{1 \text{ mol } H_2}{2.016 \text{ g } H_2} \times \frac{1 \text{ mol } O_2}{2 \text{ mol } H_2} \times \frac{32.00 \text{ g } O_2}{1 \text{ mol } O_2} = 7.9 \text{ g } O_2$$

$$\text{Grams of } H_2O = 1.0 \text{ g } H_2 \times \frac{1 \text{ mol } H_2}{2.016 \text{ g } H_2} \times \frac{2 \text{ mol } H_2O}{2 \text{ mol } H_2} \times \frac{18.02 \text{ g } H_2O}{1 \text{ mol } H_2O} = 8.9 \text{ g } H_2O$$

(c) If only 4.0 g O_2 are mixed with 1.0 g H_2, O_2 will be the limiting reactant (7.9 g of O_2 are required to react with 1.0 g of H_2) and 4.5 g H_2O should form. This is the theoretical yield of the reaction.

$$\text{Grams of } H_2O = 4.0 \text{ g } O_2 \times \frac{1 \text{ mol } O_2}{32.00 \text{ g } O_2} \times \frac{2 \text{ mol } H_2O}{1 \text{ mol } O_2} \times \frac{18.02 \text{ g } H_2O}{1 \text{ mol } H_2O} = 4.5 \text{ g } H_2O$$

6.117 The equation relating energy and specific heat is:

$q = m \times C \times \Delta T$ where $\Delta T = T_f - T_i$

$q = 155\ J$ $C = 0.222\ J/(g\ {}^\circ C)$ (Table 6.2)

$m = 125\ g$ $\Delta T = ?$ $T_f = ?$ $T_i = 26.2\,{}^\circ C$

Because we don't know either the change of temperature or the final temperature, we first solve the specific heat equation for ΔT:

$$\Delta T = \frac{q}{m \times C} = \frac{155 \text{ J}}{125 \text{ g} \times 0.222\ \dfrac{J}{g\ {}^\circ C}} = 5.59\,{}^\circ C$$

Because the tin absorbed energy, its final temperature is higher than its initial temperature:

$T_f = \Delta T + T_i = 26.2\,{}^\circ C + 5.59\,{}^\circ C = 31.8\,{}^\circ C$

The final temperature of the tin is $31.8\,{}^\circ C$.

6.119 The equation relating energy and specific heat is:

$q = m \times C \times \Delta T$ where $\Delta T = T_f - T_i$

The metal lost 220.6 J of heat, so q is -220.6 J.

$q = -220.6\ J$ $C = ?$

$m = 20.0\ g$ $\Delta T = 25.5\,{}^\circ C - 50.0\,{}^\circ C = -24.5\,{}^\circ C$

We solve the specific heat equation for C:

$$C = \frac{q}{m \times \Delta T} = \frac{-220.6 \text{ J}}{(20.0 \text{ g}) \times (-24.5^\circ C)} = 0.450\ \frac{J}{g\ {}^\circ C}$$

The specific heat is 0.450 J/(g °C) and matches most closely that of chromium (0.450 J/(g °C)).

6.121 Because energy is released we know that the heat change is -701 kJ.

$$\text{Energy per mol} = \frac{-701 \text{ kJ}}{0.250 \text{ mol}} = -2.80 \times 10^3 \text{ kJ/mol}$$

To convert from units of kJ/mol to Calories/mol, we use several steps:

$$\text{Energy kJ} \xrightarrow{1000 \text{ J} = 1 \text{ kJ}} \text{J} \xrightarrow{4.184 \text{ J} = 1 \text{ cal}} \text{cal} \xrightarrow{1000 \text{ cal} = 1 \text{ Cal}} \text{Cal}$$

$$\text{Energy in Cal} = -701 \text{ kJ} \times \frac{1000 \text{ J}}{1 \text{ kJ}} \times \frac{1 \text{ cal}}{4.184 \text{ J}} \times \frac{1 \text{ Cal}}{1000 \text{ cal}} = -168 \text{ Cal}$$

$$\text{Energy per mol} = \frac{-168 \text{ Cal}}{0.250 \text{ mol}} = -6.70 \times 10^2 \text{ Cal/mol}$$

6.123 We know that the water and the lead piece will have to reach the same temperature. The first step in finding this temperature is to look at the heat equations for both the water and the lead. For the lead we have:

$q_{Pb} = m_{Pb} \times C_{Pb} \times \Delta T_{Pb}$

$q = ?$ $C = 0.129$ J/(g °C) (Table 6.2)

$m = 20.0$ g $\Delta T = T_f - T_i = x - 24.5°C$

For the water we have:

$q_w = m_w \times C_w \times \Delta T_w$

$q = ?$ $C = 4.184$ J/(g °C) (Table 6.2)

$m = 105$ g $\Delta T = T_f - T_i = x - 55.0°C$

If you look carefully, you will see that there are two unknowns in this problem. We don't know either the energy change or the final temperature. However, from the law of conservation of energy we know that:

$q_w + q_{Pb} = 0$

This means we can write:

$m_w \times C_w \times \Delta T_w + m_{Pb} \times C_{Pb} \times \Delta T_{Pb} = 0$

Substituting known values into the expression gives us:

$$20.0 \text{ g} \times 0.129 \frac{\text{J}}{\text{g °C}} \times \left(x - 24.5°\text{C}\right) + 105 \text{ g} \times 4.184 \frac{\text{J}}{\text{g °C}}\left(x - 55.0°\text{C}\right) = 0$$

Before solving we simplify the units as much as we can. We can cancel grams and the units of specific heat because they appear in both terms.

$$20.0 \times 0.129 \times \left(x - 24.5°\text{C}\right) + 105 \times 4.184\left(x - 55.0°\text{C}\right) = 0$$

Then we solve for x, the final temperature of the mixture.

$$\left(2.58x - 63.21°\text{C}\right) + \left(439.32x - 24162.6°\text{C}\right) = 0$$

$$442x - 24226°\text{C} = 0$$

$$x = 54.8°$$

The final temperature is 54.8°C.

6.125 The law of conservation of energy tells us that the amount of energy absorbed by the water must be equal to the amount of energy released by the pipe. If we separate the known values for the pipe and the water, we see we have the following information:

Pipe Data:

$q = ?$ $C = ?$

$m = 175$ g $\Delta T = 33.43°C - 78.24°C = -44.81°C$

Water Data:

$q = ?$ $C = 4.184$ J/(g °C)

$m = 100.0$ g $\Delta T = 33.43°C - 25.00°C = 8.43°C$

With respect to the pipe, we do not have sufficient data to calculate specific heat. We do have sufficient data to calculate the energy absorbed by the water. According to the law of conservation of energy:

$q_{water} + q_{pipe} = 0$

$q_{pipe} = -q_{water}$

Once we know the amount of energy absorbed by the water, we can calculate the specific heat of the pipe.

$$q_{water} = 100.0 \text{ g} \times 4.184 \frac{J}{\text{g °C}} \times 8.43 \text{ °C} = 3.53 \times 10^3 \text{ J}$$

$$C_{pipe} = \frac{-3.53 \times 10^3 \text{ J}}{175 \text{ g} \times -44.81°C} = 0.450 \frac{J}{\text{g °C}}$$

The pipe could be made of chromium.

6.127 We begin by writing an equation that describes the dissolving of magnesium chloride:

$$MgCl_2(s) \xrightarrow{\text{H}_2\text{O}} Mg^{2+}(aq) + 2Cl^-(aq)$$

This equation tells us that for each mole of $MgCl_2$ that dissolves one mole of Mg^{2+} forms and two moles of Cl^- ions form. We use the mole relationships to calculate the quantity of Mg^{2+} ions and Cl^- ions that form when 0.250 mol $MgCl_2$ dissolve:

$$0.250 \text{ mol MgCl}_2 \times \frac{1 \text{ mol Mg}^{2+}}{1 \text{ mol MgCl}_2} = 0.250 \text{ mol Mg}^{2+}$$

$$0.250 \text{ mol MgCl}_2 \times \frac{2 \text{ mol Cl}^-}{1 \text{ mol MgCl}_2} = 0.500 \text{ mol Cl}^-$$

6.129　We begin by writing an equation that describes the dissolving of potassium sulfate:

$$K_2SO_4(s) \xrightarrow{H_2O} 2K^+(aq) + SO_4^{2-}(aq)$$

This equation tells us that for each mole of K_2SO_4 that dissolves two moles of K^+ form and one mole of SO_4^{2-} ions forms. We use the molar mass of K_2SO_4 and the mole relationships in the balanced equation to calculate the quantity of K^+ ions and SO_4^{2-} ions that form when 50.0 g K_2SO_4 dissolve:

$$50.0 \text{ g } K_2SO_4 \times \frac{1 \text{ mol } K_2SO_4}{174.26 \text{ g } K_2SO_4} \times \frac{2 \text{ mol } K^+}{1 \text{ mol } K_2SO_4} = 0.574 \text{ mol } K^+$$

$$50.0 \text{ g } K_2SO_4 \times \frac{1 \text{ mol } K_2SO_4}{174.26 \text{ g } K_2SO_4} \times \frac{1 \text{ mol } SO_4^{2-}}{1 \text{ mol } K_2SO_4} = 0.287 \text{ mol } SO_4^{2-}$$

This equation tells us that for each mole of $Al(NO_3)_3$ that dissolves one mole of Al^{3+} forms and three moles of NO_3^- ions form. We use the molar mass of $Al(NO_3)_3$ and the mole relationships in the balanced equation to calculate the quantity of Al^{3+} ions and NO_3^- ions that form when 145.0 g $Al(NO_3)_3$ dissolve:

$$145.0 \text{ g } Al(NO_3)_3 \times \frac{1 \text{ mol } Al(NO_3)_3}{213.01 \text{ g } Al(NO_3)_3} \times \frac{1 \text{ mol } Al^{3+}}{1 \text{ mol } Al(NO_3)_3} = 0.6807 \text{ mol } Al^{3+}$$

$$145.0 \text{ g } Al(NO_3)_3 \times \frac{1 \text{ mol } Al(NO_3)_3}{213.01 \text{ g } Al(NO_3)_3} \times \frac{3 \text{ mol } NO_3^-}{1 \text{ mol } Al(NO_3)_3} = 2.042 \text{ mol } NO_3^-$$

6.131　We want to determine the mass of ethanol that can be obtained when 5.0 kg of glucose reacts according to the equation:

$$C_6H_{12}O_6(s) \rightarrow 2CH_3CH_2OH(l) + 2CO_2(g)$$

The problem solving map begins with 5.0 kg of glucose, $C_6H_{12}O_6$, and ends with the mass of C_2H_5OH:

$$\text{kg } C_6H_{12}O_6 \xrightarrow{1 \text{ kg} = 1000 \text{ g}} \text{g } C_6H_{12}O_6 \xrightarrow{MM \ C_6H_{12}O_6} \text{mol } C_6H_{12}O_6 \xrightarrow{\text{mole ratio}}$$

$$\text{mol } C_2H_5OH \xrightarrow{MM \ C_2H_5OH} \text{g } C_2H_5OH$$

The molar masses of $C_6H_{12}O_6$ and C_2H_5OH are 180.16 g/mol and 46.07 g/mol, respectively. We find the mole ratio from the balanced chemical equation.

Mass of C_2H_5OH =

$$5.0 \text{ kg } C_6H_{12}O_6 \times \frac{1000 \text{ g}}{1 \text{ kg}} \times \frac{1 \text{ mol } C_6H_{12}O_6}{180.16 \text{ g } C_6H_{12}O_6} \times \frac{2 \text{ mol } C_2H_5OH}{1 \text{ mol } C_6H_{12}O_6} \times \frac{46.07 \text{ g } C_2H_5OH}{1 \text{ mol } C_2H_5OH} =$$

$$= 2.6 \times 10^3 \text{ g } C_2H_5OH = 2.6 \text{ kg } C_2H_5OH$$

6.133　We are given the amounts of both reactants mixed together so this is a limiting reactant problem. To determine the identity of the limiting reactant, we can begin by calculating the number of moles of K_2CrO_4 required to react with 0.45 mol $BaCl_2$:

$$\text{Moles of } K_2CrO_4 = 0.45 \ \cancel{\text{mol } BaCl_2} \times \frac{1 \text{ mol } K_2CrO_4}{1 \ \cancel{\text{mol } BaCl_2}} = 0.45 \text{ mol } K_2CrO_4$$

Since the initial reaction mixture contains fewer than 0.45 mol K_2CrO_4, we know that K_2CrO_4 is the limiting reactant. It determines the amount of product that forms.

$$0.20 \ \cancel{\text{mol } K_2CrO_4} \times \frac{1 \ \cancel{\text{mol } BaCrO_4}}{1 \ \cancel{\text{mol } K_2CrO_4}} \times \frac{253.3 \text{ g } BaCrO_4}{1 \ \cancel{\text{mol } BaCrO_4}} = 51 \text{ g } BaCrO_4$$

6.135　We begin by writing a balanced chemical equation:

$$2Na(s) + Br_2(l) \rightarrow 2NaBr(s)$$

To determine the limiting reactant, we can calculate the mass of Br_2 required to react with the specified mass of Na. If the mass of Br_2 required is greater than the available mass, then Br_2 is the limiting reactant. If the mass of Br_2 required is less than the available mass, then Na is the limiting reactant. If the calculated and available masses are the same, then both reactants will be completely consumed at the same time and both limit the amount of product produced.

Molar Masses: Na (22.99 g/mol), Br_2 (159.8 g/mol)

$$\text{Mass of } Br_2 = 20.0 \ \cancel{\text{g Na}} \times \frac{1 \ \cancel{\text{mol Na}}}{22.99 \ \cancel{\text{g Na}}} \times \frac{1 \ \cancel{\text{mol } Br_2}}{2 \ \cancel{\text{mol Na}}} \times \frac{159.8 \text{ g } Br_2}{1 \ \cancel{\text{mol } Br_2}} = 69.5 \text{ g } Br_2$$

Since the mass of Br_2 required is less than the available mass (100.0 g), Na is the limiting reactant. As the limiting reactant, we would expect all the Na to be consumed:

$$\text{Moles of Na} = 20.0 \ \cancel{\text{g Na}} \times \frac{1 \text{ mol Na}}{22.99 \ \cancel{\text{g Na}}} = 0.870 \text{ mol Na}$$

Chapter 7 – Electron Structure of the Atom

7.1   (a) electromagnetic radiation; (b) frequency; (c) ionization energy; (d) Hund's rule; (e) electron configuration; (f) core electron; (g) orbital; (h) continuous spectrum; (i) isoelectronic

7.3   Refer to Figure 7.6. Radio frequency, microwave, and infrared radiation all have longer wavelengths than visible light.

7.5   If the frequency of one wave is twice the frequency of the other, its wavelength will be half as long as the wavelength of the wave with the lower frequency. This means that for every cycle of the wave of the lower frequency, you should see two cycles of the wave of the higher frequency.

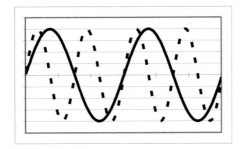

Lower frequency (——)

Higher frequency (- - - -)

7.7   The order is blue, yellow, orange, and red (use Figure 7.6 or the mnemonic ROYGBIV). ROYGBIV (Red, Orange, Yellow, Green, Blue, Indigo, Violet) lists the colors of the visible portion of the spectrum according to increasing energy or decreasing wavelength.

7.9   An inverse proportionality between two variables means that as the value of one variable increases, the value of the other decreases. With respect to the wavelength and frequency of a wave, this means that a wave with a longer wavelength has a lower frequency than a wave with a shorter wavelength. Similarly, a wave with a smaller, or shorter, wavelength has a higher frequency than a wave with a longer wavelength.

7.11   infrared radiation (Figure 7.6) (the prefix infra means below or beneath)

7.13   Gamma photons have the highest energy and the highest frequency. Radio frequency waves have the longest wavelengths.

7.15   To calculate the frequency of light from its wavelength we use the relationship:

$$c = \upsilon \lambda$$

where c is the speed of light (3.00×10^8 m/s), υ is the frequency in Hz, and λ is the wavelength in meters.

We are trying to determine frequency so we solve the equation for υ.

$$\upsilon = \frac{c}{\lambda}$$

Because we are given the wavelength in nanometers we must convert the units to meters before we solve the equation for frequency. The relationship between meters and nanometers is:

$$1 \text{ m} = 10^9 \text{ nm}$$

As a conversion factor we can write this expression in two different ways:

$$\frac{1 \text{ m}}{10^9 \text{ nm}} \quad \text{or} \quad \frac{10^9 \text{ nm}}{1 \text{ m}}$$

To convert 75.0 nm to meters we must cancel units of nm and be left with units of m.

$$\text{Wavelength in m} = 7.50 \text{ nm} \times \frac{1 \text{ m}}{10^9 \text{ nm}} = 7.50 \times 10^{-8} \text{ m}$$

Now that we have the appropriate units we can calculate the frequency.

$$\upsilon = \frac{c}{\lambda} = \frac{3.00 \times 10^8 \text{ m/s}}{7.50 \times 10^{-8} \text{ m}} = 4.00 \times 10^{15}/\text{s or } 4.00 \times 10^{15} \text{ Hz}$$

A photon with a frequency of 4.00×10^{15} Hz is in the ultraviolet region of the spectrum (see Figure 7.6).

7.17   When we know the wavelength, λ, of light we can calculate its energy using the equation:

$$E_{\text{photon}} = \frac{hc}{\lambda}$$

where $h = 6.626 \times 10^{-34}$ J·s, $c = 3.00 \times 10^8$ m/s, and λ is the wavelength in meters.

Because we are given the wavelength in nanometers, we must convert the wavelength to units of meters before we calculate frequency. The relationship between meters and nanometers is:

$1 \text{ m} = 10^9 \text{ nm}$

As a conversion factor we can write this expression two different ways:

$$\frac{1 \text{ m}}{10^9 \text{ nm}} \quad \text{or} \quad \frac{10^9 \text{ nm}}{1 \text{ m}}$$

To convert 465 nm to meters we must cancel units of nm and be left with units of m.

$$\text{Wavelength in m} = 465 \text{ nm} \times \frac{1 \text{ m}}{10^9 \text{ nm}} = 4.65 \times 10^{-7} \text{ m}$$

Now that we have the appropriate units, we can calculate the energy.

$$E_{\text{photon}} = \frac{6.626 \times 10^{-34} \text{ J} \cdot \text{s} \times 3.00 \times 10^8 \text{ m/s}}{4.65 \times 10^{-7} \text{ m}} = 4.27 \times 10^{-19} \text{ J}$$

A photon with a wavelength of 465 nm is in the visible region of the spectrum and would appear blue.

7.19   White light gives a continuous spectrum.

7.21   No. Atoms of different elements emit different line spectra because their atomic energy levels are different.

7.23   No. Bohr's model had fixed orbit radii to account for the fact that hydrogen atoms produce a line spectrum and not a continuous spectrum. If electrons could exist between orbits then elements would emit continuous light (i.e. all wavelengths of light) and we would not observe line spectra.

7.25   In the Bohr model, the electron would have to absorb energy to move from a lower-energy orbit to a higher-energy orbit. To move from a higher-energy orbit to a lower-energy orbit, the electron would have to release energy.

7.27   In the Bohr model, when an electron moves between two orbits, all the energy is released as a single photon. The wavelength of the photon will depend on the energy difference between the two orbits.

7.29   Because both electrons end up at the $n = 3$ level, the electron that falls from the highest energy level will release the highest-energy photon (Figure 7.12). The electron falling from $n = 6$ to $n = 3$ releases the highest-energy photon.

7.31   The electron that releases light with the longest wavelength will be the electron that releases the smallest amount of energy. Since both electrons end up in the $n = 3$ level, the electron that falls the shortest distance will release the lowest-energy photon (Figure 7.12). The $n = 5$ to $n = 3$ transition gives the lowest-energy photon and, therefore, the longest wavelength.

7.33   See Figure 7.10. The four lines are a result of four different transitions in the hydrogen atom. These transitions are:

$n = 6$ to $n = 2$ violet (highest energy)

$n = 5$ to $n = 2$ blue

$n = 4$ to $n = 2$ green

$n = 3$ to $n = 2$ red (lowest energy)

Because electrons making each of these four transitions produce photons with a different amount of energy, and the energy released from each transition is in the visible region of the spectrum, we observe four different colors in the visible spectrum of hydrogen.

7.35   When we know the wavelength we can calculate the photon energy using the equation:

$$E_{photon} = \frac{hc}{\lambda}$$

where $h = 6.626 \times 10^{-34}$ J·s, $c = 3.00 \times 10^8$ m/s, and λ is the wavelength in meters.

We are given the wavelength in nanometers, so we must convert the wavelength to units of meters before we calculate frequency. The relationship between meters and nanometers is:

$1\ m = 10^9\ nm$

As a conversion factor, we can write this expression in two different ways:

$$\frac{1\ m}{10^9\ nm} \quad \text{or} \quad \frac{10^9\ nm}{1\ m}$$

To convert 434.1 nm to meters we must cancel units of nm and be left with units of m.

Wavelength in m = $434.1\ \cancel{nm} \times \dfrac{1\ m}{10^9\ \cancel{nm}} = 4.341 \times 10^{-7}\ m$

Now that we have the appropriate units, we can calculate the energy.

$$E_{photon} = \frac{6.626 \times 10^{-34}\ J \cdot \cancel{s} \times 3.00 \times 10^8\ \cancel{m}\,/\,\cancel{s}}{4.341 \times 10^{-7}\ \cancel{m}} = 4.58 \times 10^{-19}\ J$$

7.37   Bohr's orbits were tracks located at fixed distances from the nucleus. The modern concept of atomic orbitals refers to regions of space surrounding the nucleus where we are most likely to find electrons.

7.39   A *p*-orbital is shown in B. One distinguishing feature of a *p*-orbital is that it has one *node*. We visualize a node as the point of space where the lobes (the round ends) come together. An *s*-orbital (D) has no nodes; the *d*-orbital shown in A has two (both at the same point); the *f*-orbital shown in C has three nodes (all at the same point).

7.41   Their primary differences are their relative sizes and their energies. The 3*p* orbital is larger and higher in energy than the 2*p* orbital.

7.43   The number of orbitals is determined by the sublevel letter.

Sublevel	Number of orbitals
s	1
p	3
d	5
f	7

(a) 1; (b) 3; (c) 5; (d) 7; (e) 1; (f) 3

7.45   For each element, we determine the atomic number and then fill the orbital diagram.

In drawing the ground state orbital diagrams be sure that you follow the following three rules:

Aufbau principle – electrons fill the lowest energy levels first

Pauli exclusion principle – no more than two electrons can occupy an orbital and if there are two electrons in an orbital they must have opposite spins

Hund's rule – electrons are distributed in an unfilled sublevel so as to give the maximum number of unpaired electrons

Si – 14 electrons

	1s	2s	2p			3s	3p	
Si	↑↓	↑↓	↑↓	↑↓	↑↓	↑↓	↑	↑

B – 5 electrons

	1s	2s	2p			3s	3p	
B	↑↓	↑↓	↑					

P – 15 electrons

	1s	2s	2p			3s	3p		
P	↑↓	↑↓	↑↓	↑↓	↑↓	↑↓	↑	↑	↑

7.47   To obtain the electron configuration, we distribute the electrons into an orbital diagram (Figure 7.18). Next we group the electrons by sublevel.

   (a) Si: 14 electrons

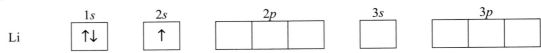

   Ground state electron configuration of Si: $1s^2 2s^2 2p^6 3s^2 3p^2$

   (b) Li: 3 electrons

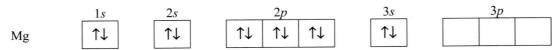

   Ground state electron configuration of Li: $1s^2 2s^1$

   (c) Mg: 12 electrons

   Ground state electron configuration of Mg: $1s^2 2s^2 2p^6 3s^2$

7.49   All orbitals, including a $2p$ orbital, can accommodate a maximum of 2 electrons.

7.51   The second principle energy level can accommodate 8 electrons: 2 in the s orbital and 6 in the p orbitals.

7.53   We use the orbital diagram for the atom to determine the number of unpaired electrons .

   P – 15 electrons

   The three electrons in the $3p$ sublevel are unpaired.

7.55   All elements in group VII A (17) have five electrons in their highest-energy level p sublevel (see Figure 7.23).

7.57   The transition elements except for Group IIB (12) have partially filled d sublevels (see Figure 7.23). The elements in Group IIB (12) have completely filled d sublevels.

7.59   Atoms in the second group of elements in the p-block, Group IVA (14), each have two electrons in their highest-energy level p sublevel. Silicon, Si, is the element at the intersection of Group IVA and the third period; it has two electrons in its $3p$ sublevel.

7.61   For this problem, refer to Figure 7.21 and your periodic table.

(a) To build the electron configuration of sodium, we take note of its location on the periodic table. Sodium, Na, is in the third period and Group IA (1). This means that its configuration ends as $3s^1$ because it is in the first group of the s-block. All of the sublevels which occur before $3s$ are filled. From this information we can build the remainder of the electron configuration:

Period 1: $1s^2$

Period 2: $2s^2 2p^6$

Period 3: $3s^1$

The complete configuration for sodium is: $1s^2 2s^2 2p^6 3s^1$

(b) Manganese, Mn, is in the fourth period and Group VIIB (7). This means its configuration ends as $3d^5$ because it is in the fifth group of the d-block. Remember that the principle energy level of the d-block elements is one less than the period number. All sublevels which occur before $3d$ are filled. From this information we can build the remainder of the electron configuration:

Period 1: $1s^2$

Period 2: $2s^2 2p^6$

Period 3: $3s^2 3p^6$

Period 4: $4s^2 3d^5$

The complete configuration for manganese is: $1s^2 2s^2 2p^6 3s^2 3p^6 4s^2 3d^5$

(c) Selenium, Se, is in the fourth period and Group VIA (16). This means its configuration ends as $4p^4$ because it is in the fourth group of the p-block. All sublevels which occur before $4p$ are filled. From this information we can build the remainder of the electron configuration:

Period 1: $1s^2$

Period 2: $2s^2 2p^6$

Period 3: $3s^2 3p^6$

Period 4: $4s^2 3d^{10} 4p^4$

The complete configuration of selenium is: $1s^2 2s^2 2p^6 3s^2 3p^6 4s^2 3d^{10} 4p^4$

7.63   There are two errors in the bromine, Br, configuration $1s^2 2s^2 2p^6 3s^2 3p^6 4s^2 4d^{10} 4p^6$. Bromine is in the fifth group of the p-block so the $4p$ orbital has one too many electrons. In addition, the $4d$ orbital should be $3d$. The correct electron configuration for bromine is $1s^2 2s^2 2p^6 3s^2 3p^6 4s^2 3d^{10} 4p^5$.

7.65   The highest principle energy level indicates the period of the element. We identify the group from the last portion of the electron configuration. (a) third period ($3s$) and fifth group in the p-block (p^5); chlorine (Cl); (b) fourth period ($4s$ is the highest energy level) and the seventh group in the d-block (d^7); cobalt (Co); (c) sixth period ($6s$) and the first group in the s-block (s^1); cesium (Cs).

7.67   To write the abbreviated electron configuration we locate the noble gas that occurs in the period above the element on the periodic table. Write the symbol for that noble gas in brackets. This is always the noble gas, with a lower atomic number, closest to the element. For example, for any element in the fifth period we would use [Kr] from the fourth period. Next, we determine the remainder of the electron configuration from the periodic table:

(a) Na: The noble gas in the period above sodium is neon (Ne). The remainder of the configuration outside the noble gas configuration is: $3s^1$. The abbreviated configuration is: [Ne] $3s^1$.

(b) Mn: The noble gas in the period above manganese is argon (Ar). The remainder of the configuration outside the noble gas configuration is: $4s^2 3d^5$. The abbreviated configuration is: [Ar] $4s^2 3d^5$.

(c) Se: The noble gas in the period above selenium is argon (Ar). The remainder of the configuration outside the noble gas configuration is: $4s^2 3d^{10} 4p^4$. The abbreviated configuration is: [Ar] $4s^2 3d^{10} 4p^4$.

7.69   The element is bromine ($1s^2 2s^2 2p^6 3s^2 3p^6 4s^2 3d^{10} 4p^5$). If the element has two filled p sublevels and one partially filled sublevel, it is in the fourth period (there are no p orbitals in the first period). In addition, because it has five p electrons the element must be in the fifth group in the p block.

7.71   Valence electrons are the electrons in the highest principle energy level.

7.73   No. The d-orbital electrons are always one energy level lower than the valence electrons. For example, calcium has two valence electrons (Ca [Ar] $4s^2$). Scandium (Sc [Ar] $4s^2 3d^1$) also has two electrons because the d electrons are in principle energy level 3.

7.75   One electron is added each time the group number increases, moving from left to right across the periodic table. The group number is the number of electrons in the s and p orbitals of the highest energy level (for the main group elements). Electrons in the d-orbitals are not included because they are always one energy level below the highest energy level (i.e. $4s3d4p$, where the 4 and 3 represent the energy levels of the electrons).

7.77   The valence level is the same as the period number of the element in the periodic table. The number of valence electrons for the main group elements is the same as the Roman numeral designating the group number.

	Element	Period	Valence level	Group Number	Valence electrons
(a)	Al	3	3	IIIA	3
(b)	S	3	3	VIA	6
(c)	As	4	4	VA	5

7.79   They are different because cations always have fewer electrons than the elements from which they are formed. They are similar in that they possess the same electron core (for example, [Ne], [Ar], or [Kr]).

7.81   First, we write the complete configuration for the atom. Next, we add or subtract the number of electrons specified by the ionic charge.

(a)   Magnesium: Mg $1s^2 2s^2 2p^6 3s^2$. To form a 2+ ion, a magnesium atom loses 2 valence electrons. The electron configuration of Mg^{2+} is $1s^2 2s^2 2p^6$. This is the same as the electron configuration of neon, so it can be abbreviated as [Ne]. To find an ion that is isoelectronic with Mg^{2+}, we need to determine which ions have 10 electrons. Fluorine has 9 electrons. A fluoride ion, F^-, has 10 electrons, so F^- is isoelectronic with Mg^{2+}.

(b)   Oxygen: O $1s^2 2s^2 2p^4$. To form a 2– ion, an oxygen atom gains 2 valence electrons. The electron configuration of O^{2-} is $1s^2 2s^2 2p^6$. This is the same as the electron configuration of neon, so it can be abbreviated as [Ne]. Nitrogen has 7 electrons. The nitride ion forms when a nitrogen atom gains three electrons. N^{3-} is isoelectronic with O^{2-}. It is also isoelectronic with Mg^{2+} and F^- (see part (a)).

(c)   Gallium: Ga $1s^2 2s^2 2p^6 3s^2 3p^6 4s^2 3d^{10} 4p^1$. Gallium atoms lose three valence electrons to form Ga^{3+} ions. As a consequence, the electron configuration of Ga^{3+} is $1s^2 2s^2 2p^6 3s^2 3p^6 3d^{10}$ ($4s$ and $4p$ electrons have been removed). The electronic configuration is abbreviated as [Ar] $3d^{10}$. Zinc atoms that have lost two electrons to form Zn^{2+} ions are isoelectronic with Ga^{3+}.

Summary of answers:

	Ion	complete	abbreviated	isoelectronic
(a)	Mg^{2+}	$1s^2 2s^2 2p^6$	[Ne]	F^-
(b)	O^{2-}	$1s^2 2s^2 2p^6$	[Ne]	N^{3-}
(c)	Ga^{3+}	$1s^2 2s^2 2p^6 3s^2 3p^6 3d^{10}$	[Ar] $3d^{10}$	Zn^{2+}

7.83   The common ions of the main group elements are all isoelectronic with the closest noble gas. O^{2-}, N^{3-}, and Na^+ are each isoelectronic with Ne (all have 10 electrons).

7.85   The valence electrons of potassium atoms are in a higher energy level than those of sodium atoms. Therefore they are farther away from the nucleus. The farther the valence electrons are from the nucleus, the less they are subject to its attraction and the easier they are to remove. As a result, potassium (with $4s$ valence electrons) has a lower ionization energy than sodium (with $3s$ valence electrons). This means that potassium holds its valence electrons less tightly than sodium does. As a result, potassium is more reactive than sodium.

7.87   Ionization energy of the elements increases as you go from left to right on the periodic table. This means that the first ionization energy of calcium is higher than that of potassium. As a result, potassium atoms hold their valence electrons less tightly than calcium holds its valence electrons. As a result, potassium is more reactive than calcium. In addition, calcium atoms must lose a second electron before they become stable. Even more energy is expended to remove the second electron.

7.89   Ionization energy increases from the bottom to the top of a group and from left to right on the periodic table. The order for increasing ionization energy is: $P < S < O$. Oxygen is above sulfur, and sulfur is to the right of phosphorus on the periodic table.

7.91   The electron configurations of lithium and sodium atoms are:

Li $1s^2 2s^1$

Na $1s^2 2s^2 2p^6 3s^1$

The valence electron of a lithium atom is in a lower energy level ($2s^1$) than that of a sodium atom ($3s^1$). It is closer to the nucleus making it more subject to the attraction of the nucleus. Therefore, more energy is required to remove an electron from a lithium atom.

7.93   Electron configurations

F $1s^2 2s^2 2p^5$

O $1s^2 2s^2 2p^4$

Two factors are important in explaining why fluorine has a higher ionization energy than oxygen. The first is that a fluorine atom has one more proton in its nucleus than an oxygen atom. Secondly, the valence electrons of fluorine and oxygen are in the same energy level. Since the electrons are in the same energy level and fluorine has a higher positive charge in its nucleus attracting the electrons, the ionization energy of fluorine is higher.

7.95   The product of the first ionization (Mg^+) is the reactant for the second ionization. In each step of the ionization only one electron is removed.

First ionization $Mg(g) \rightarrow Mg^+(g) + 1e^-$ IE_1

Second ionization $Mg^+(g) \rightarrow Mg^{2+}(g) + 1e^-$ IE_2

7.97   The third ionization of magnesium corresponds to removing an electron from the noble gas core:

$$Mg(g) \quad \rightarrow \quad Mg^+(g) \quad \rightarrow \quad Mg^{2+}(g) \quad \rightarrow \quad Mg^{3+}(g)$$

$$1s^22s^22p^63s^2 \quad \rightarrow \quad 1s^22s^22p^63s^1 \quad \rightarrow \quad 1s^22s^22p^6 \text{ or [Ne]} \quad \rightarrow \quad 1s^22s^22p^5$$

Because noble gas configurations are especially stable, a large amount of energy is required to remove the third electron from a Mg atom. The third ionization of aluminum results in the formation of a noble gas configuration.

$$Al(g) \quad \rightarrow \quad Al^+(g) \quad \rightarrow \quad Al^{2+}(g) \quad \rightarrow \quad Al^{3+}(g)$$

$$1s^22s^22p^63s^23p^1 \quad \rightarrow \quad 1s^22s^22p^63s^2 \quad \rightarrow \quad 1s^22s^22p^63s^1 \quad \rightarrow \quad 1s^22s^22p^6 \text{ or [Ne]}$$

This is the stable ion of aluminum. Thus magnesium has a higher third ionization energy than aluminum.

7.99   (a) The first ionization energy of fluorine is very high (Figure 7.26). As a result, it does not normally lose an electron. Because it has such a high ionization energy, fluorine does not form positive ions under normal conditions. (b) When sodium atoms form 1+ ions, they are isoelectronic with the noble gas neon. However, to lose a second electron, Na^+ would have to lose an electron from a noble gas configuration. Because noble gas configurations are especially stable, sodium does not normally lose a second electron. (c) Neon is a noble gas. The electron configurations of noble gases are very stable, so the first ionization energy is very high (see Figure 7.26).

7.101   Atomic radii increase as you go from right to left across a period, and down a group (Figure 7.30). The order of increasing radius is: O < S < P. Phosphorus is to the left of sulfur, and sulfur is below oxygen.

7.103   There are two factors that determine the size of an atomic radius: nuclear charge and the principle energy level of the valence electrons. Nuclear charge is more important when we compare elements in the same period (i.e. left to right). The principle energy level of the valence electrons is more important when we compare elements in the same group. The valence electrons of both chlorine and sulfur atoms are in principal energy level 3 but chlorine has more protons in its nucleus to attract the electrons. This stronger nuclear attraction causes the atomic radius of chlorine to be smaller than that of sulfur.

7.105   When an atom loses electrons, its radius decreases; when an atom gains electrons, its radius increases. This means that cations are always smaller than the atoms they are formed from, and anions are always larger than the atoms they are formed from. (a) Magnesium atoms are larger than magnesium ions, Mg^{2+}. (b) Phosphide ions, P^{3-}, are larger than phosphorus atoms.

7.107 K^+ and Ca^{2+} ions are both isoelectronic with argon, but K^+ is larger because it has fewer protons in its nucleus to attract the electrons.

7.109   Hydrogen has only one electron, and the energies of the sublevels are all the same. In multielectron atoms, the energies of the sublevels vary (compare Figure 7.17 to Figure 7.18). This happens because, besides interacting with the nucleus, the electrons interact with each other (they all have negative charge).

7.111   The f electrons are two energy levels below the valence electrons of the atom, and the d orbitals are one level below the valence level.
             Configuration

        (a) Bi [Xe] $6s^24f^{14}5d^{10}6p^3$

        (b) Rn [Rn] or [Xe]$6s^24f^{14}5d^{10}6p^6$

        (c) Ra [Rn] $7s^2$

7.113   Xenon is in energy level five, so there are 4 sets of p-orbitals (the sublevels 2p, 3p, 4p, 5p) which are filled. Each p-orbital set contains 3 individual orbitals (p_x, p_y, p_z). This means that there are 12 filled p-orbitals in a xenon atom.

7.115 A cadmium atom has two sets of *d*-orbitals which are filled (the sublevels 3*d* and 4*d*). Each *d*-orbital set is composed of five different orbitals (see figure 7.16). This means that there are 10 filled *d*-orbitals in a cadmium atom.

7.117 Both are related to the attraction of electrons to the nucleus. As this attraction increases, electrons are more tightly held. This results in a greater ionization energy, and a smaller atomic radius.

7.119 NaCl is an ionic compound composed of Na^+ and Cl^- ions. Sodium ions are smaller than sodium atoms. When atoms lose electrons, there is less repulsion among the remaining electrons, and the radius shrinks slightly. In addition, Na^+ forms when sodium atoms lose their 3*s* valence electrons. Because the 3*s* electrons are further away from the nucleus than the 2*p* electrons, there is a large decrease in the radius (compare Figures 7.30 and 7.33). Chloride ions are larger than chlorine atoms because they have picked up an additional electron. The increased repulsion of the electrons in the orbitals causes the radii to increase. The radii of sodium ions are smaller than the radii of sodium atoms; the radii of chloride ions are greater than the radii of chlorine atoms.

7.121 Neon's red color indicates lower energy, lower frequency, and longer wavelength. Krypton's blue color indicates higher energy, higher frequency, and shorter wavelength.

7.123

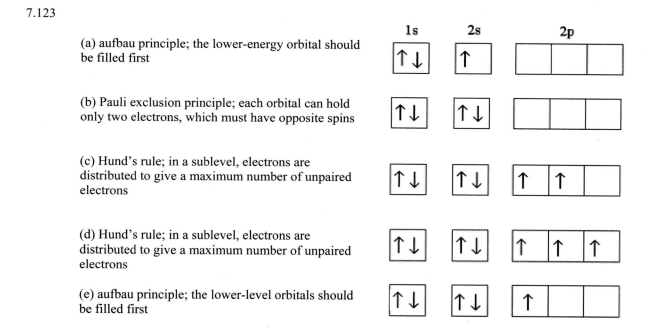

(a) aufbau principle; the lower-energy orbital should be filled first

(b) Pauli exclusion principle; each orbital can hold only two electrons, which must have opposite spins

(c) Hund's rule; in a sublevel, electrons are distributed to give a maximum number of unpaired electrons

(d) Hund's rule; in a sublevel, electrons are distributed to give a maximum number of unpaired electrons

(e) aufbau principle; the lower-level orbitals should be filled first

7.125 (a) [Ne]3s^1: elements that have one valence electron are found in group IA (1) and are called alkali metals
(b) [Ne]3$s^2$3p^3: elements that have the highest energy valence electrons in *p* orbitals in the third period are nonmetals
(c) [Ar]4$s^2$3d^{10}4p^5: elements with seven valence electrons are found in group VIIA (17) and are called halogens; they are also nonmetals
(d) [Kr]5$s^2$4d^1: elements with partially-filled *d* orbitals are transition metals
(e) [Kr]5$s^2$4d^{10}5p^6: elements with eight valence electrons are found in group VIIIA (18) and are called noble gases; they are also nonmetals

7.127 Different elements have different numbers of protons and neutrons, resulting in different atomic masses. These ions all have the same number of electrons (36) and the electron configuration of the noble gas krypton, Kr: [Ar] 4s^2 3d^{10} 4p^6. This is a set of isoelectronic ions.

7.129   Atomic radius decreases across a period in the periodic table, so strontium is smaller than rubidium. Number of valence electrons increases across a period, so strontium has more valence electrons (2) than rubidium (1). Ionization energy increases from group IA (1) to group IIA (2), so strontium has a larger ionization energy than rubidium.

7.131   The ionization energy of any element increases as each successive valence electron is removed, but it increases dramatically when a core electron is removed. Boron has three valence electrons. Removal of the fourth electron, a core electron, requires significantly more energy than removal of the first three electrons.

Chapter 8 – Chemical Bonding

8.1   (a) single bond; (b) alkane; (c) covalent bonding; (d) ionic crystal; (e) octet rule; (f) polar covalent bond; (g) alkyne; (h) electronegativity; (i) triple bond; (j) crystal lattice; (k) Lewis formula; (l) chemical bond

8.3   A chemical bond is an attractive force between atoms or ions in a substance.

8.5   Metals are most likely to form ionic bonds with nonmetals.

8.7   To determine whether bonding will be covalent or ionic, we look for the presence of a metal or ammonium (NH_4^+) ion in the chemical formula. If one (or more) of these is present with a nonmetal, the compound will have ionic bonding. If only nonmetals are present, the substance bonds covalently. Note: For the purpose of establishing whether bonding in a substance is ionic or covalent, we consider ammonium ions (NH_4^+) as though they were metal cations. Covalent bonding occurs in HF, NCl_3, and CF_4. (a) nonmetals, covalent; (b) metal-nonmetal, ionic; (c) nonmetals, covalent; (d) metal-nonmetal, ionic; (e) nonmetals, covalent

8.9   To determine whether bonding will be covalent or ionic, we look for the presence of a metal or ammonium (NH_4^+) ion in the chemical formula. If one (or more) of these is present with a nonmetal, the compound will have ionic bonding. If only nonmetals are present, the substance bonds covalently. Note: For the purpose of establishing whether bonding in a substance is ionic or covalent, we consider ammonium ions (NH_4^+) as though they were metal cations. (a) metal-nonmetal, ionic; (b) nonmetal, covalent; (c) nonmetals, covalent; (d) metal-nonmetal, ionic

8.11   In ionic compounds, each ion is surrounded by, and attracted to, several ions of opposite charge. Ionic bonds are the attractions of oppositely charged ions to one another. We can think of the ions (represented by the spheres in the diagram) as being separate from, but strongly attracted to, one another. Covalent bonding occurs when two atoms are mutually attracted to a pair (or pairs) of electrons. Because the atoms share the electrons, we can think of the atoms as being joined together where the bond occurs. Ionic bonding is shown in A and D. In A, ionic bonding is indicated because of the presence of the metal cesium Cs and nonmetal Cl. In D, the electron sharing is not indicated, so we assume this is an ionic compound.

8.13   Covalent compounds tend to have lower boiling points than ionic compounds and the boiling points of many small, covalently bonded compounds are lower than room temperature. These compounds are in the gas phase at room temperature. The covalently bonded compounds (a) CH_4, (c) SF_4, and (e) HCl will likely be gases at room temperature. It is important to note that not all low molecular weight compounds are gases at room temperature. For example, water (H_2O) and methanol (CH_3OH) are both liquids at room temperature because the molecules are very strongly attracted to each other.

8.15   Ionic compounds have relatively high boiling points because of the strong attractions of the ions in their crystal lattices. $TiCl_3$, CsCl, and CaO are ionic compounds and, therefore, have relatively high boiling points. Generally, covalently bonded substances have lower boiling points than ionically bonded substances. Therefore, we would expect the boiling points of CO_2 and O_2 to be relatively low. Note: because both CO_2 and O_2 are atmospheric gases, we know that their boiling points are below room temperature. (a) metal-nonmetal, ionic, relatively high boiling point; (b) metal-nonmetal, ionic, relatively high boiling point; (c) nonmetals, covalent, relatively low boiling point; (d) metal-nonmetal, ionic, relatively high boiling point; (e) nonmetal, covalent, relatively low boiling point.

8.17   Electronegativity follows the same trend as ionization energy. The highest electronegativity is at the top of a group and decreases as you go down.

8.19   The bond between two atoms of the same element (for example a bond between two chlorine atoms) is always nonpolar.

8.21   Generally, electronegativity increases from the bottom to the top of a group, or from left to right across a period (Figure 8.6). Note: these trends are most predictable for main group elements. (a) There are two different trends observed with these elements. The halogens, Group VIIA (17), increase in electronegativity from Br to F, so Br < Cl < F (highest electronegativity). Moving across the second period we see the trend N < O < F. Oxygen and fluorine have the highest electronegativities of all the elements, so we can write: Br < Cl < O < F (highest electronegativity). The exact placement of N is difficult, but from Figure 8.5 we see that the trend is N = Br < Cl < O < F (highest electronegativity). (b) The electronegativity of hydrogen (2.2) is between those of B (2.0) and C (2.4). The remainder of the elements are in the same period, so we expect the order of electronegativities to be H < C < N < O < F.

8.23   Bonds are polar when there is a difference in the electronegativities of the bonded atoms. If there is no electronegativity difference, the bond is nonpolar. When two atoms of the same element are bonded to each other, the bond is always nonpolar (e.g. H bonded to H). The negative end of the dipole is toward, or near, the more electronegative atom.

	Polar	Nonpolar	Reason
(a)	$^{\delta+}$H – F$^{\delta-}$	H–H	F is more electronegative than H
(b)	$^{\delta-}$Cl – I$^{\delta+}$	Cl–Cl	Cl is more electronegative than I
(c)	$^{\delta+}$B – H$^{\delta-}$	H–H	H is more electronegative than B

8.25   (a) We determine the polarity of a bond by calculating the difference in the electronegativities of the atoms sharing the electrons. Because hydrogen is involved in each bond, and hydrogen has the lowest electronegativity of the elements involved, we determine that the polarity of the bond increases as the electronegativity of the element bonded to the hydrogen atom increases. From periodic trends we know that electronegativity increases as H < C < O < F. This means that the bond polarity increases as H–H (nonpolar) < C–H < O–H < F–H.

    (b) To determine the exact order of bond polarity, we calculate the difference in electronegativity between the elements.

Bond	Electronegativity Difference
O–Cl	3.4 – 3.2 = 0.2
C–Cl	2.6 – 3.2 = –0.6
H–Cl	2.2 – 3.2 = –1.0
F–Cl	4.0 – 3.2 = 0.8

    If we ignore the sign (which indicates the direction of the dipole), we see that increasing bond polarity is given by O–Cl < C–Cl < F–Cl < H–Cl

8.27   From a Lewis symbol we can determine the number of valence electrons an atom has and the number of electrons that atom must gain or lose to achieve an octet configuration.

8.29   To draw the Lewis symbol of an atom, we first determine its number of valence electrons. For main group elements, the quickest way to do this is to look at the Roman numeral group numbers. For these elements, the Roman numerals are equal to the numbers of valence electrons. To draw the Lewis symbol we write the elemental symbol which represents the nucleus and all the inner (core) electrons. Next, we position one dot per valence electron around the elemental symbol, keeping the electrons unpaired as long as possible.

(a) ·C̈·   (b) :Ï: (c) :S̈e: (d) S̈r·   (e) C̈s   (f) :Är̈:

8.31   To draw the Lewis symbol for an ion, we first draw the Lewis symbol for the atom and then convert it to the ionic symbol by removing one electron for each positive charge, or adding one electron for each negative charge. The charge on the ion is also indicated.

	Atom	Ion
(a)	$\cdot\ddot{C}l\!:$	$:\ddot{\ddot{C}l}\!:^-$
(b)	$\cdot\dot{S}c\cdot$	Sc^{3+}
(c)	$\cdot\ddot{\dot{S}}\!:$	$:\ddot{\ddot{S}}\!:^{2-}$
(d)	$\cdot\dot{B}a$	Ba^{2+}
(e)	$\cdot\dot{B}\cdot$	B^{3+}

8.33   Draw the Lewis symbols for the ions and combine using brackets to separate the ions.

(a) $[Li^+][:\ddot{\ddot{C}l}\!:^-]$

(b) $[:\ddot{\ddot{C}l}\!:^-][Ba^{2+}][:\ddot{\ddot{C}l}\!:^-]$

(c) $[Ba^{2+}][:\ddot{\ddot{S}}\!:^{2-}]$

8.35   From the Lewis symbols for the K atom and the K^+ ion, we see that there are no valence electrons remaining to remove on the K^+ ion, giving K^+ an octet of electrons. To remove a second electron would mean taking an electron from the core. As we saw from the ionization energy trends, these core electrons are very strongly held and are not easily removed.

Atom: $\dot{K}$ Ion: K^+

8.37   When a sodium atom reacts with a fluorine atom, the sodium atom loses one electron to the more electronegative fluorine atom. As a result of the electron transfer, both ions that form satisfy the octet rule. The ionic bond formed between the two resulting ions is very strong, and the compound is electrically neutral. We can summarize this process with the equation below:

$$\overset{\frown}{Na} \quad \cdot\ddot{F}\!: \quad \longrightarrow \quad \left[Na^+\right]\left[:\ddot{\ddot{F}}\!:^-\right]$$

Any other combination than a one-to-one ratio of Na^+ and F^- ions would not produce an electrically neutral compound.

8.39   A crystal lattice is the repeating pattern observed in all crystal structures created by the attraction of oppositely charged ions for one another. An ionic crystal is the structure that forms when ions minimize the repulsive energies of like-charged ions and maximize the contact of oppositely charged ions.

8.41   Six chloride ions surround each sodium ion, as shown below:

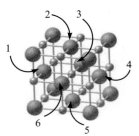

8.43   The crystal structure of any ionic compound depends on the relative sizes of the ions and on the chemical formula of the compound. The ratio of cations to anions in CaF_2 (1:2) is different than for $NaCl$ (1:1), so different ionic crystals form.

8.45   (a) Most elements try to bond to have eight electrons in their valence shell. Since an oxygen atom has six valence electrons, each will share two electrons to form a double bond.

$$:\ddot{O}\cdot\ \cdot\ddot{O}: \longrightarrow (:\ddot{O}:::\ddot{O}:) \longrightarrow :\ddot{O}=\ddot{O}:$$
$$8\ \ 8$$

Fluorine has seven valence electrons so each will share one electron to form a single bond.

$$:\ddot{F}\cdot\ \cdot\ddot{F}: \longrightarrow (:\ddot{F}:\ddot{F}:) \longrightarrow :\ddot{F}-\ddot{F}:$$
$$8\ \ 8$$

(b) Oxygen forms a double bond and fluorine forms a single bond.

8.47   Each hydrogen atom needs only one electron to satisfy its valence shell. Therefore, when two hydrogen atoms form a covalent bond, both atoms achieve a share in an electron pair. Neither of the two H atoms has another electron available to be shared with one or more additional H atoms.

$$H\cdot\ \ +\ \ \cdot H \longrightarrow (H:H)$$

8.49   The number of bonds that an atom usually forms depends on the number of unpaired electrons in the Lewis symbol. Each unpaired electron makes a bond:

	Lewis	Bonds
(a)	$H\cdot$	1 unpaired electron, 1 bond
(b)	$\cdot\ddot{N}\cdot$	3 unpaired electrons, 3 bonds
(c)	$\cdot\ddot{F}:$	1 unpaired electron, 1 bond
(d)	$:\ddot{Ne}:$	0 unpaired electrons, no bonds

8.51 By counting the bonds and lone pairs, we can usually determine the number of valence electrons the central atom must have. For example, in (b) the central atom has formed two bonds, each of which likely results when the central atom shares one electron. In addition, the central atom has two lone electron pairs. The total $2 + 4 = 6$ means that the central atom likely has six valence electrons.

(a) The central atom has formed three bonds and has no lone electron pairs indicating that it has three valence electrons. The central atom could be any nonmetal from Group IIIA (e.g. boron).

(b) The central atom has formed two bonds and has two lone electron pairs indicating that it has six valence electrons. The central atom could be any nonmetal from Group VIA (e.g. oxygen).

(c) The central atom has formed three bonds and has one lone electron pair indicating that it has five valence electrons. The central atom could be any nonmetal from Group VA (e.g. nitrogen).

8.53 The general steps for writing Lewis formulas are: 1) Count the number of available valence electrons. If the substance is charged, add one electron for each negative charge or subtract one for each positive charge. 2) Draw a reasonable skeleton. Usually, the least electronegative element is the element with the largest number of unpaired valence electrons in its Lewis symbol and it goes in the center. Note that while hydrogen may be the least electronegative element in a compound, it cannot be a central atom. 3) Place one electron pair between the central atom and each of the surrounding atoms. 4) Distribute the remaining electrons to the outside atoms so that each of these surrounding atoms has an octet of electrons. Place any remaining electrons on the central atom. 5) If any of the atoms in the resulting formula do not have an octet of electrons, rearrange the unshared electrons into double and triple bonds.

(a) HCN

		Comments
Valence electrons	$1 + 4 + 5 = 10 \text{ e}^-$	Add up valence electrons.
Skeleton	H C N	Carbon in center: least electronegative (besides H)
Bond atoms once	H $\vdots$ C $\vdots$ N	Make one bond between carbon and other atoms. 6 electrons remain.
Distribute remaining electrons	H $\vdots$ C $\vdots$ $\ddot{\text{N}}$ $\vdots$	Hydrogen's valence satisfied. Put remaining electrons on nitrogen.
Check valence (octet rule)	 2 4 8	H and N both satisfied (H has 2 valence electrons, N has 8), but carbon needs more electrons.
Multiple bonds	 2 8 8	Move two electron pairs *from the nitrogen* so that they are shared with carbon.
Replace dots with lines	H—C≡N:	*Alternate to Lewis dot formula*

(b) H₃CCN

		Comments
Valence electrons	$3 \times 1 + 2 \times 4 + 5 = 16\ e^-$	Add up valence electrons.
Skeleton	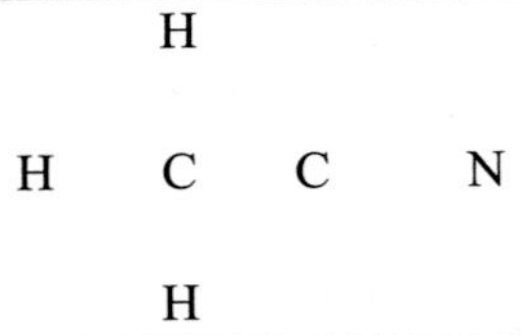 H H C C N H	Carbon atoms are in the center: least electronegative (not counting H). Hydrogen atoms can each form only one bond, so they go on the outside of the structure. Formula indicates they are all bonded to the first carbon.
Bond atoms once	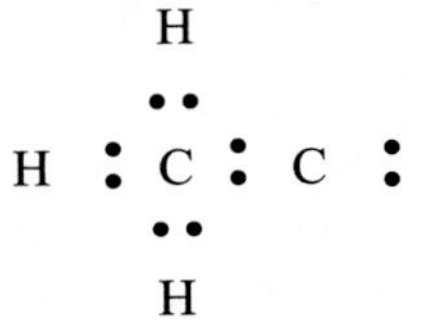 H $\cdot\cdot$ H $\ \vdots\ $ C $\ \vdots\ $ C $\ \vdots\ $ N $\cdot\cdot$ H	Connect the atoms to each other with single bonds. 6 electrons remain.
Distribute remaining electrons	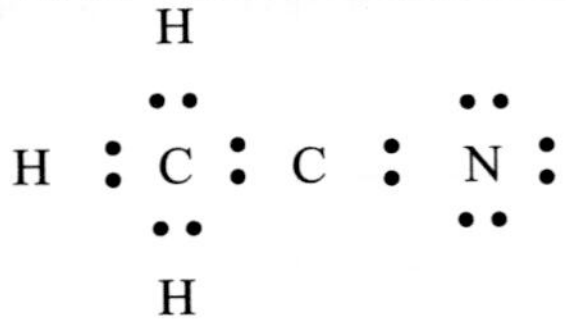 H $\cdot\cdot$ $\cdot\cdot$ H $\ \vdots\ $ C $\ \vdots\ $ C $\ \vdots\ $ N $\vdots$ $\cdot\cdot$ $\cdot\cdot$ H	Put remaining electrons on nitrogen.
Check valence (octet rule)	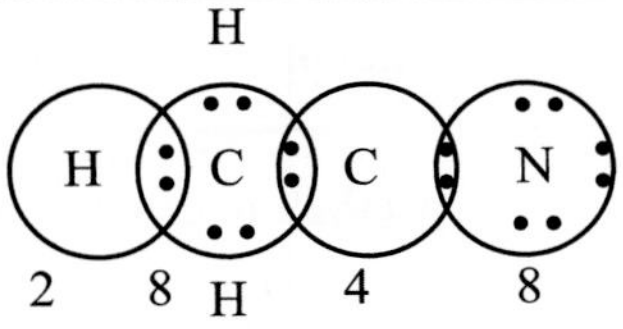 2 8 H 4 8	Valences are satisfied for all H atoms, carbon atom on the left, and N, but the carbon atom on the right needs four more electrons.
Multiple bonds	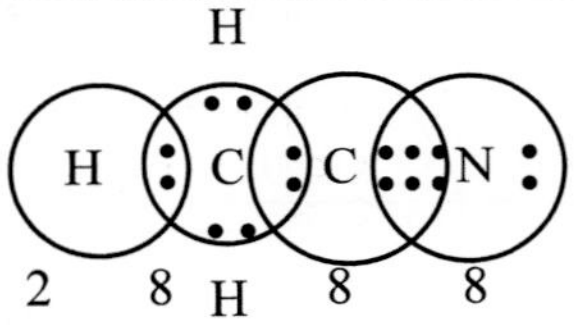 2 8 H 8 8	Move electron pairs *from the nitrogen* so that they are shared with carbon.
Replace dots with lines	H | H—C—C≡N: | H	*Alternate to Lewis dot formula*

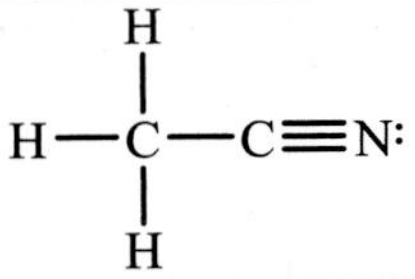

8–6

(c) C_2H_2

		Comments
Valence electrons	$2 \times 1 + 2 \times 4 = 10\ e^-$	Add up valence electrons.
Skeleton	H C C H	Place the carbon atoms in the center: they have the largest number of unpaired valence electrons. Hydrogen atoms can form only one bond each, so they go on the outside of the structure.
Bond atoms once	H $:$ C $:$ C $:$ H	Connect the atoms with single bonds. 4 electrons remain.
Distribute remaining electrons	H $:$ C $:$ C $:$ H	Distribute remaining electrons evenly on carbons.
Check valence (octet rule)	H $:$ C $:$ C $:$ H 2 6 6 2	Valences satisfied for H but both carbons need more electrons.
Multiple bonds	H $:$ C $:$ C $:$ H 2 8 8 2	Move lone pair electrons between carbons to satisfy octets.
Replace dots with lines	H—C≡C—H	*Alternate to Lewis dot formula*

(d) C_2H_4

		Comments
Valence electrons	$4 \times 1 + 2 \times 4 = 12\ e^-$	Add up valence electrons.
Skeleton	H H H C C H	Place the carbon atoms in the center: they have the largest number of unpaired valence electrons in their Lewis symbols. Hydrogen atoms can form only one bond each, so they go on the outside of the structure.
Bond atoms once	H H H : C : C : H	Connect the atoms with single bonds. 2 electrons remaining.
Distribute remaining electrons	H H H : C : C : H	Distribute remaining electrons on one of the carbon atoms.
Check valence (octet rule)	H H H C C H 2 8 6 2	Valences satisfied for all H atoms but the carbon on the right needs more electrons.
Multiple bonds	H H H C C H 2 8 8 2	Move lone pair electrons between carbons to satisfy octets. Valence satisfied.
Replace dots with lines	H H | | H—C=C—H	*Alternate to Lewis dot formula*

(e) C_2H_6

		Comments
Valence electrons	$6 \times 1 + 2 \times 4 = 14\ e^-$	Add up valence electrons.
Skeleton	H H H C C H H H	Place the carbon atoms in the center: they have the largest number of unpaired valence electrons in their Lewis symbols. Hydrogen atoms can form only one bond each, so they go on the outside of the structure.
Bond atoms once	H : C : C : H structure with H above and below each C	Connect the atoms with single bonds. No electrons remaining.
Check valence (octet rule)	H (: C (:) C :) H structure with octet circles, labeled 2 8 H H 8 2	Check octets. All valences satisfied.
Replace dots with lines	H—C—C—H structure with H above and below each C	*Alternate to Lewis dot formula*

8.55   We write Lewis formulas for ions using the same general process as for writing formulas for molecules, except that for anions, we add one electron for each negative charge, and for cations, we remove one electron for each positive charge.

a) NO_3^-

		Comments
Valence electrons	$5 + 3 \times 6 + 1 = 24\ e^-$	Add up valence electrons and add one electron for the 1– charge.
Skeleton	O O N O	Place nitrogen in the center; it has the largest number of unpaired electrons.
Bond atoms once	O O : N : O	Connect the atoms with single bonds; 18 electrons remain.
Distribute remaining electrons	:O: :O : N : O:	Distribute the remaining electrons to the oxygen atoms first.
Check valence (octet rule)		Check octets. The nitrogen atom does not have an octet.
Multiple bonds		Move one pair of electrons from one of the oxygen atoms, to form a double bond with N. Valence satisfied.
Replace dots with lines	$\left[\ \overset{\displaystyle :\ddot{O}:}{\underset{\displaystyle}{\ddot{O}\!=\!N\!-\!\ddot{O}\!:}}\ \right]^-$	*Alternate to Lewis dot formula*

b) SO_4^{2-}

		Comments
Valence electrons	$6 + 4 \times 6 + 2 = 32 \; e^-$	Add up valence electrons and add two electrons for the 2– charge.
Skeleton	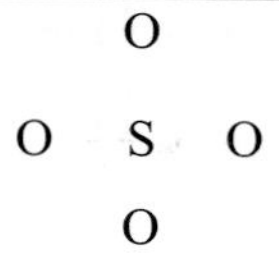 O O S O O	Locate sulfur in the middle because there are fewer sulfur atoms than oxygen and sulfur is less electronegative than oxygen. In all oxoanions, the non-oxygen atom is central.
Bond atoms once	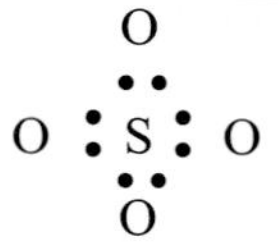 O $\cdot\cdot$ O $\cdot\cdot$ S $\cdot\cdot$ O $\cdot\cdot$ O	Connect the atoms with single bonds. Notice that sulfur already has an octet. 24 electrons remain $(32 - 8 = 24)$.
Distribute remaining electrons	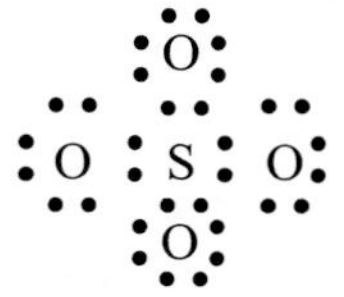 :O: :O : S : O: :O:	Distribute remaining electrons around the oxygen atoms first. All electrons distributed.
Check valence (octet rule)	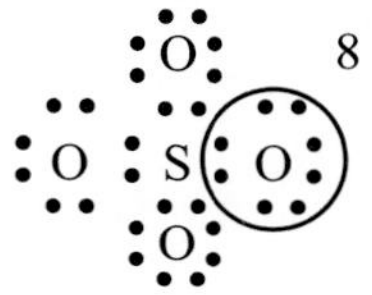 :O: 8 :O : S (O:) :O:	Sulfur already had an octet. Since all the oxygen atoms look the same, just check one. All octets are satisfied. Double and triple bonds are not needed.
Replace dots with lines	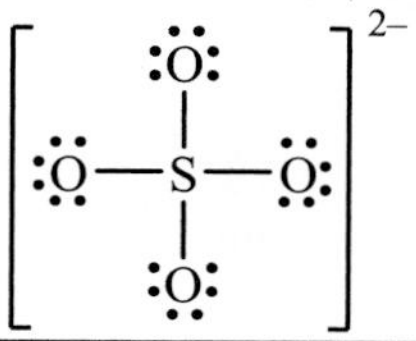 $\left[\begin{array}{c} :\ddot{O}: \\ \vert \\ :\ddot{O}-S-\ddot{O}: \\ \vert \\ :\ddot{O}: \end{array} \right]^{2-}$	*Alternate to Lewis dot formula*

8–11

c) SO_3^{2-}

		Comments
Valence electrons	$6 + 3 \times 6 + 2 = 26\ e^-$	Add up valence electrons; add two electrons because of ionic charge.
Skeleton	O O S O	Locate sulfur in the middle because there are fewer sulfur atoms than oxygen and sulfur is less electronegative than oxygen. In all oxoanions, the non-oxygen atom is central.
Bond atoms once	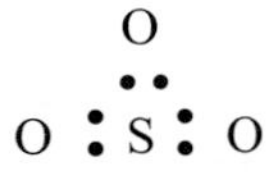 O $\cdot\cdot$ S $\cdot\cdot$ O	Connect the atoms with single bonds. 20 electrons remain (26 − 6 = 24).
Distribute remaining electrons	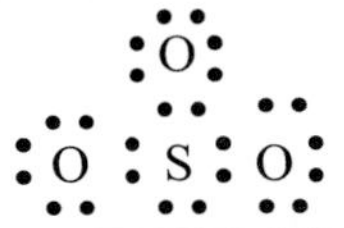	Distribute electrons to oxygen atoms first, and then to sulfur. All electrons distributed.
Check valence (octet rule)	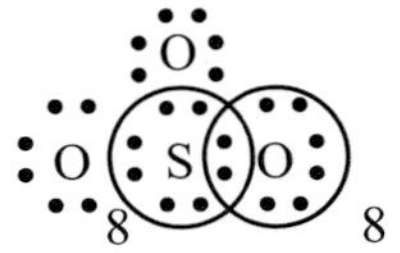	Since all the oxygen atoms look the same, just check one. All octets are satisfied. Double and triple bonds are not needed.
Replace dots with lines	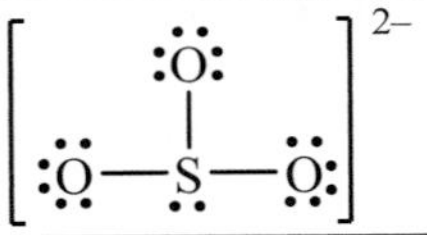	*Alternate to Lewis dot formula*

d) NO_2^-

		Comments
Valence electrons	$5 + 2 \times 6 + 1 = 18\ e^-$	Add up valence electrons and add one electron for the negative charge.
Skeleton	O N O	Locate the nitrogen atom in the center; it has the largest number of unpaired electrons, and is less electronegative than oxygen.
Bond atoms once	 O $\cdot\cdot$ N $\cdot\cdot$ O	Connect atoms with single bonds. 14 electrons remain.
Distribute remaining electrons		Distribute remaining electrons on the oxygen atoms first, then on the nitrogen atom.
Check valence (octet rule)	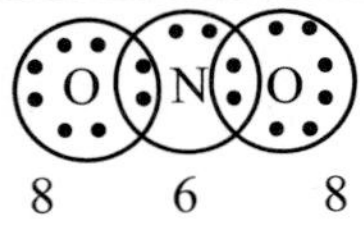 8 6 8	Valences satisfied for oxygen, but the nitrogen needs more electrons.
Multiple bonds	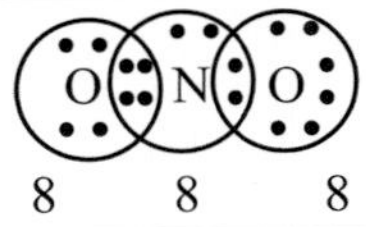 8 8 8	Move lone pair from oxygen to form a double bond with nitrogen. Valence satisfied.
Replace dots with lines	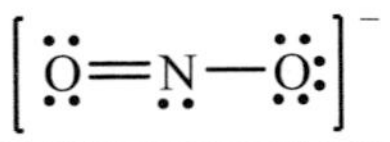	*Alternate to Lewis dot formula*

e) NO⁺

		Comments
Valence electrons	$5 + 6 - 1 = 10 \ e^-$	Add up valence electrons and subtract one electron for the positive charge.
Skeleton	N O	Arrange the atoms.
Bond atoms once	N $:$ O	Connect the atoms with a single bond. 8 electrons remain.
Distribute remaining electrons	$: \text{N} : \overset{..}{\underset{..}{\text{O}}} :$	Distribute remaining electrons on oxygen first, then nitrogen. Notice that oxygen already has an octet.
Check valence (octet rule)		Check valences. Nitrogen atom needs additional electrons.
Multiple bonds		Move electrons from oxygen to form a triple bond between the nitrogen and the oxygen. Valence satisfied.
Replace dots with lines	$[: \text{N} \equiv \text{O} :]^+$	*Alternate to Lewis dot formula*

8.57   Draw the Lewis structures of each and count the number of bonded electrons. The structure of cyanide ion is:

$$[: \text{C} \equiv \text{N} :]^-$$ three bonded pairs; 6 bonding electrons

(a) $: \overset{..}{\text{O}} = \overset{..}{\text{O}} :$ two bonded pairs; 4 bonding electrons; different

(b) $[: \text{N} \equiv \text{O} :]^+$ three bonded pairs; 6 bonding electrons; same

(c) $: \text{C} \equiv \text{O} :$ three bonded pairs; 6 bonding electrons; same

(d) $: \text{N} \equiv \text{N} :$ three bonded pairs; 6 bonding electrons; same

(e) $\text{H} - \overset{..}{\text{N}} - \text{H}$, with H below   three bonded pairs; 6 bonding electrons; same

8.59   Resonance descriptions are needed when we can represent a molecule or ion with two or more reasonable Lewis structures that differ only in the positions of the bonding and lone pairs of electrons. Note: When we draw resonance structures, we cannot change the positions of the atomic nuclei; we only change the positions of the electrons.

8.61   To determine whether a molecule exhibits resonance, we draw the Lewis formula of the molecule and then determine whether we can distribute the electrons differently without moving any atoms. Many times, resonance is observed in polyatomic oxoanions. Resonance is exhibited in (c) and (d).

(a) 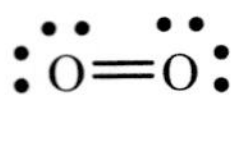   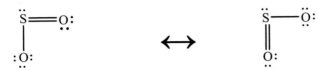   No resonance

(b)

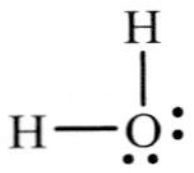

H
|
H—O:
••

No resonance

(c) SO_2 has resonance structures.

(d) NO_2 has resonance structures.

(e) 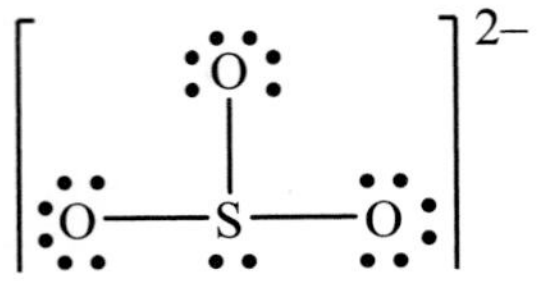   No resonance

8.63   (a) resonance structures for NO_2^-

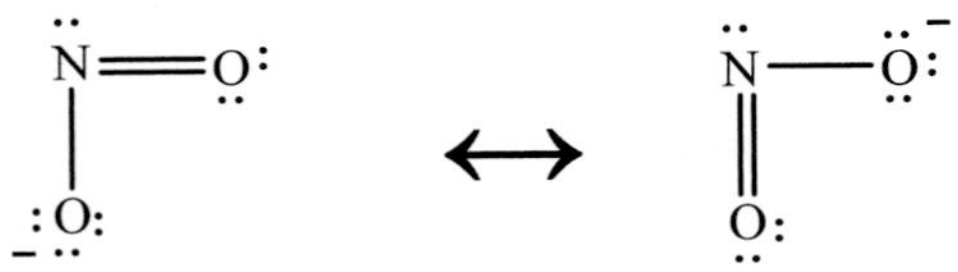

(b) resonance structures for SO_3

(c) resonance structures for CO_2

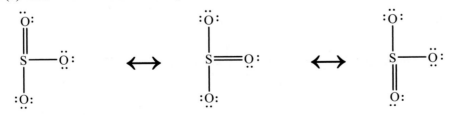

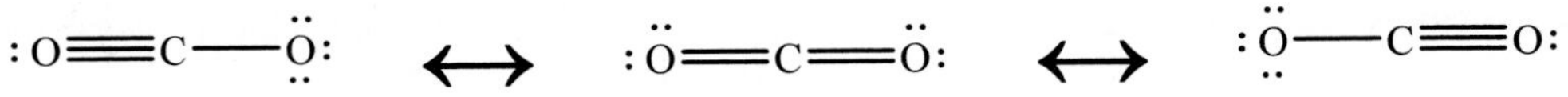

(d) resonance structures for CO_3^{2-}

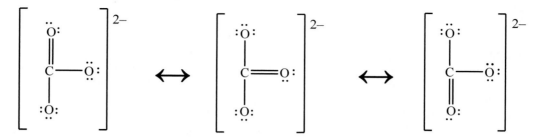

(e) resonance structures for HNO_3

8.65 Hydrogen's valence shell can hold a maximum of two electrons (a 1s orbital holds only 2 electrons). By sharing a pair of electrons, a hydrogen atom fills its valence shell. The octet rule is a simplification of the idea that configurations are more stable when the s and p orbitals of a given energy level are filled. For example, if the configuration of an atom (or ion) ends in $2s^2 2p^6$, the atom (or ion) has eight (an octet) electrons in its valence shell. In fluorine, the valence shell is also filled by sharing an electron with hydrogen:

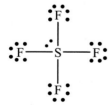

8.67 First, we draw the Lewis structures and count the electrons associated with each atom. Generally, if the octet rule is not obeyed, we will find the discrepancy on the central atom. Atoms in some compounds show one of three types of modification of the octet rule: expanded octets (where more than eight electrons surround the central atom); odd-electron species (where one atom has an odd number of electrons and, therefore, cannot satisfy the octet rule); and incomplete octet species (where the central atom has three or fewer valence electrons and so cannot make the octet through normal bonding - boron is an example).

(a) Oxygen atom obeys the octet rule.

$$H \!-\! \overset{\displaystyle ..}{\underset{\displaystyle ..}{O}} \!-\! H$$

(b) Sulfur atom does not obey the octet rule. Sulfur has an expanded octet.

(c) Fluorine atoms obey the octet rule (see (b) for structure).

(d) Sulfur atom obeys the octet rule.

$$:\!F\!-\!S\!-\!F\!:$$

8.69   First, we must draw the structures and count the electrons on each atom. Generally, if the octet rule is not obeyed, we will find the discrepancy on the central atom. Atoms in some compounds show one of three types of modification of the octet rule: expanded octets (where more than eight electrons surround the central atom); odd-electron species (where one atom has an odd number of electrons and, therefore, cannot satisfy the octet rule); and incomplete octet species (where the central atom has three or fewer valence electrons and, so, cannot make the octet through normal bonding - boron is an example).

(a)   The sulfur atom does not obey the octet rule because it is surrounded with 10 electrons.

(b)   Boron has an incomplete octet; it is surrounded by only 6 electrons.

(c)   Xenon has an expanded octet; it is surrounded by 14 electrons.

(d)   ClO$_2$ is an odd electron species; the chlorine atom has an unpaired electron.

8.71   We can divide hydrocarbons into two primary groups: aliphatic hydrocarbons and aromatic hydrocarbons. We can divide aliphatic hydrocarbons into three groups: alkanes, with only single bonds; alkenes, with double bonds; and alkynes, with triple bonds. We can divide aromatic hydrocarbons into two primary groups: benzene and its derivatives; and polynuclear aromatic compounds.

8.73   See Figure 8.22.

8.75 Functional groups are listed in table 8.4. Notice that there is a subtle difference between the structures of ketones and aldehydes. In aldehydes, the carbon atom that is doubly bonded to oxygen is also bonded to a hydrogen atom. (a) alcohol; (b) alkane; (c) ether; (d) alkene; (e) ketone

8.77 Drawing the Lewis structure helps us to identify the functional group. Once drawn, we can compare the structure to the functional groups shown in table 8.4.
(a)

		Comments
Draw the skeleton and identify missing bonds and lone pairs.	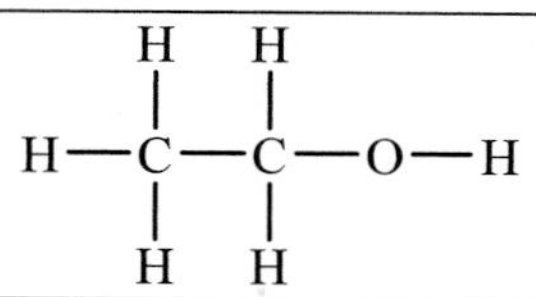 H H | | H—C—C—O—H | | H H	Carbons are central. Hydrogens are on the outside of the compound. This structure is complete except that oxygen needs two unshared pairs of electrons.
Add missing bonds and lone pairs.	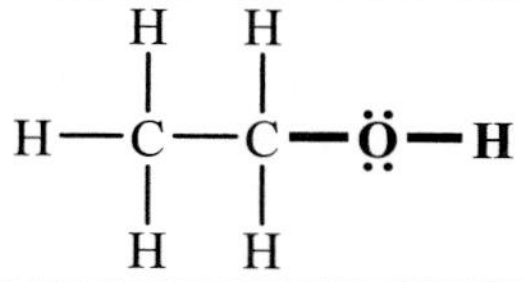 H H | | H—C—C—Ö—H | | H H	CH_3CH_2OH is an alcohol. When we draw organic compounds, we often omit the unshared electron pairs.

(b)

		Comments
Draw the skeleton and identify missing bonds and lone pairs.	H | H—C—C—C—H | H	Carbon atoms are central. Hydrogen atoms are on the outside of the compound. Look for atoms with too few bonds. Two carbon atoms are missing 2 bonds each.
Add missing bonds and lone pairs.	H | H—C—C≡C—H | H	CH_3CCH is an alkyne.

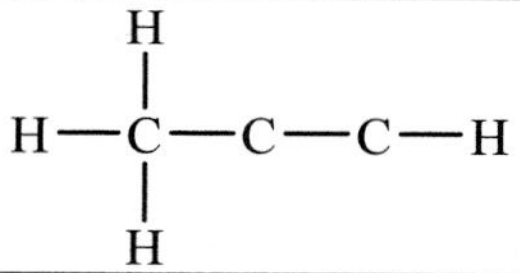

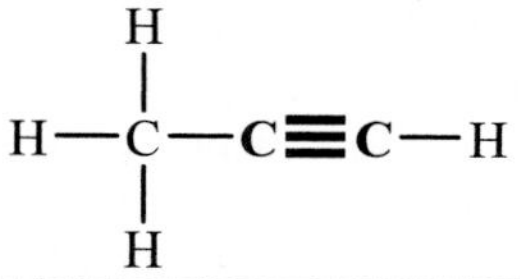

(c)

		Comments
Draw the skeleton and identify missing bonds and lone pairs.	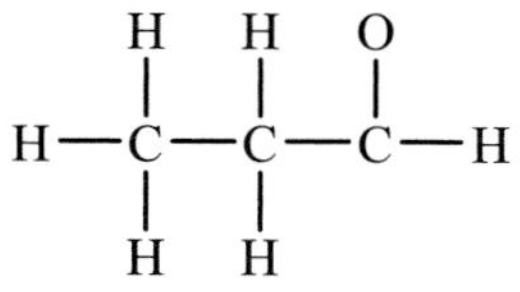 H H O | | | H—C—C—C—H | | H H	Carbons are central. The H is written before the O in the chemical formula (CHO) because both are attached to the carbon. The oxygen atom and the carbon atom to which it is attached are each missing one bond. Oxygen is missing two unshared electron pairs.
Add missing bonds and lone pairs.	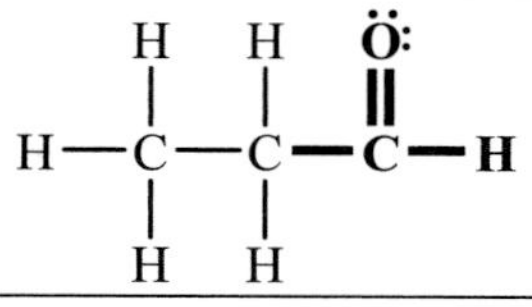 H H Ö: | | || H—C—C—C—H | | H H	CH_3CH_2CHO is an aldehyde. When we draw organic compounds, we often omit the unshared electron pairs.

			Comments
Draw the skeleton and identify missing bonds and lone pairs.	H—C—C—O—C—C—H (with H atoms above and below each carbon)		Carbons are central. The oxygen atom connects the two carbon chains. Each of the carbon atoms has four bonds, but the oxygen is missing its unshared electron pairs.
Add missing bonds and lone pairs.	H—C—C—Ö—C—C—H (with H atoms above and below each carbon)		$CH_3CH_2OCH_2CH_3$ is an ether.

8.79   (a) Because the compound contains only carbon and hydrogen atoms and no double or triple bonds this is an alkane. (b) The blue atom represents nitrogen. The functional group $-NH_2$ indicates that this is an amine.

(c) The red atoms represent oxygen atoms. This is the ester functional group ($-\overset{\overset{\displaystyle O}{\|}}{C}-O-$).

8.81

H—C—C—C—C—H (with H atoms on first three carbons, and a double-bonded O on the fourth carbon)

8.83   We predict molecular shapes by finding the geometry that gives the greatest distance between the electron groups around one or more central atoms (Table 8.5). To do this, we draw the Lewis structure and count the bonded groups and unshared electron pairs on the atom around which we want to determine the geometry and shape. The geometry and the arrangement of the bonded atoms is determined by the configuration that will allow these bonded groups and unshared electron pairs to be as separated from each other as possible (Table 8.6).

8.85   To determine the shape at a particular atom within a molecule we need to know the number of electron domains (bonded atoms and unshared electron pairs). We determine the number of bonded atoms and unshared electron pairs by drawing the Lewis structure. The molecular shapes, which describe the arrangement of atoms, are shown in Table 8.6.

8.87   The shapes are found in Table 8.6.
  (a) A solid wedge indicates that the group is coming out in front of the plane of the paper, and a hashed wedge
       represents a group that is extending behind the plane of the paper.

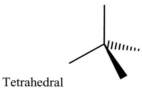

        Tetrahedral

  (b) Trigonal planar arrangements do not require perspective drawings, because all of the groups are in the same
       plane, as the name trigonal *planar* implies.

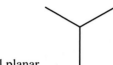

        Trigonal planar

  (c) Note that the bent shape can derive from either a trigonal planar or a tetrahedral structure. These two
       structures differ in the number of unshared electron pairs on the central atom (trigonal planar bent has one
       unshared electron pair and tetrahedral bent has two unshared electron pairs). Regardless of the parent
       geometry, however, the bent shape looks approximately the same for both, because the unshared electron
       pairs are not part of the shape.

        Bent

  (d) The trigonal pyramid comes from the tetrahedral geometry. It is drawn in a similar fashion but leaving out
       the top bond gives it the pyramidal appearance:

8.89 We first draw the structures and then determine the number of electron domains attached to the central atom. From this, we can determine which parent structure or geometric arrangement exists. For example, in $BeCl_2$ there are two electron domains, so the parent structure is linear. In SO_2, the central S atom is attached to two atoms (the oxygen atoms) and one unshared electron pair. This corresponds to a trigonal planar arrangement. Note that we count a double bond or triple bond as one electron domain.

		Electron Domains	Parent Structure	Lewis structure
(a)	$BeCl_2$	2	Linear	$:\!\ddot{C}l\!-\!Be\!-\!\ddot{C}l\!:$
(b)	PH_3	4	Tetrahedral	H—P̈—H with H above
(c)	SCl_2	4	Tetrahedral	$:\!\ddot{C}l\!-\!\ddot{S}\!-\!\ddot{C}l\!:$
(d)	SO_2	3	Trigonal planar	$:\!\ddot{O}\!=\!\ddot{S}\!-\!\ddot{O}:$
(e)	H_2Te	4	Tetrahedral	H—T̈e—H
(f)	SiH_4	4	Tetrahedral	H—Si—H with H above and H below
(g)	BBr_3	3	Trigonal planar	Br—B—Br with Br above
(h)	H_2O	4	Tetrahedral	H—Ö—H

8.91   First we draw the Lewis structure of the compound and then determine its parent structure, as shown in Table 8.5. Then we can consult Tables 8.6 and 8.5 for the information we need to determine molecular shape and approximate bond angles. For example, in $BeCl_2$ the Be atom is bonded to two atoms and no unshared electron pairs. Because there are two electron domains, the molecule's parent geometry is linear, with a bond angle of 180°. Because there are no unshared electron pairs, the shape of the molecule is also linear, as shown on Table 8.6.

		Parent Structure	Atoms Bonded to Central Atom	Unshared Electron Pairs on Central Atom	Shape	Approximate Bond Angle	Lewis Structure
(a)	$BeCl_2$	Linear	2	0	Linear	180°	:Cl—Be—Cl:
(b)	PH_3	Tetrahedral	3	1	Trigonal pyramidal	109.5°	H—P—H with H above
(c)	SCl_2	Tetrahedral	2	2	Bent	109.5°	:Cl—S—Cl:
(d)	SO_2	Trigonal planar	2	1	Bent	120°	:O=S—O:
(e)	H_2Te	Tetrahedral	2	2	Bent	109.5°	H—Te—H
(f)	SiH_4	Tetrahedral	4	0	Tetrahedral	109.5°	H—Si—H with H above and below
(g)	BBr_3	Trigonal planar	3	0	Trigonal planar	120°	Br—B—Br with Br above
(h)	H_2O	Tetrahedral	2	2	Bent	109.5°	H—O—H

8–21

8.93 For each compound, we draw the Lewis structure and use it to determine the parent geometry. The parent geometry determines the bond angle.

	Structure	Electron Domains Around Central Atom, Approximate Bond Angle
(a)	H——N̈——H | H	four electron domains, 109.5°
(b)	H——Ö——H	four electron domains, 109.5°
(c)	H——C̈l:	A bond angle is defined by three atoms. Since there are only two atoms, there is no angle defined for HCl.
(d)	H——C≡N:	two electron domains, 180°
(e)	:F̈——B——F̈: | :F̈:	three electron domains, 120°
(f)	Ö: ‖ H——C——H	three electron domains, 120°
(g)	:C̈l——P——C̈l: | :C̈l:	four electron domains, 109.5°

8.95 We can best approach this problem by thinking, "What does the central atom of this structure look like?" In (b) for example, the Lewis symbol for the central atom must have five electrons (one pair, and three unpaired). This means it is in group VA. The central atom could be N, P, or As. The other atoms only need to have a single unpaired electron (pick something easy such as H, or an element in Group VII A).
(a) BF_3; (b) NH_3; (c) SCl_2

8.97 We use the bond angles to determine the parent geometry. In (c) for example, the parent geometry is linear. The central atom needs to have only two atoms and no unshared electron pairs (otherwise it would be bent). We can deduce that the central atom is in group IIA, so $BeCl_2$ will work. The information on Table 8.6 is also very helpful. Notice that carbon could also be the central atom. The shape of CO_2 works because there are two double bonds and no unshared electron pairs on the central atom. (a) BCl_3; (b) H_2O; (c) $BeCl_2$; (d) NH_4^+

8.99   The structure represents ClO_3^-. If we draw the Lewis structure of ClO_3^-, we find three oxygen atoms and one unshared electron pair attached to the chlorine atom.

$$\left[:\ddot{O}\!-\!\overset{\displaystyle\cdot\cdot}{Cl}\!-\!\ddot{O}: \atop \hspace{1.2em} \underset{:\ddot{O}:}{|} \right]^{-}$$

This means that the molecule is trigonal pyramidal in shape. NO_3^- is trigonal planar.

8.101   Image B has one unshared electron pair on the central atom. Even though we don't draw them, unshared electron pairs take up space. When we look at the shape, we know there is an unshared electron pair on the central atom because the structure bends away from the unshared electron pair. If the unshared electron pair were not present, the structure would be linear.

8.103   From the Lewis structure we find that each nitrogen atom has one unshared electron pair.

$$H\!-\!\overset{\displaystyle\cdot\cdot}{N}\!-\!\overset{\displaystyle\cdot\cdot}{N}\!-\!H \atop \hspace{0.3em}\underset{H}{|}\hspace{0.6em}\underset{H}{|}$$

8.105   One possible Lewis structure for chloropicrin is:

$$\begin{array}{c} :\ddot{Cl}: \hspace{2em} \overset{\cdot\cdot}{O}{\cdot\cdot} \\ | \hspace{2.5em} \| \\ :\ddot{Cl}\!-\!C\!-\!N \\ | \hspace{2.5em} \backslash \\ :\ddot{Cl}: \hspace{1.5em} \cdot\ddot{O}\cdot \end{array}$$

The bond angles around the carbon atom are approximately 109.5° (four electron groups, and no unshared electron pairs). The bond angles around the nitrogen are approximately 120° (three electron groups, and no unshared electron pairs).

8.107   A bond is polar if there is a difference in the electronegativities of the two bonding atoms. A molecule is polar if the polarities of the bonds around the central atom do not cancel one another.

8.109   The chlorine atoms are all attracting the electrons in their bonds with the central carbon atom with equal force (the electronegativity difference between C and Cl is the same in all cases). Because the molecule is geometrically symmetrical, the polarity of the bonds cancels. You could imagine that you were having a four way tug-of-war and each group had the same strength. If the forces are pulling equally in opposite directions, no group can win the tug-of-war.

8.111 For a dipole to exist in a molecule at least one of the bonds must be polar and the geometry must be such that the dipoles do not cancel each other. Molecules that are symmetrical (linear, trigonal planar, tetrahedral from Table 8.6) are nonpolar.

			Comments
(a)	H——Ï:	Polar	Polar bond; not symmetrical
(b)	:F: | H——C——F: | :F:	Polar	Symmetrical geometry, but bond polarities have different magnitudes and do not cancel Tetrahedral
(c)	:Ö: | :Cl——S——Cl: | :O:	Polar	Symmetrical geometry, but bond polarities have different magnitudes and do not cancel Tetrahedral
(d)	:F——P——F: | :F:	Polar	Molecule not symmetrical Trigonal pyramidal

8.113 For a dipole to exist in a molecule at least one of the bonds must be polar and the geometry must be such that the dipoles do not cancel one another. Molecules that are symmetrical (linear, trigonal planar, tetrahedral from Table 8.6) are nonpolar.

			Comments
(a)	:Ö==S——Ö:	Polar	Polar bonds; not symmetrical Bent
	:Ö==C==Ö:	Nonpolar	Polar bonds; symmetrical geometry cancels out bond dipoles Linear
(b)	:Ö==S——Ö:	Polar	Polar bonds; not symmetrical Bent
	:Ö: | :Ö——S==Ö:	Nonpolar	Polar bonds; symmetrical geometry cancels out bond dipoles Trigonal planar
(c)	:Cl——Se——Cl:	Polar	Polar bonds; not symmetrical Bent
	:Cl——Be——Cl:	Nonpolar	Polar bonds; symmetrical geometry cancels out bond dipoles Linear
(d)	H | H——C——H | H	Nonpolar	Polar bonds; symmetrical geometry cancels out bond dipoles Tetrahedral
	H | H——C——Ï: | H	Polar	Polar bonds, but magnitudes of bond polarities are different and do not cancel Tetrahedral

8.115 Polar molecules will align with the electric field. The only molecule in this grouping that could be polar is B. This is because the other three are all symmetrical.

8.117 Polar molecules are generally soluble in water. CCl_2F_2 is most likely to be soluble in water because it is polar. CF_4 is nonpolar, so it is not likely to be soluble in water.

8.119 Because molecule A is symmetrical, it is not polar. Molecule in B is not symmetrical and is likely to be polar.

8.121   To draw the Lewis symbols, we first determine the number of valence electrons for each atom. The quickest way to do this for main group elements is to look at the Roman numeral group numbers. The Roman numeral is equal to the number of valence electrons. Next, we place the appropriate number of dots (one dot per valence electron) around the elemental symbol keeping the electrons unpaired as long as possible.

(a) $:\ddot{Br}\cdot$; (b) $\cdot Pb\cdot$; (c) $:\ddot{S}\cdot$; (d) $Ca\cdot$; (e) $\ddot{Be}\cdot$; (f) $:\ddot{Xe}:$

8.123   Electronegativity increases from the bottom to the top of a group: F > Cl > Br > I (lowest)

8.125   If a compound contains a metal or ammonium ion, it is probably an ionic compound. Substances formed from nonmetals are covalently bonded. (a) covalent; (b) ionic; (c) covalent; (d) covalent; (e) ionic

8.127   (a) N_2H_2

		Comments
Valence electrons	$2 \times 5 + 2 \times 1 = 12$ e	Add up valence electrons.
Skeleton	H N N H	Nitrogen atoms are central. Most unpaired electrons and hydrogen must be at ends of molecule.
Bond atoms once	H : N : N : H	Add one bond between each atom. 6 electrons remaining.
Distribute remaining electrons and check valence shell	H : $\ddot{N}$ (: $\ddot{N}$:) H	Distribute remaining electrons onto nitrogen atoms. Circled nitrogen does not have an octet. Move one pair of electrons from the other nitrogen to form a double bond.
Multiple bonds	H : $\ddot{N}$ :: $\ddot{N}$: H	Octets are satisfied; Lewis structure complete.
Replace dots with lines	H——$\ddot{N}$══$\ddot{N}$——H	*Alternate to Lewis dot formula*

(b) CS_2

		Comments
Valence electrons	$4 + 2 \times 6 = 16$ e	Add up valence electrons.
Skeleton	S C S	Carbon atom is central because it has the most unpaired electrons.
Bond atoms once	S : C : S	Add one bond between each atom. 12 electrons remaining.
Distribute remaining electrons and check valence shell	:S̈ : C : S̈ :	Distribute remaining electrons onto sulfur atoms first. Carbon does not have an octet. Move one pair of electrons from the sulfur to form a double bonds.
Multiple bonds	:S̈ :: C :: S̈ :	Octets are satisfied; Lewis structure complete.
Replace dots with lines	:S̈＝C＝S̈:	*Alternate to Lewis dot formula*

(c) AsF_3

		Comments
Valence electrons	$5 + 3 \times 7 = 26$ e	Add up valence electrons.
Skeleton	F As F F	Arsenic is central because it has the most unpaired electrons.
Bond atoms once	F : As : F F	Add one bond between each atom. 20 electrons remaining.
Distribute remaining electrons and check valence shell	:F̈ : As : F̈ : :F̈ :	Octets are satisfied; Lewis structure complete.

Replace dots with lines	:F̈——As——F̈:
	        |
	      :F̈:
		Alternate to Lewis dot formula

(d) CO_2

		Comments
Valence electrons	$4 + 2 \times 6 = 16\ e$	Add up valence electrons.
Skeleton	O C O	Carbon atom is central because it has the most unpaired electrons.
Bond atoms once	O : C : O	Add one bond between each atom. 12 electrons remaining.
Distribute remaining electrons and check valence shell	$:\overset{..}{\underset{..}{O}} : C : \overset{..}{\underset{..}{O}} :$	Distribute remaining electrons onto oxygen atoms first. Carbon does not have an octet. Move one pair of electrons from the sulfur to form a double bonds.
Multiple bonds	$:\overset{..}{O} :: C :: \overset{..}{O} :$	Octets are satisfied; Lewis structure complete.
Replace dots with lines	$:\overset{..}{O}=\!=C=\!=\overset{..}{O}:$	*Alternate to Lewis dot formula*

(e) CO

		Comments
Valence electrons	$4 + 6 = 10\ e$	Add up valence electrons.
Skeleton	C O	
Bond atoms once	C : O	Add one bond between each atom. 8 electrons remaining.
Distribute remaining electrons and check valence shell	$: C : \overset{..}{\underset{..}{O}} :$	Distribute remaining electrons onto oxygen atom first. Carbon does not have an octet. Move two pairs of electrons from the oxygen to form a triple bond.
Multiple bonds	$: C ::: O :$	Octets are satisfied; Lewis structure complete.
Replace dots with lines	$:C\equiv O:$	*Alternate to Lewis dot formula*

8.129   When drawing the Lewis structure, check to see if the octets can be satisfied by contributing pairs of electrons from different atoms. The different electron arrangements are resonance structures.

(a) ClO_4^-

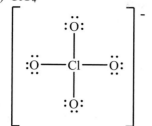

(b) NO_2^-

| :N : O: :O: | The nitrogen needs one more bond to satisfy its octet. Either oxygen atom can contribute a bond to the carbon. Two resonance structures are possible. |

(c) NCO^-

| :N : C : O: | The carbon needs two more bonds (electron pairs) to satisfy its octet. Three resonance structures are possible. |

(d) HCO_2^-

| H : C : O: :O: | The carbon needs one more bond (electron pair) to satisfy its octet. Either oxygen atom can contribute a bond to the carbon. Two resonance structures are possible. |

(e) BF_3

8.131 To determine the molecular geometry, first draw the Lewis structure and then determine the parent geometry and shape (Table 8.5 and Table 8.6).

	Lewis Structure	Geometry Shape Bond Angle		Lewis Structure	Geometry Shape Bond Angle
(a) $SiCl_4$		Tetrahedral Tetrahedral 109.5°	(b) $GaCl_3$		Trigonal planar Trigonal planar 120°
(c) NCl_2^+		Trigonal planar Bent 120°	(d) IO_3^-		Tetrahedral Trigonal pyramid 109.5°
(e) PCl_4^+		Tetrahedral Tetrahedral 109.5°	(f) OF_2		Tetrahedral Bent 109.5°
(g) GeH_4		Tetrahedral Tetrahedral 109.5°	(h) $SOCl_2$		Tetrahedral Trigonal pyramid 109.5°
(i) Br_2O		Tetrahedral Bent 109.5°	(j) ClO_2^-		Tetrahedral Bent 109.5°

8.133 If the bonds are all equivalent and the molecule is symmetrical, the bond polarities cancel and the molecule is nonpolar. If the bonds are not equivalent or the molecule does not exhibit symmetry, the molecule is likely to be polar.

(a) $BeCl_2$ is nonpolar because the molecule is symmetrical and the dipoles of the bonds cancel. OCl_2 is polar because the molecule is bent. The bend does not allow for the cancellation of the bond dipoles.

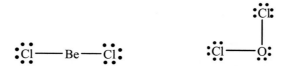

(b) PH_3 is polar. The bonds are polar and the molecule is not symmetrical (trigonal pyramidal). BH_3 is nonpolar because the molecule is symmetrical (trigonal planar) and the bonds are equivalent.

(c) BCl_3 is nonpolar because the molecule is symmetrical (trigonal planar) and the bonds are equivalent. $AsCl_3$ is polar because the bonds are polar and the molecule is not symmetrical (trigonal pyramidal).

(d) SiH_4 nonpolar because the molecule is symmetrical (tetrahedral) and the bonds are equivalent. NH_3 polar because the bonds are polar the molecule is not symmetrical (trigonal pyramidal).

8.135

			Comments
(a)	H ‖ :Cl—C—Cl: ‖ H	More polar	Polar bonds; not symmetrical Tetrahedral
	:Cl: ‖ :Cl—C—Cl: ‖ :Cl:		Bond polarities cancel because of symmetrical geometry. Tetrahedral
(b)	H ‖ H—C—F: ‖ H	More polar	Fluorine is more electronegative than bromine, so the dipole is greater in this molecule.
	H ‖ H—C—Br: ‖ H		
(c)	:F—N—F: ‖ :F:	More polar	Both molecules have the same geometry, but fluorine is more electronegative than hydrogen, so the dipole is greater in this molecule.*
	H—N—H ‖ H		
(d)	:F—O—F:		
	H—O—H	More polar	Both molecules have the same geometry, but the electronegativity difference of hydrogen and oxygen is greater than that of fluorine and oxygen.

*If you continue in your studies of chemistry, you will eventually learn that the polarity of this molecule is mostly cancelled by the lone pair of electrons on the nitrogen.

8.137 Ionic compounds have relatively high boiling points because of the strong attractions of the ions in their crystal lattices. KCl and $MgBr_2$ are ionic compounds and, therefore, have relatively high boiling points. Generally, covalently bonded substances have lower boiling points than ionically bonded substances. Therefore, we would expect the boiling points of N_2, NO_2, and NH_3 to be relatively low. Note: because both N_2 and NO_2 are atmospheric gases, we know that their boiling points are below room temperature.
(a) nonmetal, covalent, relatively low boiling point; (b) nonmetals, covalent, relatively low boiling point; (c) metal-nonmetal, ionic, relatively high boiling point; (d) metal-nonmetal, ionic, relatively high boiling point; (e) nonmetals, covalent, relatively low boiling point

8.139 To determine whether bonding will be covalent or ionic, we look for the presence of a metal or ammonium (NH_4^+) ion in the chemical formula. If one (or more) of these is present with a nonmetal, the compound will have ionic bonding. If only nonmetals are present, the substance bonds covalently. Note: For the purpose of establishing whether bonding in a substance is ionic or covalent, we consider ammonium ions (NH_4^+) as though they were metal cations. (a) nonmetals, covalent; (b) metal-polyatomic ion, ionic bonding between the ions, covalent bonding within the polyatomic ion; (c) metal-polyatomic ion, ionic bonding between the ions, covalent bonding within the polyatomic ion

8.141 As alcohols the two compounds have an –OH functional group. From the formula, we know that these alcohols have a three-carbon chain. One of these compounds will have the –OH group at the end of the three-carbon chain, $CH_3CH_2CH_2OH$.

		Comments
Draw the skeleton and identify missing bonds and lone pairs.	H H H | | | H—C—C—C—O—H | | | H H H	Carbons are central. Hydrogens are on the outside of the compound. This structure is complete except that oxygen needs two unshared pairs of electrons.
Add missing bonds and lone pairs.	H H H | | | H—C—C—C—Ö—H | | | H H H	This is an alcohol. When we draw organic compounds, we often omit the unshared electron pairs.

The other alcohol will have the OH group on the central carbon.

Draw the skeleton and identify missing bonds and lone pairs.	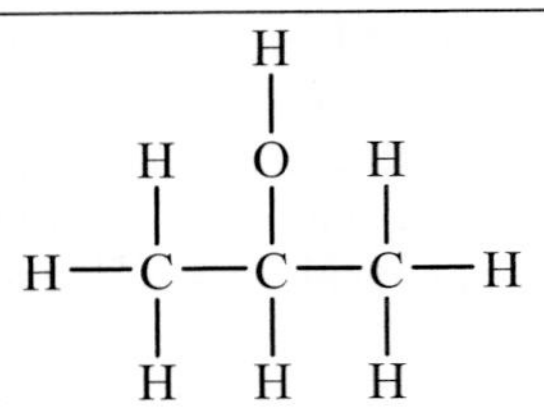 　　　　　　H 　　　　　　| 　　　H　　O　　H 　　　|　　|　　| 　H—C—C—C—H 　　　|　　|　　| 　　　H　　H　　H
Add missing bonds and lone pairs.	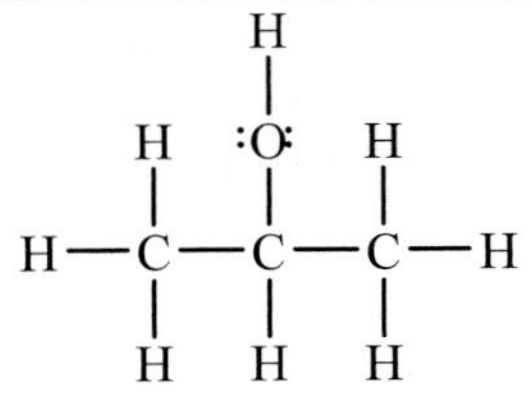 　　　　　　H 　　　　　　| 　　　H　:O:　H 　　　|　　|　　| 　H—C—C—C—H 　　　|　　|　　| 　　　H　　H　　H

8.143   The complete Lewis structure for HCN is

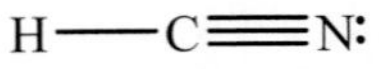

H——C≡≡N:

There is one single bond, no double bonds, and one triple bond. There are two unshared electrons.

8.145

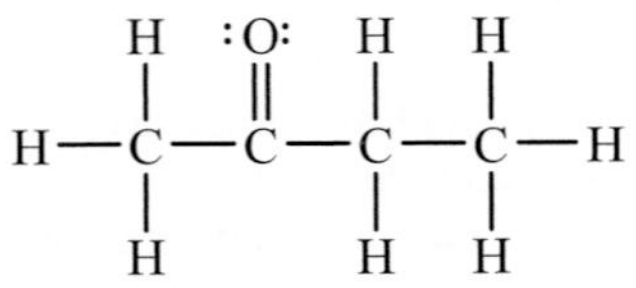

　　　H　:O:　H　　H
　　　\|　　\|\|　　\|　　\|
　H—C—C—C—C—H
　　　\|　　　　\|　　\|
　　　H　　　　H　　H

Chapter 9 – The Gaseous State

9.1   (a) effusion; (b) Boyle's law; (c) combined gas law; (d) ideal gas; (e) pressure; (f) Dalton's law of partial pressures; (g) molar volume; (h) ideal gas constant, R

9.3   The y-axis units and label are missing (in both the graph and the data tables), x-axis units are missing, the points should not be connected by lines (a straight "best fit" line that comes closest to all points should be used), gridlines should be uniformly labeled (for example, 10, 20, 30, 40, 50, etc.), the x-axis maximum should be 60 instead of 120, a graph title is missing.

9.5   There is a trend, but it is not linear because the data clearly are curved in a downward direction.

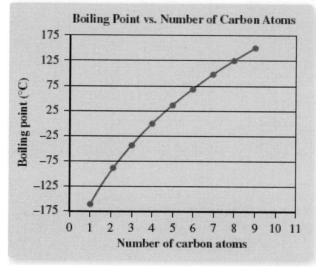

9.7   The values for Δh represent the pressure applied to the gas in excess of 1 atm. The length of the gas column is proportional to the volume of the gas. When the pressure is high, the volume is small. As the pressure decreases, the volume increases. Gas pressure and volume are inversely related.

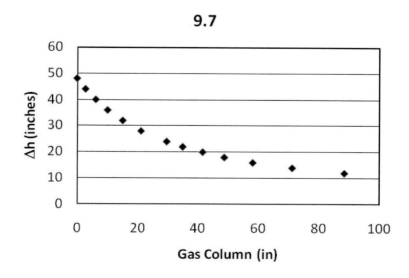

9.9 You can estimate the vapor pressure by drawing a line vertically from the temperature and seeing where the intersection point lies on the vapor pressure axis (horizontal line).

22 °C   21 torr
38 °C   50 torr

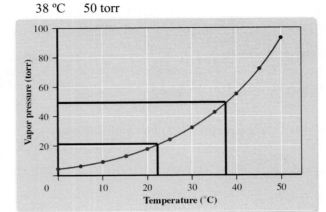

9.11 (a) $5x + 1 = 3$; subtract 1 from both sides
$5x = 2$; divide both sides by five
$$x = \frac{2}{5} = 0.4$$

(b) $0.412 / x = 2.00$; multiply both sides by x
$0.412 = 2.00 \cdot x$; divide both sides by 2
$$x = \frac{0.412}{2.00} = 0.206$$

(c) $x^2 = 32 - x^2$; add x^2 to both sides
$2 \cdot x^2 = 32$; divide both sides by 2
$x^2 = 16$; take the square root of both sides
$$x = \sqrt{x^2} = \sqrt{16} = \pm 4$$

(d) $2x = 6 - x$; add x to both sides
$3x = 6$; divide both sides by 3
$$x = \frac{6}{3} = 2$$

9.13 For temperature conversions we use the equation:
$F = 1.8 \cdot C + 32$
To convert from Fahrenheit to Celsius we rearrange the equation:
$$C = \frac{(F - 32)}{1.8}$$
It is worth noting that 1.8 and 32 are exact in this equation and that the significant figures are determined by the precision of the temperature you are converting:

(a) $C = \dfrac{(212 - 32)}{1.8} = \dfrac{180.}{1.8} = 100.°C$

(b) $C = \dfrac{(80.0 - 32)}{1.8} = \dfrac{48.0}{1.8} = 26.7°C$

(c) $\quad C = \dfrac{(32.0 - 32)}{1.8} = \dfrac{0.0}{1.8} = 0.0°C$

(d) $\quad C = \dfrac{(-40.0 - 32)}{1.8} = \dfrac{-72.0}{1.8} = -40.0°C$

9.15   Solve the equation algebraically and then calculate for the unknown value.

(a) $\quad T = \dfrac{PV}{nR} = \dfrac{(1.00)(1.00)}{(.500)(0.08206)} = 24.4$

(b) $\quad n = \dfrac{PV}{RT} = \dfrac{(0.750)(3.00)}{(0.08206)(237)} = 0.116$

(c) $\quad V = \dfrac{nRT}{P} = \dfrac{(1.50)(0.08206)(455)}{(3.25)} = 17.2$

(d) $\quad P = \dfrac{nRT}{V} = \dfrac{(2.67)(0.08206)(322)}{(15.0)} = 4.70$

9.17   When compared to other states of matter, gases have low densities and are very compressible. They also take the shapes of their containers. We describe a gas in terms of its pressure, temperature, volume, and the number of moles of the gas in the sample.

9.19   The density of warm air is lower than the density of cool air.

9.21   As temperature decreases, gas molecules move more slowly. As a result, they collide with less force, and a larger number of molecules occupy a unit volume. As shown in the figure, at lower temperatures the gas density is higher, but the molecular velocity is lower.

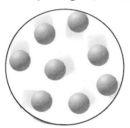

9.23   Gas pressure is the amount of force exerted by the gas particles divided by the area over which the force is exerted. $P = \dfrac{\text{Force}}{\text{Area}}$

9.25   We measure absolute pressure with a barometer (Figure 9.11). Essentially, a barometer allows us to compare the pressure exerted by a column of a liquid (usually mercury) to the pressure exerted by the atmosphere. A tire gauge compares the pressure of the air inside a tire to atmospheric pressure.

9.27   If the temperature doesn't change, the velocity of the gas molecules remains constant. Because the volume is larger, the density of molecules is lower.

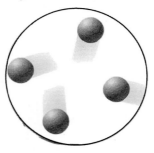

9.29   We use the following relationships to convert between the various pressure units:

1 atm = 29.9 in Hg = 76 cm Hg = 760 mm Hg = 760 torr = 101,325 Pa = 101.325 kPa = 14.7 lb/in^2

For example, to convert between pascals and mm Hg, use the relationship: 760 mm Hg = 101,325 Pa.

(a) Pressure in atm = 745 torr $\times \dfrac{1\ \text{atm}}{760\ \text{torr}}$ = 0.980 atm

(b) Pressure in torr = 1.23 atm $\times \dfrac{760\ \text{torr}}{1\ \text{atm}}$ = 935 torr

(c) Pressure in atm = 90.1 mm Hg $\times \dfrac{1\ \text{atm}}{760\ \text{mm Hg}}$ = 0.119 atm

(d) Pressure in Pascal = 0.643 kPa $\times \dfrac{1000\ \text{Pa}}{1\ \text{kPa}}$ = 643 Pa

(e) Pressure in mm Hg = 1.35 $\times 10^5$ Pa $\times \dfrac{760\ \text{mm Hg}}{101,325\ \text{Pa}}$ = 1.01 $\times 10^3$ mm Hg

(f) Pressure in torr = 7.51 $\times 10^4$ Pa $\times \dfrac{760\ \text{torr}}{101,325\ \text{Pa}}$ = 563 torr;

(g) Pressure in Pa = 798 torr $\times \dfrac{101,325\ \text{Pa}}{760\ \text{torr}}$ = 1.06 $\times 10^5$ Pa;

(h) Pressure in mm Hg = 29.3 cm Hg $\times \dfrac{10\ \text{mm}}{1\ \text{cm}}$ = 293 mm Hg

9.31   The conversion from inches of mercury to pascals is quite lengthy. The first half of the conversion involves converting from inches to millimeters (English to metric conversion). The second half involves using the relationships among pressure units to convert from millimeters of mercury to pascals. The problem solving map looks like:

Pressure in Hg $\xrightarrow{\ 2.54\ \text{cm} = 1\ \text{in}\ }$ cm Hg $\xrightarrow{\ 10\ \text{mm} = 1\ \text{cm}\ }$ mm Hg

mm Hg $\xrightarrow{\ 760\ \text{mm Hg} = 1\ \text{atm}\ }$ atm $\xrightarrow{\ 1\ \text{atm} = 101,325\ \text{Pa}\ }$ Pa

Pressure in Pa = 30.24 in $\times \dfrac{2.54\ \text{cm}}{1\ \text{in}} \times \dfrac{10\ \text{mm}}{1\ \text{cm}} \times \dfrac{1\ \text{atm}}{760\ \text{mm}} \times \dfrac{101,325\ \text{Pa}}{1\ \text{atm}}$ = 1.024 $\times 10^5$ Pa

An alternate, but important, conversion method involves using the definitions of pressure given in the text. We know that 1 atm = 29.9 in Hg and also that 1 atm = 101,325 Pa. This means that 29.9 in Hg = 101,325 Pa. Using this factor, we can convert from inches to pascals in one step (although with less precision):

$$\text{Pressure in Pa} = 30.24 \text{ in Hg} \times \frac{101{,}325 \text{ Pa}}{29.9 \text{ in Hg}} = 1.02 \times 10^5 \text{ Pa}$$

9.33 Boyle's law tells us that, at constant temperature, as the pressure of a gas increases the volume decreases. This is an inverse relationship, so that if the pressure increases by a factor of three, the volume decreases to one third (1/3) of its original volume, assuming the temperature remains constant.

9.35 If the container volume decreases, the particles will collide with the container walls more frequently, causing an increase in pressure. If the container size is decreased to half its original volume, and the sample size remains constant, we expect to find twice as many molecules in the same space. The velocity of the molecules will not change as long as the temperature remains constant, so the pressure on the container walls will double.

9.37 (a) 3.60 L; (b) 26.7 mL; (c) 0.392 mL. Boyle's law describes the relationship of pressure and volume changes (P & V) on a gas sample, assuming that the sample size (number of moles) and temperature are kept constant. Boyle's law is an inverse relationship: As the volume of a gas sample increases, the pressure of the gas decreases, assuming that the temperature is kept constant. For example, in part (a) because the pressure increases by a factor of 5.00 atm/2.00 atm (units must be the same), we conclude that the volume will decrease by a factor of 2.00 atm/5.00 atm. We can compute the final volume as follows:

$$\text{Final volume} = \frac{3.00 \text{ atm}}{5.00 \text{ atm}} \times 6.00 \text{ L} = 3.60 \text{ L}$$

Boyle's law also states that $P_1V_1 = P_2V_2$. For part (a) we have:

V_1	P_1	P_2	V_2
6.00 L	3.00 atm	5.00 atm	?

We can rearrange Boyle's law and solve for V_2 as follows:

$$V_2 = \frac{P_1 V_1}{P_2}$$

$$V_2 = \frac{(3.00 \text{ atm})(6.00 \text{ L})}{5.00 \text{ atm}} = 3.60 \text{ L}$$

By either applying the proportionality, or using Boyle's law, we obtain the same result. After you complete your calculation, you should make sure to evaluate your answer to ensure that it is reasonable. The effect of the pressure increase should be a volume decrease.

(b) The pressure increases, so we should see a volume decrease, and we do.

$$V_2 = \frac{(60.0 \text{ torr})(40 \text{ mL})}{90.0 \text{ torr}} = 26.7 \text{ mL}$$

(c) The pressure increases, so we should see a volume decrease, and we do.

$$V_2 = \frac{(40.0\ torr)(2.50\ mL)}{255\ torr} = 0.392\ mL$$

9.39   According to Boyle's law, if the volume of a gas sample increases by a factor of 1512 mL/405 mL, its pressure decreases by a factor of 205 mL/1512 mL. For part (a) we can write:

$$P_2 = \frac{405\ mL}{1512\ mL} \times 602\ torr = 161\ torr$$

We can obtain the same results using the mathematical expression of Boyle's law, $P_1V_1 = P_2V_2$, solving for P_2:

$$P_2 = \frac{P_1V_1}{V_2}$$

Note: When we solve these problems, the volume units must agree so that they cancel properly.

(a) $P_2 = \dfrac{(602\ torr)(405\ mL)}{1512\ mL} = 161\ torr$

Because the sample volume increases we expect that the pressure will decrease, and it does.

(b) Before we can use Boyle's law, we must convert 1.50 L into mL.

Volume in mL = $1.50\ L \times \dfrac{1000\ mL}{1\ L} = 1.50 \times 10^3\ mL$

$$P_2 = \frac{(0.00100\ torr)(1.50 \times 10^3\ mL)}{15.0\ mL} = 0.100\ torr$$

Because the volume decreases we expect the pressure to increase, and it does.

(c) $P_2 = \dfrac{(0.832\ atm)(805\ L)}{37.5\ L} = 17.9\ atm$

Because the volume decreases we expect pressure to increase, and it does.

9.41   It is helpful to organize the data in table form.

P_1	V_1	P_2	V_2
1.25 atm	925 L	?	6.35 L

Based on Boyle's law, we predict that if the volume of the gas sample decreases by a factor of 6.35 L/925 L, the pressure should increase by a factor of 925 L/6.35 L, assuming the sample size and temperature are kept constant. Solving Boyle's law for P_2 we have:

$$P_2 = \frac{P_1V_1}{V_2}$$

$$P_2 = \frac{1.25\ atm \times 925\ L}{6.35\ L} = 182\ atm$$

As we predicted, the pressure is higher by a factor of 925 L/6.35 L.

9.43   Organize the data in table form.

P_1	V_1	P_2	V_2
1.00 atm	0.550 L	725 torr	?

Notice that the units of pressure for P_1 and P_2 differ. Because 1 atm = 760 torr, we can substitute 1.00 atm with 760 torr in the Boyle's law expression:

$$V_2 = \frac{P_1 V_1}{P_2} = \frac{(760 \text{ torr})(0.550 \text{ L})}{725 \text{ torr}} = 0.577 \text{ L}$$

The volume of H_2 required is larger than 0.550 L because it is being collected at a lower pressure.

9.45   Charles's law states that if the pressure and sample size of a gas are kept constant, the volume and temperature (in kelvins) of the gas are directly proportional to each other (volume increases when temperature increases; volume decreases when temperature decreases). If the temperature increases, the gas will occupy a larger volume.

9.47   The velocity of the gas particles increases as temperature increases. This means they strike the walls of the container with greater force. If the container volume does not increase, the gas pressure increases. To maintain a constant pressure, as stated in the problem, the volume of the container will increase.

9.49   In Charles's law problems, we must always express temperature in kelvins. We can state Charles's law mathematically as:

$$\frac{V_1}{T_1} = \frac{V_2}{T_2}$$

Solving for V_2: $V_2 = \frac{V_1}{T_1} \times T_2$

We see that if the temperature, T_2, increases, the volume, V_2 must also increase.

(a) Both temperatures are given in Celsius. The conversion is: $T_K = T_{°C} + 273.15$ K.

$T_1 = 30.0°C + 273.15 = 303.2$ K

$T_2 = 0.0°C + 273.15 = 273.2$ K

$$V_2 = \frac{6.00 \text{ L}}{303.2 \text{ K}} \times 273.2 \text{ K} = 5.41 \text{ L}$$

Because the temperature decreases, the volume also decreases.

(b) Both temperatures are given in Celsius. The conversion is: $T_K = T_{°C} + 273.15$ K.

$T_1 = -60.0°C + 273.15 = 213.15$ K (4 sig figs)

$T_2 = 401.0°C + 273.15 = 674.15$ K (4 sig figs)

$$V_2 = \frac{212 \text{ mL}}{213.15 \text{ K}} \times 674.15 \text{ K} = 671 \text{ mL}$$

Because temperature increases, the volume also increases.

(c) $V_2 = \dfrac{47.5 \text{ L}}{212 \text{ K}} \times 337 \text{ K} = 75.5 \text{ L}$

Because the temperature increases, the volume also increases.

9.51   When we solve Charles's law problems, we must always express temperature in kelvins. We can state Charles's as:

$$\frac{V_1}{T_1} = \frac{V_2}{T_2}$$

Solving for T_2: $T_2 = V_2 \times \dfrac{T_1}{V_1}$

(a) Although we are asked to give the final temperature in Celsius, we must solve for T_2 in kelvins and then express that temperature in degrees Celsius.

$T_1 = 0.0°C + 273.15 = 273.2 \text{ K}$

$T_2 = 140.0 \text{ mL} \times \dfrac{273.2 \text{ K}}{70.0 \text{ mL}} = 546 \text{ K}$

Because the volume doubles, the temperature also doubles. Convert T_2 to Celsius:

$T_{°C} = T_K - 273.15 \text{ K} = 546 \text{ K} - 273.15 = 273°C$

(b) The units for volume must be the same so they will cancel. First, we convert 85 mL to liters and −37°C to kelvins before we apply Charles's law:

Volume in liters = $85 \text{ mL} \times \dfrac{1 \text{ L}}{1000 \text{ mL}} = 0.085 \text{ L}$

$T_1 = -37°C + 273.15 = 236 \text{ K}$

$T_2 = 0.085 \text{ L} \times \dfrac{236 \text{ K}}{2.55 \text{ L}} = 7.9 \text{ K}$

Because the volume decreases, the temperature also decreases.

$T_{°C} = T_K - 273.15 \text{ K} = 7.9 \text{ K} - 273.15 = -265.2°C$

(c) $T_2 = 135 \text{ L} \times \dfrac{165 \text{ K}}{87.5 \text{ L}} = 255 \text{ K}$

$T_{°C} = T_K - 273.15 \text{ K} = 255 \text{ K} - 273.15 = -19°C$

9.53   It is often helpful to create a table like the one shown below and input the data from the problem.

V_1	T_1	V_2	T_2
0.150 mL	24.2°C	?	62.5°C

Because only temperature and volume are changing, we use Charles's law to solve for the new volume:

$$V_2 = \frac{V_1}{T_1} \times T_2$$

Before doing the calculation, we must convert the temperatures to kelvins:

$T_1 = 24.2°C + 273.15 = 297.4$ K

$T_2 = 62.5°C + 273.15 = 335.7$ K

$$V_2 = \frac{0.150 \text{ mL}}{297.4 \cancel{K}} \times 335.7 \cancel{K} = 0.169 \text{ mL}$$

Because the temperature increases, the volume of the bubble also increases.

9.55   Nothing happens to the particles if the temperature, pressure, and volume are constant. If the tank is sealed so that no gas molecules can escape, then the pressure does not change. If the tank is opened, then some of the gas molecules will leave the tank to maintain an equilibrium pressure with the air outside the tank.

9.57   We know that the temperature and pressure of a fixed volume of gas are directly related, so we can write an equation relating the change in pressure that occurs when the temperature of a gas sample changes. The equation is similar to that for Charles's law, except that we substitute pressure for volume (recall that volume and temperature are also directly proportional).

$$\frac{P_1}{T_1} = \frac{P_2}{T_2}$$

Solving for P_2: $P_2 = T_2 \times \dfrac{P_1}{T_1}$

(a)   Convert temperatures to kelvins:

$T_1 = 0.0°C + 273.15 = 273.2$ K

$T_2 = 105.0°C + 273.15 = 378.2$ K

$$P_2 = 378.2 \cancel{K} \times \frac{302 \text{ torr}}{273.2 \cancel{K}} = 418 \text{ torr}$$

The temperature increases, so the pressure also increases.

(b)   Convert temperatures to kelvins:

$T_1 = 25.0°C + 273.15 = 298.2$ K

$T_2 = 0.0°C + 273.15 = 273.2$ K

$$P_2 = 273.2 \cancel{K} \times \frac{735 \text{ torr}}{298.2 \cancel{K}} = 673 \text{ torr}$$

The temperature decreases, so the pressure also decreases.

(c) $P_2 = 373 \cancel{K} \times \dfrac{3.25 \text{ atm}}{273 \cancel{K}} = 4.44 \text{ atm}$

Because the temperature increases, the pressure also increases.

9.59   We know that temperature and pressure of a fixed volume of a gas are directly related, so we can write an equation that relates the change in pressure of a gas sample with a change in temperature. The equation is similar to Charles's law, except that we substitute pressure for volume (recall that volume and temperature are also directly proportional).

$$\frac{P_1}{T_1} = \frac{P_2}{T_2}$$

Solving for T_2: $T_2 = P_2 \times \dfrac{T_1}{P_1}$

(a) Convert temperature to kelvins:

$T_1 = 30.0°C + 273.15 = 303.2$ K

$$T_2 = 915 \text{ torr} \times \frac{303.2 \text{ K}}{1525 \text{ torr}} = 182 \text{ K}$$

Convert T_2 to degrees Celsius:

$T_{°C} = T_K - 273.15 \text{ K} = 182 \text{ K} - 273.15 = -91°C$

(b) Convert temperature to kelvins:

$T_1 = 250.0°C + 273.15 = 523.2$ K

The pressures need to be expressed in the same units. Convert 1042 torr to atm:

$$\text{Pressure in atm} = 1042 \text{ torr} \times \frac{1 \text{ atm}}{760 \text{ torr}} = 1.371 \text{ atm}$$

$$T_2 = 1.371 \text{ atm} \times \frac{523.2 \text{ K}}{0.70 \text{ atm}} = 1.0 \times 10^3 \text{ K}$$

Convert T_2 to degrees Celsius:

$T_{°C} = T_K - 273.15 \text{ K} = 1.0 \times 10^3 \text{ K} - 273.15 = 750°C$

(c) $T_2 = 1000.0 \text{ torr} \times \dfrac{355 \text{ K}}{500.0 \text{ torr}} = 7.10 \times 10^2 \text{ K}$

Convert T_2 to degrees Celsius:

$T_{°C} = T_K - 273.15 \text{ K} = 7.10 \times 10^2 \text{ K} - 273.15 = 437°C$

9.61   We know that the temperature and pressure of a fixed volume of a gas are directly related, so we can write an equation relating the pressure change associated with a change in the temperature of a gas sample. The equation is similar to Charles's law, except that pressure replaces volume (recall that volume and temperature are also directly proportional).

$$\frac{P_1}{T_1} = \frac{P_2}{T_2}$$

Solving for P_2: $P_2 = T_2 \times \dfrac{P_1}{T_1}$

P_1	T_1	P_2	T_2
7.25 atm	18.5°C	?	37.2°C

Convert temperatures to Celsius and calculate P_2:

$T_1 = 18.5°C + 273.15 = 291.7 \text{ K}$

$T_2 = 37.2°C + 273.15 = 310.4 \text{ K}$

$P_2 = 310.4 \, \cancel{K} \times \dfrac{7.25 \text{ atm}}{291.7 \, \cancel{K}} = 7.71 \text{ atm}$

9.63   When we use the combined gas law, it helps if we solve the equation for the desired variable before we substitute numbers into the expression. When we rearrange the equation, we try to keep the state 1 and state 2 variables separate (it helps us to keep from mixing up the variables). We must also make sure that the units are consistent (i.e. that the pressures are in the same units and the temperatures are in kelvins).

Combined gas law: $\dfrac{P_1 V_1}{T_1} = \dfrac{P_2 V_2}{T_2}$

(a) The missing value on the data table is V_2:

$V_2 = \dfrac{P_1 V_1}{T_1} \times \dfrac{T_2}{P_2}$

P_1	V_1	T_1	P_2	V_2	T_2
0.50 atm	2.50 L	20.0°C	760.0 torr	?	0.0°C

We need to express the pressures in common units and convert the temperatures to kelvins. Since 760.0 torr is 1.000 atm, we can simply substitute 1.000 atm for P_2. First, we convert temperatures to kelvins and then solve for V_2:

$T_1 = 20.0°C + 273.15 = 293.2 \text{ K}$

$T_2 = 0.0°C + 273.15 = 273.2 \text{ K}$

$V_2 = \dfrac{(0.50 \, \cancel{\text{atm}})(2.50 \text{ L})}{293.2 \, \cancel{K}} \times \dfrac{273.2 \, \cancel{K}}{1.000 \, \cancel{\text{atm}}} = 1.2 \text{ L}$

(b) The value missing from the data table is T_2:

$T_2 = \dfrac{T_1}{P_1 V_1} \times P_2 V_2$

P_1	V_1	T_1	P_2	V_2	T_2
0.250 atm	125 L	25.0°C	100.0 torr	62.0 L	?

Temperature and pressure both need to be converted to appropriate units:

$T_1 = 25.0°C + 273.15 = 298.2 \text{ K}$

$P_2 \text{ (atm)} = 100.0 \, \cancel{\text{torr}} \times \dfrac{1 \text{ atm}}{760 \, \cancel{\text{torr}}} = 0.1316 \text{ atm}$

$T_2 = \dfrac{298.2 \text{ K}}{(0.250 \, \cancel{\text{atm}})(125 \, \cancel{L})} \times (0.1316 \, \cancel{\text{atm}})(62.0 \, \cancel{L}) = 77.8 \text{ K (or } -195.3°C)$

(c) The value missing from the data table is P_2:

$$P_2 = \frac{P_1 V_1}{T_1} \times \frac{T_2}{V_2}$$

P_1	V_1	T_1	P_2	V_2	T_2
200.0 torr	455 mL	300.0 K	?	200.0 mL	327°C

First, we convert T_2 to kelvins:

$$T_2 = 327°C + 273.15 = 6.00 \times 10^2 \text{ K}$$

$$P_2 = \frac{(200.0 \text{ torr})(455 \text{ mL})}{(300.0 \text{ K})} \times \frac{6.00 \times 10^2 \text{ K}}{200.0 \text{ mL}} = 9.10 \times 10^2 \text{ torr}$$

9.65   We want to calculate the volume of oxygen at STP (V_1). Standard temperature and pressure are 273.15 K (0°C) and 1 atm (by definition). The tank will hold 0.500 L at 3.50 atm and 24.5°C. This can be considered the final state of the gas.

P_1	V_1	T_1	P_2	V_2	T_2
1 atm	?	273.15 K	3.50 atm	0.500 L	24.5°C

Convert 24.5°C to kelvins:

$$T_2 = 24.5°C + 273.15 = 297.7 \text{ K}$$

Solve the combined gas law for V_1.

$$V_1 = \frac{P_2 V_2}{T_2} \times \frac{T_1}{P_1} = \frac{(3.50 \text{ atm})(0.500 \text{ L})}{297.7 \text{ K}} \times \frac{273.15 \text{ K}}{1 \text{ atm}} = 1.61 \text{ L}$$

9.67   Gay-Lussac's Law states that volumes of gases react in simple, whole number ratios when the volumes of the reactants and products are measured at the same temperature and pressure. These ratios correspond to the coefficients in the balanced chemical equation for the reaction.

9.69   The molar volume of all gases is approximately 22.414 L/mol at STP.

9.71   At STP, 1.00 mol of any gas occupies 22.414 L. We determine the mass of each gas sample from the number of moles and the molar mass of the gas.

(a)   Moles $CH_4 = 8.62 \text{ L} \times \frac{1 \text{ mol}}{22.414 \text{ L}} = 0.385 \text{ mol}$

Given that the molar mass of CH_4 (16.04 g/mol) we can calculate the mass of CH_4:

Mass $CH_4 = 0.385 \text{ mol} \times \frac{16.04 \text{ g}}{1 \text{ mol}} = 6.17 \text{ g}$

(b) Convert mL to L:

$$\text{Volume in L} = 350.0 \text{ mL} \times \frac{1 \text{ L}}{1000 \text{ mL}} = 0.3500 \text{ L}$$

$$\text{Moles Xe} = 0.3500 \text{ L} \times \frac{1 \text{ mol}}{22.414 \text{ L}} = 1.562 \times 10^{-2} \text{ mol}$$

We use the molar mass of Xe (131.3 g/mol) to calculate the mass of Xe gas in the sample:

$$\text{Mass Xe} = 1.562 \times 10^{-2} \text{ mol} \times \frac{131.3 \text{ g}}{1 \text{ mol}} = 2.050 \text{ g}$$

(c) $\text{Moles CO} = 48.1 \text{ L} \times \dfrac{1 \text{ mol}}{22.414 \text{ L}} = 2.15 \text{ mol}$

We use the molar mass of CO (28.01 g/mol) to calculate the mass of CO gas in the sample:

$$\text{Mass CO} = 2.15 \text{ mol} \times \frac{28.01 \text{ g}}{1 \text{ mol}} = 60.1 \text{ g}$$

9.73   Avogadro's law tells us that because both balloons have the same volume they contain the same number of gas particles (and, therefore, the same number of moles). Because argon atoms are more massive than helium atoms, the balloon containing argon has the greater mass, and, therefore, the greater density (the volumes of the balloons are the same).

9.75   We use the molar mass of each gas to calculate the number of moles of gas in each sample. Because the gases are at STP, one mole of each gas occupies 22.414 L.
(a) 5.8 g NH_3 (17.03 g/mol)

$$\text{Moles NH}_3 = 5.8 \text{ g} \times \frac{1 \text{ mol}}{17.03 \text{ g}} = 0.34 \text{ mol}$$

$$\text{Volume NH}_3 = 0.34 \text{ mol} \times \frac{22.414 \text{ L}}{1 \text{ mol}} = 7.6 \text{ L}$$

(b) 48 g O_2 (32.00 g/mol)

$$\text{Moles O}_2 = 48 \text{ g} \times \frac{1 \text{ mol}}{32.00 \text{ g}} = 1.5 \text{ mol}$$

$$\text{Volume O}_2 = 1.5 \text{ mol} \times \frac{22.414 \text{ L}}{1 \text{ mol}} = 34 \text{ L}$$

(c) 10.8 g He (4.003 g/mol)

$$\text{Moles He} = 10.8 \text{ g} \times \frac{1 \text{ mol}}{4.003 \text{ g}} = 2.70 \text{ mol}$$

$$\text{Volume He} = 2.7 \text{ mol} \times \frac{22.414 \text{ L}}{1 \text{ mol}} = 60.5 \text{ L}$$

9.77   We convert the mass of helium (425 g, given in the figure) to the equivalent number of moles, and then to volume in liters, using the molar mass of helium (4.003 g/mol) and the molar volume of a gas at STP (22.414 L/mol). Once we have determined the volume, we can determine approximately how many balloons we can fill from the tank.

$$\text{Moles He} = 425 \, \cancel{g} \times \frac{1 \, \text{mol}}{4.003 \, \cancel{g}} = 106 \, \text{mol}$$

$$\text{Volume He} = 106 \, \cancel{mol} \times \frac{22.414 \, \text{L}}{1 \, \cancel{mol}} = 2.38 \times 10^3 \, \text{L}$$

Because each balloon holds 1.0 L of gas, we can estimate that we can fill approximately 2380 balloons.

9.79   Convert 25.0°C to kelvins: $T_1 = 25.0°C + 273.15 = 298.2 \, \text{K}$. For STP, $T_2 = 273.15 \, \text{K}$ and $P_2 = 1 \, \text{atm}$. Solve the combined gas law for V_2.

$$V_2 = \frac{P_1 V_1}{T_1} \times \frac{T_2}{P_2} = \frac{(1.00 \, \cancel{atm})(12.0 \, \text{L})}{298.2 \, \cancel{K}} \times \frac{273.15 \, \cancel{K}}{1 \, \cancel{atm}} = 11.0 \, \text{L}$$

9.81   An ideal gas is any gas whose behavior is described by the five postulates of kinetic-molecular theory. Ideal gases have elastic collisions (they bounce and/or collide without losing energy), travel in straight lines (because they are not attracted to other molecules), and occupy zero volume. In addition, the average kinetic energy of an ideal gas is directly proportional to the temperature of the gas.

9.83   To calculate the volume occupied by each gas, we first determine the number of moles of each gas and convert the temperature of the gases to kelvins. Whenever we use the ideal gas law, we must make certain that the units of P, V, n, and T match the units of R that we are using (R = 0.08206 L·atm/(mol·K)).

For each gas, the temperature is $T = 100.0°C + 273.15 \, \text{K} = 373.15 \, \text{K}$ (four significant figures)

Rearranging the ideal gas law for volume: $V = \dfrac{nRT}{P}$

(a) 5.8 g NH$_3$ (17.03 g/mol)

$$\text{Moles NH}_3 = 5.8 \, \cancel{g} \times \frac{1 \, \text{mol}}{17.03 \, \cancel{g}} = 0.34 \, \text{mol}$$

$$V = \frac{\left(0.34 \, \cancel{mol}\right)\left(0.08206 \dfrac{\text{L} \cdot \cancel{atm}}{\cancel{mol} \cdot \cancel{K}}\right)\left(373.15 \, \cancel{K}\right)}{15.0 \, \cancel{atm}} = 0.69 \, \text{L}$$

(b) 48 g O$_2$ (32.00 g/mol)

$$\text{Moles O}_2 = 48 \, \cancel{g} \times \frac{1 \, \text{mol}}{32.00 \, \cancel{g}} = 1.5 \, \text{mol}$$

$$V = \frac{\left(1.5 \, \cancel{mol}\right)\left(0.08206 \dfrac{\text{L} \cdot \cancel{atm}}{\cancel{mol} \cdot \cancel{K}}\right)\left(373.2 \, \cancel{K}\right)}{15.0 \, \cancel{atm}} = 3.1 \, \text{L}$$

(c) 10.8 g He (4.003 g/mol)

$$\text{Moles He} = 10.8 \ \cancel{g} \times \frac{1 \ \text{mol}}{4.003 \ \cancel{g}} = 2.70 \ \text{mol}$$

$$V = \frac{(2.70 \ \cancel{mol})\left(0.08206 \ \dfrac{L \cdot atm}{\cancel{mol} \cdot \cancel{K}}\right)(373.2 \ \cancel{K})}{15.0 \ \cancel{atm}} = 5.51 \ \text{L}$$

9.85   Rearranging the ideal gas law for the number of moles in the sample: $n = \dfrac{PV}{RT}$

Because $R = 0.08206$ L·atm/(mol·K), we must express pressure in atmospheres and, of course, temperature in kelvins.

$$\text{Pressure in atm} = 722 \ \cancel{torr} \times \frac{1 \ \text{atm}}{760 \ \cancel{torr}} = 0.950 \ \text{atm}$$

$$T = 87.5°C + 273.15 = 360.7 \ \text{K}$$

(a) 7.62 L CH_4 (16.04 g/mol)

$$n = \frac{PV}{RT} = \frac{(0.950 \ \cancel{atm})(7.62 \ \cancel{L})}{\left(0.08206 \ \dfrac{\cancel{L} \cdot \cancel{atm}}{mol \cdot \cancel{K}}\right)(360.7 \ \cancel{K})} = 0.245 \ \text{mol}$$

$$\text{Mass } CH_4 = 0.245 \ \cancel{mol} \times \frac{16.04 \ g}{1 \ \cancel{mol}} = 3.92 \ g$$

(b) 135 mL H_2 (2.016 g/mol)
First convert mL to L:

$$\text{Volume in L} = 135 \ \cancel{mL} \times \frac{1 \ L}{1000 \ \cancel{mL}} = 0.135 \ L$$

$$n = \frac{PV}{RT} = \frac{(0.950 \ \cancel{atm})(0.135 \ \cancel{L})}{\left(0.08206 \ \dfrac{\cancel{L} \cdot \cancel{atm}}{mol \cdot \cancel{K}}\right)(360.7 \ \cancel{K})} = 4.33 \times 10^{-3} \ \text{mol}$$

$$\text{Mass } H_2 = 4.33 \times 10^{-3} \ \cancel{mol} \times \frac{2.016 \ g}{1 \ \cancel{mol}} = 8.74 \times 10^{-3} \ g$$

(c) 8.96 L N_2 (28.02 g/mol)

$$n = \frac{PV}{RT} = \frac{(0.950 \ \cancel{atm})(8.96 \ \cancel{L})}{\left(0.08206 \ \dfrac{\cancel{L} \cdot \cancel{atm}}{mol \cdot \cancel{K}}\right)(360.7 \ \cancel{K})} = 0.288 \ \text{mol}$$

$$\text{Mass } N_2 = 0.288 \ \cancel{mol} \times \frac{28.02 \ g}{1 \ \cancel{mol}} = 8.06 \ g$$

9.87   The balloons sink or float depending on how their densities compare to the density of air. Recall that if the balloons (or any containers) have equal volume, pressure, and temperature, they contain the same number of particles (Avogadro's Law or ideal gas law). Because each CO_2 molecule is heavier than each He atoms, and there are equal numbers of each, the CO_2-filled balloon will have more mass and, therefore, higher density. Why doesn't CO_2 float in air? Air is less dense than CO_2. Air is primarily a mixture of N_2 and O_2, both of which have lower masses than CO_2. The CO_2-filled balloon sinks because CO_2 gas is more dense than air.

9.89   At STP, one mole of any gas occupies 22.414 L. If we know the molar mass of a gas, we can calculate its density ($d = m/V$, the mass of one mole divided by the volume of one mole).
(a) NH_3 (17.03 g/mol).

$$d = \frac{17.03 \text{ g}}{22.414 \text{ L}} = 0.7598 \text{ g/L}$$

(b) N_2 (28.02 g/mol)

$$d = \frac{28.02 \text{ g}}{22.414 \text{ L}} = 1.250 \text{ g/L}$$

(c) N_2O (44.02 g/mol)

$$d = \frac{44.02 \text{ g}}{22.414 \text{ L}} = 1.964 \text{ g/L}$$

9.91   Density is the mass of a substance divided by its volume. For gas samples, it is convenient for us to determine density by dividing the molar mass of the gas by its volume, which we calculate using the ideal gas law.

$T = 25°C + 273.15 = 298.2 \text{ K}$

Pressure in atm $= 735 \text{ torr} \times \dfrac{1 \text{ atm}}{760 \text{ torr}} = 0.967 \text{ atm}$

$n = 1.00$ mol (we choose this for convenience)

$$V = \frac{nRT}{P} = \frac{(1.00 \text{ mol})\left(0.08206 \dfrac{\text{L} \cdot \text{atm}}{\text{mol} \cdot \text{K}}\right)(298.2 \text{ K})}{0.967 \text{ atm}} = 25.3 \text{ L}$$

To calculate the density, we divide the molar mass by the volume it occupies:

(a) NH_3 (17.03 g/mol) – One mole of NH_3 has a mass of 17.03 g and occupies a volume of 25.3 L. The density is:

$$d = \frac{m}{V} = \frac{17.03 \text{ g}}{25.3 \text{ L}} = 0.673 \text{ g/L}$$

(b) N_2 (28.02 g/mol) – One mole of N_2 has a mass of 28.02 g and occupies a volume of 25.3 L. The density is:

$$d = \frac{m}{V} = \frac{28.02 \text{ g}}{25.3 \text{ L}} = 1.11 \text{ g/L}$$

(c) N_2O (44.02 g/mol) – One mole of N_2O has a mass of 44.02 g and occupies a volume of 25.3 L. The density is:

$$d = \frac{m}{V} = \frac{44.02 \text{ g}}{25.3 \text{ L}} = 1.74 \text{ g/L}$$

9.93   According to the ideal gas law, equal-molar amounts of every gas occupy the same volume at the same pressure and temperature. Using this law, we can calculate the number of moles of gas in the container. First, we convert pressure and temperature to units consistent with the units in the gas constant:

Pressure in atm = $840 \text{ torr} \times \dfrac{1 \text{ atm}}{760 \text{ torr}} = 1.1 \text{ atm}$

Temperature in K = $50.0°C + 273.15 = 323.2 \text{ K}$

$$n = \frac{PV}{RT} = \frac{(1.1 \text{ atm})(5.00 \text{ L})}{\left(0.08206 \dfrac{\text{L} \cdot \text{atm}}{\text{mol} \cdot \text{K}}\right)(323.2 \text{ K})} = 0.21 \text{ mol}$$

Now we can determine the mass of each gas using its molar mass:

(a) H_2 (2.016 g/mol)

Mass $H_2 = 0.21 \text{ mol} \times \dfrac{2.016 \text{ g}}{\text{mol}} = 0.42 \text{ g}$

(b) CH_4 (16.04 g/mol)

Mass $CH_4 = 0.21 \text{ mol} \times \dfrac{16.04 \text{ g}}{\text{mol}} = 3.3 \text{ g}$

(c) SO_2 (64.06 g/mol)

Mass $SO_2 = 0.21 \text{ mol} \times \dfrac{64.06 \text{ g}}{\text{mol}} = 13 \text{ g}$

9.95   Dalton's law of partial pressures states that the total pressure in a container is the sum of the pressures exerted by each of the individual gases in the container. Each gas exerts a pressure on the walls of the container as if it were the only gas in the container. Gas molecules behave in this way because they are not strongly attracted to each other.

9.97   According to Dalton's law of partial pressures, the total pressure (728 torr) is the sum of the pressure exerted by the oxygen gas and water vapor (20.0 torr). To calculate the partial pressure of oxygen gas we subtract the water vapor pressure from the total pressure. Mathematically, Dalton's law of partial pressures looks like:

$$P_T = P_{O_2} + P_{water}$$

$$P_{O_2} = P_T - P_{water} = 728 \text{ torr} - 20.0 \text{ torr} = 708 \text{ torr}$$

9.99   (a) We use molar mass to calculate the number of moles of N_2 (28.02 g/mol).

$$\text{Moles } N_2 = 78.0 \text{ g} \times \frac{1 \text{ mol}}{28.02 \text{ g}} = 2.78 \text{ mol } N_2$$

(b) We use molar mass to calculate the number of moles of Ne (20.18 g/mol).

$$\text{Moles Ne} = 42.0 \text{ g} \times \frac{1 \text{ mol}}{20.18 \text{ g}} = 2.08 \text{ mol Ne}$$

(c) To calculate the partial pressure of N_2 we use the ideal gas law ($PV = nRT$). We know the number of moles of N_2 (part (a)) and temperature (50.0°C), but the volume is missing. To calculate the volume, we take advantage of a concept from Dalton's law of partial pressures; that each gas occupies the total volume of the container. Because we know the total pressure, number of moles of N_2, and temperature, we can calculate the total volume.

$P_T = 3.75$ atm

$n_T = 2.78$ mol $+ 2.08$ mol $= 4.86$ mol

$T = 50.0°C + 273.15 = 323.2$ K

$V_T = ?$

$$V_T = \frac{nRT}{P_T} = \frac{(4.86 \text{ mol})\left(0.08206 \dfrac{\text{L} \cdot \text{atm}}{\text{mol} \cdot \text{K}}\right)(323.2 \text{ K})}{3.75 \text{ atm}} = 34.4 \text{ L}$$

Using the total volume, calculate the partial pressure of N_2 using the moles of N_2.

$$P_{N_2} = \frac{n_{N_2} RT}{V_T} = \frac{(2.78 \text{ mol})\left(0.08206 \dfrac{\text{L} \cdot \text{atm}}{\text{mol} \cdot \text{K}}\right)(323.2 \text{ K})}{34.4 \text{ L}} = 2.15 \text{ atm}$$

(d) To calculate the partial pressure of Ne, we can do exactly as we did in (c) *or* we can use Dalton's law of partial pressures. Either method gives the same result. According to Dalton's law of partial pressures, the total pressure is the sum of the pressures of Ne and N_2. Mathematically, we can express Dalton's law of partial pressure as:

$P_T = P_{N_2} + P_{Ne}$

$P_{Ne} = P_T - P_{N_2} = 3.75$ atm $- 2.15$ atm $= 1.60$ atm

9.101   Given gas density (g/L), we can calculate molar mass (g/mol) using the molar volume of a gas (L/mol). Under STP conditions, the molar volume of a gas is 22.414 L/mol. Multiplying density (g/L) by molar volume (L/mol) converts density units to molar mass units:

(a) $MM = \dfrac{1.785 \text{ g}}{1 \text{ L}} \times \dfrac{22.414 \text{ L}}{1 \text{ mol}} = 40.01 \text{ g/mol}$

(b) $MM = \dfrac{1.340 \text{ g}}{1 \text{ L}} \times \dfrac{22.414 \text{ L}}{1 \text{ mol}} = 30.03 \text{ g/mol}$

(c) $MM = \dfrac{2.052 \text{ g}}{1 \text{ L}} \times \dfrac{22.414 \text{ L}}{1 \text{ mol}} = 45.99 \text{ g/mol}$

(d) $MM = \dfrac{0.905\ \text{g}}{1\ \cancel{L}} \times \dfrac{22.414\ \cancel{L}}{1\ \text{mol}} = 20.3\ \text{g/mol}$

(e) $MM = \dfrac{0.714\ \text{g}}{1\ \cancel{L}} \times \dfrac{22.414\ \cancel{L}}{1\ \text{mol}} = 16.0\ \text{g/mol}$

9.103   The five postulates of the kinetic-molecular theory are:

1) Gases are composed of small, widely separated particles.
2) Gas particles behave independently of each other.
3) Gas particles move in straight lines.
4) Gas pressure results from the force exerted by the particles in the container. This force is the sum of the forces exerted by the particles as they bounce off the container walls.
5) The average kinetic energy of gas particles depends only on the absolute temperature.

The key to understanding the postulates 1–3 of the kinetic-molecular theory is recognizing that particles (atoms or molecules) in the gas phase are not strongly attracted to each other. Particles that are strongly attracted tend to be liquids or solids at room temperature. In addition, because they are not attracted to each other, the particles move through space on straight paths. Postulate 4 also is easy to understand. As particles strike the container walls, they exert a force on those walls. This is analogous to the force you feel when you walk into a wall (if you try this, make sure no one is looking). One person walking into a wall exerts only a small force, but if the whole class ran into the same wall at the same time, they could do some damage! The forces of individual particles are additive. The fifth postulate is most difficult to understand. When a gas particle absorbs energy, its only options are to store it as potential energy or use it as kinetic energy. Kinetic energy is energy of motion. Particles travel faster and vibrate faster when they have more energy.

9.105   As the temperature of a gas sample increases the kinetic energy of the particles increases. Recall that kinetic energy is the energy of motion. This means that increasing the temperature results in an increase in particle velocity. Faster-moving particles strike the walls of a container with greater force than slower-moving particles, so the pressure is higher. In addition, when the particles are moving faster, the frequency of their collisions with the container walls also increases.

9.107   The pressure of a gas depends on the force and frequency of collisions in a given area. The density of gas particles is inversely proportional to volume. If the volume decreases, the density of the gas sample increases. Because the density of gas particles increases with decreasing volume, we can conclude that the frequency of collisions in a given area also increases with decreasing volume. This means that the pressure increases.

9.109   Because the gases are all at the same temperature, their kinetic energies are the same. The equation for kinetic energy, KE, is: $KE = \frac{1}{2}\ mv^2$. This means that if two objects have the same kinetic energy, the heavier object will move more slowly than the lighter object. For a gas sample (with many particles), the equation is described in terms of the average kinetic energy: $KE_{ave} = \frac{1}{2}\ m(v_{ave})^2$. This equation shows us that, at a given temperature, gas particles with the largest molar mass will move the slowest. Of the substances listed, CO_2 has the largest molar mass, so it has the lowest velocity. H_2 has the smallest molar mass, so it has the highest velocity.

(lowest velocity) $CO_2 < CH_4 < He < H_2$ (highest velocity)

9.111   Gas particles with smaller molar masses have higher average velocities than gas particles with larger molar masses, when at the same temperature. As a result, gases with smaller molar masses diffuse faster than gases with larger molar masses.

9.113   The He atoms are lighter than Ne atoms, so the He atoms will have a higher average velocity, and therefore effuse faster than the Ne atoms.

9.115   The volumes of gases that react are directly proportional to the stoichiometric coefficients in the balanced chemical equation for the reaction. If the products are at the same temperature and pressure as the reactants, the gaseous reaction product volumes are also proportional to the stoichiometric coefficients of the chemical equation. For the given reaction, this means that 2 L H_2 react with 1 L O_2.

$$2H_2(g) + O_2(g) \rightarrow 2H_2O(g)$$

We can calculate the volume of H_2 required to react with 12 L of O_2 using this problem solving map:

Volume O_2 $\xrightarrow{\text{volume ratio}}$ Volume H_2

Volume of O_2 = 12 $\cancel{L\,O_2} \times \dfrac{2\text{ L }H_2}{1\ \cancel{L\,O_2}}$ = 24 L H_2

9.117   When we measure the volumes of gases in chemical reactions under the same pressure and temperature conditions, the volume ratios are equivalent to the mole ratios from the balanced chemical equation. For the reaction of hexane with oxygen, we can say that 2 L of C_6H_{14} react with 19 L of O_2 to produce 12 L of CO_2 and 14 L of H_2O.

Volume CO_2 = 8.00 $\cancel{L\,C_6H_{14}} \times \dfrac{12\text{ L }CO_2}{2\ \cancel{L\,C_6H_{14}}}$ = 48.0 L CO_2

Volume O_2 = 8.00 $\cancel{L\,C_6H_{14}} \times \dfrac{19\text{ L }O_2}{2\ \cancel{L\,C_6H_{14}}}$ = 76.0 L O_2

9.119   Use the ideal gas law to calculate the number of moles of N_2O produced in the reaction. Then use this number of moles of N_2O to calculate the number of moles, and mass, of NH_4NO_3 (80.05 g/mol) required for the reaction.

Volume N_2O $\xrightarrow{PV = nRT}$ mol N_2O $\xrightarrow{\text{mole ratio}}$ mol NH_4NO_3 $\xrightarrow{MM}$ g NH_4NO_3

Pressure in atm = 2850 $\cancel{torr} \times \dfrac{1\text{ atm}}{760\ \cancel{torr}}$ = 3.75 atm

Temperature in K = 42°C + 273.15 = 315.2 K

$$n = \dfrac{PV}{RT} = \dfrac{(3.75\ \cancel{atm})(145\ \cancel{L})}{\left(0.08206\dfrac{\cancel{L}\cdot\cancel{atm}}{mol\cdot\cancel{K}}\right)(315.2\ \cancel{K})} = 21.0\text{ mol }N_2O$$

Mass NH_4NO_3 = 21.0 $\cancel{mol\,N_2O} \times \dfrac{1\ \cancel{mol\,NH_4NO_3}}{1\ \cancel{mol\,N_2O}} \times \dfrac{80.05\text{ g }NH_4NO_3}{1\ \cancel{mol\,NH_4NO_3}}$ = 1.68×10^3 g NH_4NO_3

9.121   At higher altitudes, the pressure on the outside of the balloon is lower than the pressure the balloon experiences on the ground. The balloon expands or contracts so that the external pressure and the internal pressure are the same.

9.123   The density of a gas decreases when it is heated and its volume is allowed to expand, or if the gas sample size in a fixed volume container is reduced. A hot air balloon actually has an opening in the bottom. As the air in the balloon is heated, the gas expands and the "extra" air leaves from the opening at the bottom of the balloon. Since gas molecules are escaping and the volume is remaining fairly constant, the air in the hot air balloon becomes less dense than the surrounding air, allowing the balloon to float.

9.125   We are looking at a change of state (i.e. "Assume the pressure at the surface is 760 torr and changes to 150 torr."). Therefore, we can assume that this is a combined gas law problem, and organize the data into a table as shown below:

P_1	V_1	T_1	P_2	V_2	T_2
760 torr	nc	?	150 torr	nc	218 K

*nc = no change

Next, we write the combined gas law, cancel the variables that do not change, and solve for T_1:

Combined gas law: $\dfrac{P_1 V_1}{T_1} = \dfrac{P_2 V_2}{T_2}$

$$T_1 = P_1 \times \frac{T_2}{P_2} = 760 \text{ torr} \times \frac{218 \text{ K}}{150 \text{ torr}} = 1.10 \times 10^3 \text{ K}$$

9.127   The table below summarizes the changes on the macroscopic and microscopic levels.

What will happen if…	Macroscopic view	Microscopic view
temperature increases and pressure remains constant	balloon gets bigger	molecules move more rapidly molecules collide more forcefully collision frequency increases molecules are further apart
temperature decreases and pressure remains constant	balloon gets smaller	molecules move more slowly molecules collide less energetically collision frequency decreases molecules are closer together
10,000 feet (temperature decreases and pressure decreases)	balloon probably gets a little bigger	molecules move more slowly molecules collide less energetically collision frequency decreases

9.129   To calculate the volume of O_2, we must first determine the number of moles of O_2 produced. We can calculate the number of moles from the mass of HgO that reacted, the molar mass of HgO (216.6 g/mol), and the balanced chemical equation.

g HgO $\xrightarrow{\quad MM \quad}$ mol HgO $\xrightarrow{\text{mole ratio}}$ mol O_2 $\xrightarrow{\quad PV = nRT \quad}$ Volume O_2

Moles O_2 = 27.0 g HgO $\times \dfrac{1 \text{ mol HgO}}{216.6 \text{ g HgO}} \times \dfrac{1 \text{ mol } O_2}{2 \text{ mol HgO}} = 0.0623 \text{ mol } O_2$

Temperature in kelvins = 50.0°C + 273.15 = 323.2 K

$$V = \frac{n_{O_2} RT}{P} = \frac{\left(0.0623 \text{ mol}\right)\left(0.08206 \dfrac{L \cdot atm}{mol \cdot K}\right)\left(323.2 \text{ K}\right)}{0.947 \text{ atm}} = 1.75 \text{ L } O_2$$

9.131   If we assume that the reaction takes place at 745 torr and 25.0°C, we can use the stoichiometric coefficients to calculate the volume of the butene combusted:

$$\text{Volume } C_4H_8 = 12.0 \text{ L } O_2 \times \frac{1 \text{ L } C_4H_8}{6 \text{ L } O_2} = 2.00 \text{ L } C_4H_8$$

Then, we can use the combined gas law to calculate the volume occupied by this amount of butene at 188°C and 2.50 atm:

P_1	V_1	T_1	P_2	V_2	T_2
745 torr	2.00 L	25.0°C	2.50 atm	?	188°C

$$\text{Pressure of } C_4H_8 \text{ in atm} = 745 \text{ torr} \times \frac{1 \text{ atm}}{760 \text{ torr}} = 0.980 \text{ atm}$$

Convert temperatures to kelvins:

$T_1 = 25.0°C + 273.15 = 298.2 \text{ K}$

$T_2 = 188°C + 273.15 = 461 \text{ K}$

Solve the combined gas law for V_2:

$$V_2 = \frac{P_1 V_1}{T_1} \times \frac{T_2}{P_2} = \frac{(0.980 \text{ atm})(2.00 \text{ L})}{298.2 \text{ K}} \times \frac{461 \text{ K}}{2.50 \text{ atm}} = 1.21 \text{ L}$$

9.133   First, we use the balanced chemical equation and the molar mass of CuO (79.55 g/mol) to calculate the number of moles of H_2 required to react with 85.0 g of CuO. Then we can use the ideal gas law to calculate the volume of H_2 gas.

$$\text{Mass CuO} \xrightarrow{MM \text{ CuO}} \text{mol CuO} \xrightarrow{\text{Mole ratio}} \text{mol } H_2 \xrightarrow{PV = nRT} \text{Volume } H_2$$

$$\text{Moles } H_2 = 85.0 \text{ g CuO} \times \frac{1 \text{ mol CuO}}{79.55 \text{ g CuO}} \times \frac{1 \text{ mol } H_2}{1 \text{ mol CuO}} = 1.07 \text{ mol } H_2$$

Temperature in kelvins = 27.0°C + 273.15 = 300.2 K

$$\text{Pressure in atm} = 722 \text{ torr} \times \frac{1 \text{ atm}}{760 \text{ torr}} = 0.950 \text{ atm}$$

$$V = \frac{nRT}{P} = \frac{\left(1.07 \text{ mol}\right)\left(0.08206 \dfrac{L \cdot atm}{mol \cdot K}\right)\left(300.2 \text{ K}\right)}{0.950 \text{ atm}} = 27.7 \text{ L}$$

9.135   Because the gas volumes are all measured at the same conditions, we can use the mole ratios as volume ratios:

(a) Volume H_2S = 5.00 $L\,H_2 \times \dfrac{1\,L\,H_2S}{1\,L\,H_2}$ = 5.00 L

(b) Volume NH_3 = 5.00 $L\,H_2 \times \dfrac{2\,L\,NH_3}{3\,L\,H_2}$ = 3.33 L

(c) Volume C_6H_{12} = 5.00 $L\,H_2 \times \dfrac{1\,L\,C_6H_{12}}{3\,L\,H_2}$ = 1.67 L

9.137   From Chapter 5, we learned that, when heated, calcium carbonate decomposes to form calcium oxide and carbon dioxide gas:

$CaCO_3(s) \rightarrow CaO(s) + CO_2(g)$ *Reaction 1*

The reaction of CaO with carbon is:

$2CaO(s) + 5C(s) \rightarrow 2CaC_2(s) + CO_2(g)$ *Reaction 2*

The equation given in this problem is:

$CaC_2(s) + 2H_2O(l) \rightarrow Ca(OH)_2(aq) + C_2H_2(g)$ *Reaction 3*

If we determine the number of moles of $CaCO_3$ (limestone) that react, then we can calculate the number of moles of C_2H_2 that form using the stoichiometric coefficients from reactions 1-3.

$$\text{mol } CaCO_3 \xrightarrow[\text{reaction 1}]{\text{mole ratio}} \text{mol } CaO \xrightarrow[\text{reaction 2}]{\text{mole ratio}} \text{mol } CaC_2 \xrightarrow[\text{reaction 3}]{\text{mole ratio}} \text{mol } C_2H_2$$

Moles $CaCO_3$ = 5.00 $g\,CaCO_3 \times \dfrac{1\,\text{mol } CaCO_3}{100.09\,g\,CaCO_3}$ = 0.0500 mol $CaCO_3$

Moles C_2H_2 = 0.0500 $mol\,CaCO_3 \times \dfrac{1\,mol\,CaO}{1\,mol\,CaCO_3} \times \dfrac{2\,mol\,CaC_2}{2\,mol\,CaO} \times \dfrac{1\,mol\,C_2H_2}{1\,mol\,CaC_2}$ = 0.0500 mol C_2H_2

Finally, we use the ideal gas law to calculate the temperature of the gas:

$$T = \frac{PV}{nR} = \frac{(0.750\,atm)(1.25\,L)}{(0.0500\,mol)\left(0.08206\,\dfrac{L \cdot atm}{mol \cdot K}\right)} = 228\,K\ (-45°C)$$

9.139   (a) Assuming the temperature and pressure are kept constant, the number of moles of a gas and its volume are directly proportional to each other. Increasing the number of moles of gas increases the volume occupied by the gas.

(b) Pressure and temperature must be held constant for this relationship to be true. We can use the ideal gas law to show this:

$$V = \frac{nRT}{P} = k \times n$$

where k is constant as long as T and P are constant. This means volume is directly proportional to the number of moles of gas at constant T and P.

9.141   We can determine the molar mass of the liquid by dividing the mass of the liquid by the number of moles of liquid. Mass is given, but we need to determine the number of moles from the information given. We are given the pressure, temperature, and volume of the vaporized liquid, so we can use the ideal gas law to calculate the number of moles of vapor (which is the same as the number of moles of liquid).

$$\text{Volume in liters} = 127 \text{ mL} \times \frac{1 \text{ L}}{1000 \text{ mL}} = 0.127 \text{ L}$$

$$\text{Pressure in atm} = 691 \text{ torr} \times \frac{1 \text{ atm}}{760 \text{ torr}} = 0.909 \text{ atm}$$

$$\text{Temperature in kelvins} = 98°C + 273.15 = 371 \text{ K}$$

$$n = \frac{PV}{RT} = \frac{(0.909 \text{ atm})(0.127 \text{ L})}{\left(0.08206 \frac{\text{L} \cdot \text{atm}}{\text{mol} \cdot \text{K}}\right)(371 \text{ K})} = 3.79 \times 10^{-3} \text{ mol}$$

$$\text{Molar mass of liquid} = \frac{0.495 \text{ g}}{3.79 \times 10^{-3} \text{ mol}} = 131 \text{ g/mol}$$

9.143   If we calculate the total number of moles of gas produced by the reaction, we can calculate their total volume.

$$\text{g } C_3H_5(ONO_2)_3 \xrightarrow{MM} \text{mol } C_3H_5(ONO_2)_3 \xrightarrow{\text{mole ratio}} \text{mol gas} \xrightarrow{PV = nRT} \text{Volume gas}$$

The molar mass of nitroglycerine is 227.1 g/mol. Note: We are not concerned with the volume produced by any one gas, so we can group the stoichiometric coefficients of the gases:

1 mol $C_3H_5(ONO_2)_3$ = 12 + 10 + 6 + 1 = 29 mol gas

$$\text{Moles gas} = 1.00 \text{ g } C_3H_5(ONO_2)_3 \times \frac{1 \text{ mol } C_3H_5(ONO_2)_3}{227.1 \text{ g } C_3H_5(ONO_2)_3} \times \frac{29 \text{ mol gas}}{4 \text{ mol } C_3H_5(ONO_2)_3} = 0.0319 \text{ mol gas}$$

Temperature in K = 275°C + 273.15 = 548 K

$$V = \frac{nRT}{P} = \frac{(0.0319 \text{ mol})\left(0.08206 \frac{\text{L} \cdot \text{atm}}{\text{mol} \cdot \text{K}}\right)(548 \text{ K})}{2.00 \text{ atm}} = 0.718 \text{ L}$$

9.145   (a) Average kinetic energy does not change when volume is changed because it is constant at a given temperature.
      (b) Average molecular velocity does not change when volume is changed because for a given gas, average velocity only changes when kinetic energy changes: $KE_{ave} = m(v_{ave})^2$.
      (c) Pressure increases as the gas particles become more crowded with resulting greater collisions per unit area.

9.147   (a) Average kinetic energy decreases with temperature.
      (b) Average molecular velocity decreases because kinetic energy decreases.
      (c) Volume decreases to maintain constant pressure as gas particle velocity decreases.

9.149   Under the identical conditions of temperature and pressure, the ratio of gas densities is the same as the ratio of their molar masses because equal volumes of gases contain the same number of gas particles (Avogadro's hypothesis). Therefore, to find a gas that has 8.0 times the density we must identify a gas with a molar mass that is 8.0 times the molar mass of H_2:

Molar mass of unknown gas = molar mass $H_2 \times 8.0 = 2.016$ g/mol $\times 8.0 = 16$ g/mol

The molar mass of O_2 is 32.00 g/mol, the molar mass of CH_4 is 16.04 g/mol, and the molar mass of Ne is 20.18 g/mol. The molar mass of CH_4 is the best match, so of the three possibilities, it is most likely the unknown gas.

9.151   The pressure of a gas can be increased by reducing the volume, increasing the temperature, or adding more gas to the container. Another way would be to increase the pressure of the surrounding air if the gas is in an elastic container.

9.153   (a) $n = \dfrac{PV}{RT} = \dfrac{91 \text{ atm} \times 1.0 \text{ L}}{0.08206 \frac{\text{L} \cdot \text{atm}}{\text{K} \cdot \text{mol}} \times 740 \text{ K}} = 1.5 \text{ mol}$

(b)   Mass of carbon dioxide = $1.5 \text{ mol } CO_2 \times \dfrac{44.01 \text{ g } CO_2}{1 \text{ mol } CO_2} = 66 \text{ g } CO_2$

(c)   To calculate density we divide mass by volume. Mass was calculated in (b). The volume used to determine moles in (a) was 1.0 L so this is the volume we must use to calculate density:

$\text{Density} = \dfrac{\text{mass}}{\text{volume}} = \dfrac{66 \text{ g}}{1.0 \text{ L}} = 66 \text{ g/L}$

$\text{Density in units of g/mL} = \dfrac{66 \text{ g}}{\text{L}} \times \dfrac{1 \text{ L}}{1000 \text{ mL}} = 0.066 \text{ g/mL}$

Chapter 10 – The Liquid and Solid States

10.1   (a) vapor pressure; (b) melting point; (c) equilibrium; (d) evaporation; (e) induced dipole; (f) alloy; (g) boiling point; (h) sublimation; (i) molecular solid; (j) intermolecular force; (k) London dispersion force; (l) normal freezing point; (m) crystal

10.3   Some comparisons of the liquid and gas states are indicated below. Note that intermolecular attraction and kinetic energy are relative terms, and that while liquids and gases are fluid (conform to their containers), only gases completely occupy the available volume of their containers.

	Liquid	Gas
Particle spacing	dense or closely spaced	molecules far apart
Intermolecular attraction	moderate	weak
Kinetic energy	low	high

10.5   In a solid, the particles (molecules, atoms, or ions) do not have enough energy to overcome the strong attractions to their neighboring particles. As a result, the particles cannot easily move with respect to each other. Rather than conforming to their containers, the shapes of solids are defined by the shapes of the particle groupings. Substances in the liquid phase have relatively high energy compared to their forces of attraction. As a result, they can move past each other and conform to the shapes of their containers.

10.7   The substance is in the solid state. If the substance were a liquid or gas, it would not maintain its regular shape.

10.9   The image is a molecular level representation of a substance in the liquid state. Both the liquid and solid states have dense particle spacing (the particles are always in contact with each other), but in the image, the particles do not show any clear, ordered pattern. The relative disorder we see is a characteristic of the liquid state. In the solid state we expect to see an ordered structure at the molecular level.

10.11   Images 1, 2, and 3 represent the solid, gas, and liquid states, respectively. In Image 1, we see a high degree of order, which is characteristic of the solid state. Image 2 shows the gas phase, in which the particles have high kinetic energy and are spaced far apart. Image 3 represents the liquid phase because there is dense particle packing, but no microscopic order. The names of the phase changes are specific to the direction of the change. For example, melting is the change from solid to liquid ($1 \rightarrow 3$), while the change from liquid to solid ($3 \rightarrow 1$) is freezing. Phase change A represents deposition, the formation of a solid from a gas. Phase change B represents vaporization, the formation of the gas from the liquid. Phase change C represents melting, the formation of a liquid from a solid.

10.13   The vapor pressure of all liquids increase as temperature increases. Eventually, if the temperature increases enough, the vapor pressure of the liquid equals the atmospheric pressure above the liquid. At this point, bubbles of vapor form in the liquid (the liquid boils). Liquids boil when their vapor pressures equal the pressure above them (usually, atmospheric pressure). The temperature at which boiling occurs is the boiling point. The normal boiling point of a liquid is the temperature at which the vapor pressure of the liquid is equal to 1 atm.

10.15   Molecules of a substance in the gas state have significantly more kinetic energy than they do in the liquid state, where they are not only moving more slowly, but also feel the attractions of the molecules around them. In order to go from the liquid to the gas state, the energy of the molecules must increase enough to allow them to overcome their attractions for each other and move about independently. Because the molecules must absorb energy to go into the gas state, the process is endothermic (i.e. absorbs energy from the surroundings).

10.17

10.19 In the diagrams, the particles are changing from the solid state to the gas state (sublimation). Particles of a substance in the gas phase have much higher kinetic energies than they do in the solid phase, so energy must be absorbed in the process. Therefore sublimation is an endothermic process.

10.21 All molecules and atoms have some degree of attraction for one another. As a gas cools, the particles lose kinetic energy. Eventually, the attractive forces are greater than the kinetic energy of the particles, and the particles begin to coalesce (stick together) to form the liquid state. If you put a hot butter knife and a cold butter knife above a pot of boiling water, condensation occurs on the cold knife. This happens because the molecules lose kinetic energy as they strike the cold knife and coalesce there to form liquid water. Nothing happens on the hot knife (if it's hot enough) because the molecules pick up kinetic energy and bounce off it.

10.23 Evaporation of water is an endothermic process because molecules in the gas phase have higher energies than molecules in the liquid phase. As a liquid evaporates, it absorbs energy from its surroundings. This makes the surroundings cooler. If it is circulated by a fan, we can use the cooled air to cool a house.

10.25 Water boils when it has been heated to the point that its vapor pressure equals atmospheric pressure. At high elevations, atmospheric pressure can be significantly lower than 1 atm (atmospheric pressure is about 0.6 to 0.7 atm at 12,000 ft). As a result, the temperature at which water boils is lower. Because the water boils at a lower temperature, the pasta cooks more slowly.

10.27 The average kinetic energy of molecules in the gas phase is higher than that of the same molecules in the liquid phase. Increasing the temperature increases the kinetic energy of the molecules in the liquid. As a result, the number of molecules on the liquid surface with enough energy to go into the gas phase increases, which increases the vapor pressure of the liquid.

10.29 A temperature of 100°C is higher than the melting point, but lower than boiling point, of gallium. Therefore, gallium is a liquid at 100°C. A temperature of 15°C is below gallium's freezing point, so it is a solid.

10.31   The image shows a gas and liquid at equilibrium. This would occur on segment B–C, which is at the boiling point of the substance. We know that this occurs at the boiling point because it is taking place at constant temperature and heat is being removed. A gas-liquid equilibrium is expected when the liquid is not boiling (Segment C–D), but that process does not take place at constant temperature.

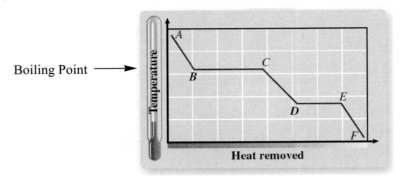

10.33   To calculate the energy required to evaporate 105.0 g of H_2O (18.02 g/mol), we use the molar heat of vaporization of water, 4.07×10^4 J/mol, and the following problem solving map:

$q_{vaporization} = n \times$ heat of vaporization

$$\text{g } H_2O \xrightarrow{\quad MM \quad} \text{mol } H_2O \xrightarrow{\text{heat of vaporization}} J$$

$$\text{Energy} = 105.0 \text{ g } H_2O \times \frac{1 \text{ mol } H_2O}{18.02 \text{ g } H_2O} \times \frac{4.07 \times 10^4 \text{ J}}{1 \text{ mol } H_2O} = 2.37 \times 10^5 \text{ J (or 237 kJ)}$$

10.35   Before we begin dealing with numbers, it is useful to consider the specific steps involved when we convert 15.0 g of liquid water at 52.5°C to steam at 238.2°C. There are actually three steps involved:

$$\text{Water } 52.5°C \xrightarrow{q_{water}} \text{Water } 100°C \xrightarrow{q_{vaporization}} \text{Steam } 100°C \xrightarrow{q_{steam}} \text{Steam } 238.2°C$$

This means we need to determine three different energies before we can determine the total energy required to complete the process.

$q_{total} = q_{water} + q_{vaporization} + q_{steam}$

Calculating q_{water}

First, we determine the amount of energy required to raise the temperature of liquid water to its boiling point, 100.0°C.

$q_{water} = m \times C \times \Delta T$

$m = 15.0$ g $C = 4.18$ J/(g·°C) $\Delta T = T_f - T_i = 100.0°C - 52.5°C = 47.5°C$

$$q_{water} = \left(15.0 \text{ g}\right) \times \left(4.18 \frac{J}{g \cdot °C}\right) \times \left(47.5 °C\right) = 2.98 \times 10^3 \text{ J}$$

Calculating $q_{vaporization}$

Next, we calculate the amount of energy required to vaporize the sample at 100°C.

$$\text{g H}_2\text{O} \xrightarrow{\quad MM \quad} \text{mol H}_2\text{O} \xrightarrow{\quad \text{heat of vaporization} \quad} \text{J}$$

The molar heat of vaporization of water is 4.07×10^5 J.

$$q_{vaporization} = 15.0 \ \text{g H}_2\text{O} \times \frac{1 \ \text{mol H}_2\text{O}}{18.02 \ \text{g H}_2\text{O}} \times \frac{4.07 \times 10^4 \ \text{J}}{1 \ \text{mol H}_2\text{O}} = 3.39 \times 10^4 \ \text{J} \ \text{(or 33.9 kJ)}$$

Calculating q_{steam}

Finally, we calculate the amount of energy required to raise the temperature of the steam from 100.0°C to 238.2°C.

$$q_{steam} = m \times C \times \Delta T$$

$m = 15.0$ g $C = 2.02$ J/(g·°C) $\Delta T = T_f - T_i = 238.2°\text{C} - 100.0°\text{C} = 138.2°\text{C}$

$$q_{steam} = \left(15.0 \ \text{g}\right) \times \left(2.02 \ \frac{\text{J}}{\text{g} \cdot °\text{C}}\right) \times \left(47.5 \ °\text{C}\right) = 4.19 \times 10^3 \ \text{J}$$

Calculating q_{total}

We calculate the total energy required for the process by adding up all the energies:

$$q_{total} = q_{water} + q_{vaporization} + q_{steam}$$

$$q_{total} = 2.98 \times 10^3 \ \text{J} + 3.39 \times 10^4 \ \text{J} + 4.19 \times 10^3 \ \text{J} = 4.10 \times 10^4 \ \text{J} \ \text{(or 41.0 kJ)}$$

Notice that, of the three processes, the process of vaporizing the water requires the largest amount of energy.

10.37   In this problem, you are taking ice at −15.0°C to steam at 145°C. There are five steps in this process. They can be summarized by the following problem map:
Warm the ice → Melt the ice → Warm the Water → Vaporize the Water →Warm the Steam
Calculate the energy for each step and then add all of them together to get the total energy change, $q_{tot.}$

Warm the ice: The ice starts at −15.0 °C and increases in temperature to 0 °C. It stops warming at this temperature because it undergoes a phase change.

$$q_{ice} = m \times C \times \Delta T$$

   $m = 542$ g $C = 2.03$ J/(g·°C) $\Delta T = 0.0 \ °\text{C} - (-15.0°\text{C}) = 15.0°\text{C}$

$$q_{water} = 542 \ \text{g} \left(2.03 \ \frac{\text{J}}{\text{g} \cdot °\text{C}}\right) \left(15.0 \ °\text{C}\right) = 16500 \ \text{J}$$

Melt the ice: Next, we calculate the amount of energy required to melt the ice at 0.0°C.

$$q_{fusion} = n \times \text{heat of fusion}$$

$$\text{g H}_2\text{O} \xrightarrow{\quad MM \quad} \text{mol H}_2\text{O} \xrightarrow{\quad \text{heat of fusion} \quad} \text{J}$$

The molar heat of fusion of water is 6.01×10^3 J.

$$q_{fusion} = 542 \ \text{g H}_2\text{O} \times \frac{1 \ \text{mol H}_2\text{O}}{18.02 \ \text{g H}_2\text{O}} \times \frac{6.01 \times 10^3 \ \text{J}}{1 \ \text{mol H}_2\text{O}} = 1.81 \times 10^5 \ \text{J}$$

Warm the Water: Next we calculate the energy required to raise the temperature of the liquid water from 0.0°C to 100.0°C. Remember, the temperature of water stops changing when it reaches its next phase transition (i.e. starts to boil).

$q_{water} = m \times C \times \Delta T$

$m = 542$ g $C = 4.18$ J/(g·°C) $\Delta T = 100.0°C - 0.0°C = 100.0°C$

$q_{water} = 542 \, \cancel{g} \times \left(4.18 \dfrac{J}{\cancel{g} \cdot \cancel{°C}}\right) \times \left(100.0 \, \cancel{°C}\right) = 2.27 \times 10^5$ J

Vaporize the water: Next, we calculate the amount of energy required to vaporize the water at 100.0°C

$q_{vap} = n \times$ heat of vaporization

$$\text{g H}_2\text{O} \xrightarrow{MM} \text{mol H}_2\text{O} \xrightarrow{\text{heat of vaporization}} \text{J}$$

The molar heat of vaporization of water is 4.07×10^4 J.

$q_{fusion} = 542 \, \cancel{\text{g H}_2\text{O}} \times \dfrac{1 \, \cancel{\text{mol H}_2\text{O}}}{18.02 \, \cancel{\text{g H}_2\text{O}}} \times \dfrac{4.07 \times 10^4 \text{ J}}{1 \, \cancel{\text{mol H}_2\text{O}}} = 1.22 \times 10^6$ J

Warm the steam: Next we calculate the energy required to raise the temperature of the steam from 100.0°C to 145°C.

$q_{steam} = m \times C \times \Delta T$

$m = 542$ g $C = 2.02$ J/(g·°C) $\Delta T = 145°C - 100.°C = 45°C$

$q_{steam} = 542 \, \cancel{g} \times \left(2.02 \dfrac{J}{\cancel{g} \cdot \cancel{°C}}\right) \times \left(45 \, \cancel{°C}\right) = 4.9 \times 10^4$ J

Calculating total energy, q_{total}: If you add all these energies together, you will have the energy for the entire process:

$q_{total} = 16500 \text{ J} + 181000 \text{ J} + 227000 \text{ J} + 1220000 \text{ J} + 49000 \text{ J} = 1.70 \times 10^6$ J

10.39   In this process, you are heating 125g of liquid ethanol at 25.0 °C to ethanol vapor at 96.0 °C. In the process, the ethanol vaporizes at its boiling point 78.4 °C. This temperature defines the end of the heating of the liquid and beginning of the heating of the vapor. The process looks like this:
Warm the Ethanol → Vaporize the Ethanol → Warm the Ethanol Vapor

Warm the Ethanol: Next we calculate the energy required to raise the temperature of the liquid ethanol from 25.0°C to 78.4 °C. Remember, the temperature of ethanol stops changing when it reaches its next phase transition (i.e. starts to boil).

$q_{ethanol} = m \times C \times \Delta T$

$m = 125$ g $C = 2.44$ J/(g·°C) $\Delta T = 78.4°C - 25.0°C = 53.4°C$

$q_{ethanol} = 125 \, \cancel{g} \times \left(2.44 \dfrac{J}{\cancel{g} \cdot \cancel{°C}}\right) \times \left(53.4 \, \cancel{°C}\right) = 1.63 \times 10^4$ J

Vaporize the ethanol: Next, we calculate the amount of energy required to vaporize the ethanol.

$q_{vap} = n \times$ heat of vaporization

$$\text{g Ethanol} \xrightarrow{MM} \text{mol Ethanol} \xrightarrow{\text{heat of vaporization}} \text{J}$$

The molar heat of vaporization of ethanol is 3.86×10^4 J/mol

$$q_{fusion} = 125 \text{ g Ethanol} \times \frac{1 \text{ mol Ethanol}}{46.06 \text{ g Ethanol}} \times \frac{3.86 \times 10^4 \text{ J}}{1 \text{ mol Ethanol}} = 1.05 \times 10^5 \text{ J}$$

Warm the ethanol vapor: Next we calculate the energy required to raise the temperature of the ethanol vapor from 78.4°C to 96.0°C.

$q_{vap} = m \times C \times \Delta T$

$$m = 125 \text{ g} \qquad C = 1.42 \text{ J/(g·°C)} \qquad \Delta T = 96.0°C - 78.4°C = 17.6°C$$

$$q_{vap} = 125 \text{ g} \times \left(1.42 \frac{\text{J}}{\text{g} \cdot °C} \right) \times \left(17.6 °C \right) = 3124 \text{ J}$$

Calculating total energy, q_{total}: If you add all these energies together, you will have the energy for the entire process:

$$q_{total} = 1.63 \times 10^4 \text{ J} + 1.05 \times 10^5 \text{ J} + 3124 \text{ J} = 1.24 \times 10^5 \text{ J}$$

10.41   An intermolecular force is an attractive force between individual particles (molecules or atoms) of a substance. An analogy would be "interpersonal" skills. This refers to the way people relate to each other, rather than to themselves.

10.43   If molecules were not attracted to one another, the predominant state of matter would be the gas state.

10.45   Argon, Ar, is a gas (a noble gas) at room temperature. None of these substances are polar, but both I_2 and Hg have very high molar masses and, therefore, experience significant London dispersion forces. The strength of these forces causes Hg to be liquid and I_2 to be solid at room temperature.

10.47   To answer this question we need to first determine which of the molecules are polar. We do this by finding the molecule(s) composed of atoms with differing electronegativities (if the electronegativities of the atoms are all the same, the molecule cannot be polar). Then, we draw the Lewis structures of those compounds that contain polar bonds, and determine if the molecule is symmetrical (and, therefore, nonpolar). Of the molecules listed, CO_2 has polar bonds, but is symmetrical and, therefore, nonpolar. NO, NF_3, and CH_3Cl are all polar and can experience dipole-dipole forces.

Molecular structures

(a)   (b)   (c)   (d)

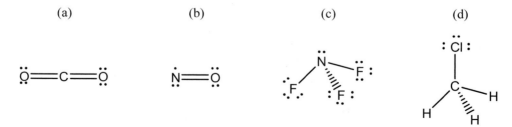

10.49   A bond is an attractive force between two atoms in a molecule. A hydrogen bond is not a true bond, because it usually occurs between two atoms from different molecules. Hydrogen bonds are much weaker than covalent bonds, and the electrons in hydrogen bonds are not shared.

10.51   Hydrogen bonding only occurs in substances in which a hydrogen atom is bonded to nitrogen, oxygen, or fluorine. Of the substances shown, only CH_3NH_2 (B) can form hydrogen bonds, because the nitrogen is bonded to a hydrogen atom. The compound shown in A contains oxygen and hydrogen, but the oxygen atom is bonded to two carbon atoms, not to a hydrogen atom. The compound shown in C does not contain one of the three highly electronegative elements (N, O, or F) necessary for hydrogen bonding.

10.53 H_2O (a) and NH_3 (b) have hydrogen atoms bonded to O and N atoms, respectively, so they can participate in hydrogen bonding.

10.55   To form hydrogen bonds with water molecules, a substance needs to have unshared electron pairs on an electronegative element. Both (a) and (b) meet this requirement and can participate in hydrogen bonding with water. In SCl_2 (a), the unshared electron pairs can only accept hydrogen bonds from water. HF (b) possesses a hydrogen atom bonded to a fluorine atom, so HF can both contribute and accept hydrogen bonds from water molecules.

10.57   The vapor pressures of pure liquids depend on the strength of the intermolecular forces between the substance particles. Liquids with stronger intermolecular forces have lower vapor pressures (at a specified temperature) than liquids with weaker intermolecular forces because the strongly attracted particles require more energy to overcome the forces and enter the gas phase.

10.59   When a molecule is polar, some portion of it carries a permanent, slightly negative charge, while another portion carries a permanent, slightly positive charge. These charges come from the unequal sharing of electrons in chemical bonds (Chapter 9). The oppositely-charged areas of different molecules attract each other (dipole-dipole interaction). Similar size polar and nonpolar molecules experience the same magnitude of induced dipole attraction (London dispersion force). Therefore, while polar and nonpolar molecules possess London dispersion forces, polar molecules experience additional attractive forces.

10.61   As molecular weight increases, the importance of dispersion forces increases. High molecular weight compounds experience greater intermolecular attractive forces than lower molecular weight compounds experience. Nonpolar molecules could have greater intermolecular forces than polar molecules if the nonpolar molecules are much greater in size.

10.63   All substances experience dispersion forces. In addition, polar molecules experience dipole-dipole interactions. Only molecules with hydrogen atoms directly bonded to either nitrogen, oxygen, or fluorine atoms can experience hydrogen bonding. We draw Lewis structures to determine whether the molecular geometry of a substance allows for polarity.

	dispersion force	Dipole-dipole	hydrogen bonding	Lewis structure
(a) C_6H_6	X			
(b) NH_3	X	X	X	

	dispersion force	dipole-dipole	hydrogen bonding	Lewis structure
(c) CS_2	X			$\ddot{S}=C=\ddot{S}$
(d) $CHCl_3$	X	X		(Lewis structure of $CHCl_3$)

10.65   Intermolecular forces must be overcome to melt or boil each of the substances. All substances experience dispersion forces. In addition, polar molecules experience dipole-dipole interactions. Only molecules with hydrogen atoms directly bonded to either nitrogen, oxygen, or fluorine atoms can experience hydrogen bonding. We draw Lewis structures to determine whether the molecular geometry of a substance allows for polarity.

	dispersion force	dipole-dipole	hydrogen bonding	Lewis structure
(a) Kr	X			$:\ddot{Kr}:$
(b) CO	X	X		$:C\equiv O:$
(c) CH_4	X			(Lewis structure of CH_4)
(d) NH_3	X	X	X	(Lewis structure of NH_3)

10.67   Because all of the substances can form hydrogen bonds and experience London dispersion forces, the total intermolecular force (IMF) increases with increasing molar mass.

(lowest IMF) $CH_3OH < CH_3CH_2OH < CH_3CH_2CH_2OH$ (highest IMF)

The boiling points of the substances follow the same trends:

(lowest boiling point) $CH_3OH < CH_3CH_2OH < CH_3CH_2CH_2OH$ (highest boiling point)

10.69   Nitrogen dioxide has polar bonds and is bent, so it will experience dipole-dipole interactions in addition to London dispersion forces. The dipoles of the C=O bonds will cancel, so CO_2 experiences only London dispersion forces. Therefore, NO_2 experiences stronger intermolecular forces than CO_2.

10.71   Hydrogen (H_2) is physically larger (it is composed of two atoms), but not more massive, than helium. The size of the hydrogen molecule allows the electrons to be moved around more easily than is the case for He, so H_2 experiences greater dispersion forces. More energy is required to vaporize liquid hydrogen than to vaporize liquid helium, so the boiling point of H_2 is greater than that of He.

10.73   When we compare the intermolecular forces of HCl and F_2 we see that both possess approximately equal London dispersion forces (based on molecular weight), but HCl is polar. Because F_2 is nonpolar, it can't participate in dipole–dipole interactions. Therefore, because the intermolecular forces in HCl are greater than those in F_2, HCl has a higher boiling point than F_2.

10.75   To choose the substance from each pair with the higher equilibrium vapor pressure, we evaluate the intermolecular forces each experiences. The substance experiencing the stronger intermolecular forces will have the lower equilibrium vapor pressure at a given temperature.
  (a) CH_3OH – higher vapor pressure
        CH_2CH_5OH – has stronger London dispersion forces
  (b) CH_3OH – higher vapor pressure
        H_2O – experiences weaker London dispersion forces and similar dipole-dipole forces, but forms much stronger hydrogen bonds. H_2O possesses two hydrogen atoms and two unshared electron pairs with which to form hydrogen bonds.
  (c) NH_3 – higher vapor pressure
        H_2O – experiences approximately the same London dispersion forces and similar dipole-dipole forces, but engages in much stronger hydrogen bonding. H_2O possesses two hydrogen atoms and two unshared electron pairs with which to form hydrogen bonds.

10.77   (lowest) $He < N_2 < H_2O < I_2$ (highest). You might have guessed that H_2O experiences the strongest intermolecular forces, but I_2 has a high molecular weight and is quite large. The induced dipole of I_2 causes it to be a solid at room temperature. He and N_2 are both nonpolar, but He experiences the weakest intermolecular forces because it has the lowest molecular weight.

10.79   Both molecules are polar, but water experiences very strong hydrogen bonding while acetone experiences dipole-dipole interactions. As a result, water has a higher boiling point.

10.81   On a molecular level, particles of a substance in the liquid state have more energy than when they are in the solid state so they have the freedom to change their positions relative to one another. In addition, particles in the liquid state do not demonstrate a long-range structure comparable to the crystal lattices of many solids. On a macroscopic level, liquids flow and can take the shapes of their containers, whereas solids do not. Liquids tend to be less dense than solids.

10.83   Viscosity is the resistance of a liquid to flow. The more viscous a liquid is, the more slowly it will flow when it is poured. For example, pancake syrup and honey are viscous liquids. Both pour much more slowly than water, which is not viscous.

10.85   All the particles in a liquid experience attractive forces with other particles in the liquid. Particles in the middle experience forces that are equal in all directions. On the surface, the particles only experience attractive forces to the sides and "downward" (toward the center of the liquid). This imbalance of forces causes several interesting phenomena. If you drop a liquid, it tends to form spheres as the particles experience a net pull inward (think, for example, of a drop of rain). A sphere (or drop) represents the lowest energy shape because it minimizes the surface area of the liquid. You can lay a sewing needle on its side on the surface of water and it will not immediately fall into the liquid, because the pin cannot break the intermolecular forces of the surface molecules. Surface tension causes liquids to bead up on solid surfaces (think of water beading up on a waxed floor). In addition, surface tension is related to capillary action and to the formation of a meniscus (curved surface) when most liquids are confined to narrow tubes.

10.87   There is an imbalance of forces experienced by the molecules at the surface of a liquid. Minimizing the surface area of the liquid minimizes the energy imbalance. Therefore, a given volume of a substance will assume the shape with the smallest surface area. As a result, liquids have a tendency to form spherical shapes to minimize their energy (see answer to question 85).

10.89   When ice forms on bodies of water such as lakes and rivers, it remains on the surface and provides an insulating layer. Energy loss through evaporation is minimized, and less freezing occurs than would be the case if the ice formed at the bottom or in the body of the lake or river. If ice were more dense than water, it would continually form and sink. Many bodies of water would freeze completely, making the persistence of life in them impossible during the coldest months.

10.91   A crystal is a solid that, on a molecular or atomic level, has regular repeating patterns of atoms. See Figure 10.37 for examples of the types of regular patterns that occur in crystals.

10.93   Knowing the crystal structure helps us understand the macroscopic properties of the solid state (i.e. conductivity, hardness, brittleness).

10.95   Many metals have hexagonal close packed structures. This arrangement is the most common for structures composed of similar-size atoms because it is the most efficient and stable (creates the smallest amount of wasted space). When oranges or apples are stacked, they also assume the hexagonal packing structure (Figure 10.34).

10.97   Some solids, such as tar, are amorphous. Amorphous solids have a tendency to flow because the molecules are not arranged in a crystal lattice. Their structures are highly disorganized.

10.99   We classify solids by the strength and type of attractions between their subunits. This allows us to broadly classify the characteristics of substances.

10.101 Ionic solids are composed of ions and have high melting points. The high melting point is the result of the strong attractive forces between the positive and negative ions in the solid. Ionic solids do not conduct electricity at low temperatures because the ions are immobilized in the crystal lattice. When ionic solids melt, the ions move and, therefore, are conductive (can carry an electric current).

10.103 There are two reasons for the difference in the properties of sodium and sodium chloride. First, the attractive forces between metal atoms are relatively weak. This allows them to be relatively easily separated from each other. In addition, the electrons in metal atoms are delocalized amongst all their neighboring atoms. Atoms can move with relatively little effect on the overall charge distribution in the sample. In ionic solids, cations and anions cannot be separated easily, so particles do not move apart easily. In addition, the electrons are not delocalized, but are localized around the individual ions, causing ionic solids to be brittle.

10.105 Both ionic and metallic solids form crystal lattices. Within a metal crystal the atoms are all the same size, and they tend to pack as close together as possible. For ionic solids, the cations and anions are usually very different in size. The anions usually pack closely in a cubic arrangement, and the cations fit into the holes between anions.
        The attractive forces between metal atoms are relatively weak. This allows them to be relatively easily separated from each other. In addition, the electrons in metal atoms are delocalized amongst all their neighboring atoms. Atoms can move with relatively little effect on the overall charge distribution in the sample. In ionic solids, the electrons are not delocalized, but are localized around the individual ions, causing ionic solids to be brittle. In ionic compounds, displaced ions cannot slip past each other to easily reestablish the crystal lattice.

10.107 Molecular solids are held together by dipole-dipole interactions, hydrogen bonding, and/or London dispersion forces. Ionic solids are held together by ionic attractions (i.e. forces of attraction between oppositely charged ions). Because ionic interactions are much stronger than intermolecular forces, the ionic solids tend to melt at much higher temperatures. Ice is a molecular solid and melts at 100°C, but sodium chloride, an ionic solid, melts at 801°C.

10.109 (a) Diamond is a network solid. This means that the entire structure is held together by a network of covalent bonds.
        (b) Carbon dioxide is a nonpolar molecular compound. In the solid state the molecules are held together by London forces.
        (c) Water molecules in ice are held together primarily by hydrogen bonds and, to a lesser extent, by London forces and dipole-dipole interactions.
        (d) Like diamonds, silicon dioxide exists as a network solid in which the atoms are held together by covalent bonds.

10.111 To answer this question, we must classify these substances by their types. The most difficult to recognize are network solids (see Table 10.3). (a) ionic; (b) molecular; (c) metallic; (d) network ; (e) molecular

10.113 Molecular solids are made up of individual molecules. Both CO_2(b) and H_2O(e) form molecular solids. SiO_2 forms a network solid, $CaCl_2$ forms an ionic solid, and Fe forms a metallic solid.

10.115 The image is of a molecular solid. We can tell it is a molecular solid because it is highly ordered and composed of molecules (Cl_2).

10.117 SiF_4 must be a molecular solid because of its extremely low melting point.

10.119 Ice sublimes below its freezing point. Ice reforms by deposition if it is in equilibrium with the vapor. However, in many cold weather situations, the rate of sublimation is greater than the rate of deposition. (If you've had a climatology/weather class, you will recognize that this condition occurs when the "relative humidity" is less than 100%.) What does this ultimately mean? The rate of ice loss is greater than the rate of deposition, so eventually all the ice "disappears."

10.121 Because all four of the compounds are composed of carbon and hydrogen, the order of increasing boiling points is based on increasing molecular weight. As molecular weight increases, the London dispersion forces experienced by the molecules increases. Dipole-dipole interactions are not important because these molecules are very symmetrical, and therefore, nonpolar. CH_4 ($-162°C$), C_2H_6 ($-88°C$), C_3H_8 ($-42°C$), C_4H_{10} ($0°C$)

10.123 A glass plate, composed of SiO_2, is very polar. Polar substances are attracted to polar substances. Water is attracted to glass, so it spreads out. Mercury is not polar and is attracted more to itself than to the polar glass, so it tends to bead up.

10.125 Ionic solids are formed from ionic compounds (metal-nonmetal combinations). This means that $CaCl_2$ (c) forms an ionic solid. Glass (SiO_2) is a common network solid; both CO_2 and H_2O form molecular solids; Fe forms a metallic solid.

10.127 Substances with high vapor pressures experience weak intermolecular attractions. (a) Cl_2 (weaker London forces); (b) CH_3SH (cannot form hydrogen bonds); (c) NF_3 (smaller dipole)

10.129 Refer to Table 10.3.

Substance (solid type)	Hardness	Electrical Conductivity	Melting Point	Vapor Pressure
NaCl (ionic)	Hard	In molten state	High	Low
SiC (network)	Hardest	Not conductive	Highest	Lowest
SiCl₄ (molecular)	Softest	Not conductive	Lowest	Highest
Fe (metallic)	Soft*	Always conductive	High	Low

*metals are soft relative to network or ionic solids

10.131   solid → liquid   melting
         liquid → solid   freezing
         solid → gas   sublimation
         gas → solid   deposition
         liquid → gas   vaporization
         gas → liquid   condensation

10.133   In the solid state, the Cl_2 molecules are highly ordered. When the solid melts, the molecules leave the crystal lattice and form a liquid which is far less ordered than the solid.

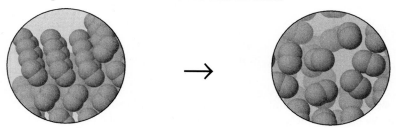

10.135   Substances that have dipole-dipole forces are composed of polar molecules. The molecules CH_4 and CCl_4 are nonpolar so they have no dipole-dipole forces (only London-dispersion forces). The molecules CH_3Cl, CH_2Cl_2, NCl_3 are all asymmetric with polar bonds, so they are polar. The compounds CH_3Cl, CH_2Cl_2, NCl_3 therefore have dipole-dipole forces.

10.137   (a) The 10W motor oil has a higher viscosity so it must consist of larger molecules which interact more strongly by London-dispersion forces.
         (b) Viscosity increases when temperature is lowered. To avoid a motor oil that is too thick and viscous at low temperatures, a lower viscosity oil (thinner oil) is used.

10.139   The substance that boils off first is the one with the lowest boiling point and therefore weaker intermolecular forces. Both methanol and ethanol have hydrogen bonding in their liquid states, but methanol, CH_3OH, has weaker London-dispersion forces than ethanol, CH_3CH_2OH (because methanol is a smaller molecule with fewer electrons than ethanol), so methanol should boil off first.

10.141   The world would definitely be very different and life as we know it would not exist. The most obvious difference is that liquid water would be virtually nonexistent except if some very cold climates exist. If there was ice it would not float because it is the hydrogen bonding that makes it less dense than water. Earth temperatures would vary much more wildly because the high specific heat of liquid water allows the oceans and lakes to absorb and release large amounts of heat providing relatively moderate air temperatures. Because much of the life on earth depends on liquid water for survival, other life forms might exist instead. Another difference would be the structure of DNA and other biomolecules, which are dependent upon hydrogen bonding. The building blocks of life might not even work or they would work very differently.

10.143   These two compounds have the same molar mass and therefore approximately the same London-dispersion force strengths. However, the *cis* compound is polar and the more symmetrical *trans* compound is nonpolar. The polar *cis* compound also has dipole-dipole forces which cause it to have a higher boiling point than the *trans* compound. The boiling point of the *trans* compound is 47.5°C and the boiling point of *cis* compound is 60.3 °C.

Chapter 11 – Solutions

11.1 (a) saturated solution; (b) aqueous solution; (c) molarity; (d) solvent; (e) entropy; (f) solubility; (g) parts per million; (h) miscible; (i) osmosis; (j) colligative property; (k) percent by volume; (l) mass/volume percent

11.3 The solute is the substance that is dissolved in the solution. The solvent is the substance that dissolves the solute. Usually, the solute is the substance that is present in the lesser amount, and the solvent is the substance present in the greater amount in the solution.

Solute	Solvent
(a) sodium chloride (and other salts)	water
(b) carbon	iron
(c) oxygen (O_2)	water

11.5 Strong electrolytes are substances that fully ionize when dissolved in solution. For example, when sodium chloride dissolves in water, it completely dissociates into sodium ions and chloride ions. Strong acids, strong bases, and soluble ionic compounds are all strong electrolytes. Most molecular compounds (with the notable exception of strong acids) do not dissociate in water and are not electrolytes (they are nonelectrolytes).

11.7 The solute appears to be a molecular compound, and is likely a nonelectrolyte. If the solute were an electrolyte, we would expect to see cations and anions formed from the ionization of the substance. However, this solution appears to contain only one type of particle.

11.9 Acids, bases, and soluble salts are electrolytes because they dissociate (or ionize) in water. (a) HBr is an acid that produces $H^+(aq)$ and $Br^-(aq)$ when it dissolves in water. (b) NH_4Cl is a soluble salt that dissociates into $NH_4^+(aq)$ and $Cl^-(aq)$ when it dissolves in water. (c) Butanol is a molecular compound, so it does not form ions but retains its molecular form ($CH_3CH_2CH_2CH_2OH(aq)$) in water. Note: Even though butanol has an –OH group, it is not a base. Most bases are ionic compounds composed of metal ions and hydroxide (OH^-) ions.

11.11 When we count the spheres, we discover a two to one ratio of Cl^- to Mg^{2+} in image A. Magnesium chloride is a soluble salt, so it will produce one Mg^{2+} (the smaller spheres) and two Cl^- (larger spheres) for each formula unit that dissolves. Images B and C are incorrect because they do not show the ions formed by $MgCl_2$ in solution.

11.13 Water-soluble ionic compounds separate into ions when they dissolve in water. Each ion is surrounded with water molecules, a situation we indicate with the (*aq*) designation, which stands for aqueous. When water-soluble molecular substances (other than strong acids) dissolve, water molecules surround them in their entirety, something we indicate with the (*aq*) designation.

(a) $Ca(OH)_2(s) \xrightarrow{H_2O} Ca^{2+}(aq) + 2\,OH^-(aq)$

(b) $N_2(g) \xrightarrow{H_2O} N_2(aq)$

(c) $CH_3OH(l) \xrightarrow{H_2O} CH_3OH(aq)$

11.15 Attractions in both the solute and the solvent must be disrupted before a solute can dissolve in a solvent. This means that because the solute is an ionic compound, the ionic bonds in the crystal lattice must break so the cations and anions can separate, and both the dipole-dipole and hydrogen bonding interactions in the water must be overcome. Finally, ion-dipole interactions form in the solution as the water molecules surround (solvate) the ions in the solution.

11.17 The London dispersion forces in both the solute and the solvent must be overcome. The forces attracting the solvent to the solute in the solution are also London dispersion forces.

11.19 Because the solution cools, we know that energy is absorbed as the solution forms. This means that more energy is required to overcome the attractions between the NH_4^+ and NO_3^- ions in the solute and between the water molecules (the solvent) than is released when the water molecules surround (solvate) the ions to form the solution. Therefore, we can say that the relative strengths of the attractive forces between the solute and solvent are weaker than the attractive forces between particles of the pure substances.

11.21 Energy and entropy changes drive the formation of solutions (Figure 11.11). When the intermolecular forces which act between the solute and solvent are stronger than the intermolecular forces within the pure solvent and solute, the formation of the solution produces energy and solution formation is energetically favorable. Solution formation is also favored when the dissolving process creates more overall disorder than order.

11.23 The intermolecular attractions between water molecules and between oil molecules are not the same because water is polar and cooking oil is nonpolar. Water molecules are attracted to substances that are ionic or polar in nature and oil molecules are attracted to nonpolar substances. Because the intermolecular forces are so dissimilar (they are not "like" in the sense that "like dissolves like") oil and water are immiscible.

11.25 The molecules of both substances (ethanol CH_3CH_2OH; water H_2O) are polar and can form hydrogen bonds. Substances that experience similar intermolecular forces tend to be soluble in each other because their molecules are mutually attracted to one another.

11.27 To apply the "like dissolves like" rule, we must first determine the types of intermolecular forces that exist between molecules or formula units of each substance. Water molecules are polar and form hydrogen bonds. Substances that are polar, form hydrogen bonds, or are soluble ionic compounds are soluble in water. (a) Benzene, a hydrocarbon, is nonpolar so it does not dissolve in water. (b) From its formula we can tell that ethylene glycol has −OH groups, which allow it to form hydrogen bonds. Because ethylene glycol and water are both polar and form hydrogen bonds, ethylene glycol is soluble in water. (c) Potassium iodide is ionic and, based on the solubility rules, dissolves in water. Potassium iodide is soluble in water.

11.29 Compared to other types of interactions, ion-ion interactions are very strong so it takes considerable energy to separate ions from one another. Because nonpolar solvents are not strongly attracted to ions, not enough energy is released, nor entropy gained, through formation of ion-solvent interactions to cause the dissolution process to be favorable.

11.31 The ions in NaCl are attracted by ionic bonding. Water molecules are attracted by dipole-dipole forces and hydrogen bonding. Although disrupting the attractive forces in this solute and solvent requires considerable energy, formation of the ion-dipole interactions between solute and solvent particles releases enough energy to allow the ionic compound to dissolve.

11.33 The solubility of gases, such as O_2, decreases as temperature increases. This happens because the increased kinetic energy of the nonpolar gas molecules in the warmer solution (i.e. $O_2(aq)$) disrupts their weak interactions with the solvent, allowing them to escape more easily into the gas phase.

11.35 Gas solubility decreases as the partial pressure of the gas over the solution decreases. At higher elevations, while the percent composition of the atmosphere is relatively unchanged, the atmospheric pressure is lower. This lower pressure results in lower partial pressures of oxygen and nitrogen. Because the solubility of these gases is directly proportional to their pressures above the solution, the solubilities of oxygen and nitrogen go down as altitude increases.

11.37 Divers get what is known as the bends when they ascend from a dive too rapidly. The rapid decrease in the solubility of their blood gases results in the formation of air bubbles in the divers' blood streams. Besides being very painful, the presence of these air bubbles leads to death. Putting divers in pressurized chambers causes the insoluble gases to re-dissolve into their blood. Then, as the chamber pressure slowly decreases (decompression), the divers can exhale the gases (primarily nitrogen) from their lungs.

11.39 At higher temperatures, the solubility of gases decreases so fewer molecules remain in solution.

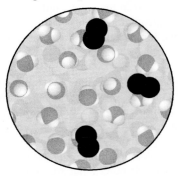

11.41 We use the water solubility of KCl, 34.0 g per 100 g water, as a conversion factor for calculating the amount of KCl that can dissolve into any mass of water at 20°C. We can express this conversion factor in two different ways:

$$\frac{34.0 \text{ g KCl}}{100 \text{ g H}_2\text{O}} \quad \text{and} \quad \frac{100 \text{ g H}_2\text{O}}{34.0 \text{ g KCl}}$$

Because we are given the mass of water, we choose the expression that allows the units 'g H₂O' to cancel and gives us the number of grams of KCl.

$$\text{Mass KCl} = 50.0 \text{ g H}_2\text{O} \times \frac{34.0 \text{ g KCl}}{100 \text{ g H}_2\text{O}} = 17.0 \text{ g KCl}$$

11.43 On a macroscopic level, we can't dissolve more solute in a saturated solution, but we can dissolve more solute in an unsaturated solution. A saturated solution contains the maximum amount of solute that can be dissolved at a particular temperature, while an unsaturated solution contains less than this amount of dissolved solute. On a molecular level, the ions in a saturated solution are in equilibrium with the solid. This means that the solid is dissolving at a rate equal to the rate at which it is forming from the ions in the solution. In an unsaturated solution, none of the solid is present.

11.45 We could add solute a little at a time to the solvent until no more solid dissolves and we see traces of undissolved solute in the solution. At this point, the dissolved solute will be in equilibrium with the undissolved solute in the saturated solution.

11.47 Because the solubility of Ca(OH)₂ is 0.15 g/100g of water, only 0.15 g of the 1.00 g sample will dissolve when we add it to the water (assuming the water temperature remains at 30°C). The resulting solution will be saturated, and 0.85 g of Ca(OH)₂(s) will remain undissolved.

11.49 Percent by mass is:

$$\text{Percent by mass} = \frac{\text{g solute}}{\text{g solution}} \times 100\%$$

Note that the mass of the solution represents the mass of the solute plus the mass of the solvent.

Grams solute = 15.0 g NaCl

Grams solution = 15.0 g NaCl + 90.0 g H₂O = 105.0 g

$$\text{Percent by mass} = \frac{15.0 \text{ g}}{105.0 \text{ g}} \times 100\% = 14.3\%$$

11.51 We can write two different relationships that represent a 15.0% KI solution:

$$\frac{15.0 \text{ g KI}}{100 \text{ g solution}} \quad \text{and} \quad \frac{100 \text{ g solution}}{15.0 \text{ g KI}}$$

Because we are given the mass of the solution, we choose the conversion that allows us to cancel 'g of solution' and gives us 'g KI.'

$$\text{Grams of KI} = 700.0 \text{ g solution} \times \frac{15.0 \text{ g KI}}{100 \text{ g solution}} = 105 \text{ g KI}$$

11.53 To calculate mass percent of acetic acid in a solution, we need to know the mass of acetic acid (54.50 g) and the mass of solution. We use the density of the solution and its volume to calculate the mass of solution, and then calculate the mass percent of acetic acid in the solution.

$$\text{Volume in mL} = 1.000 \text{ L} \times \frac{1000 \text{ mL}}{1 \text{ L}} \times \frac{1.005 \text{ g}}{1 \text{ mL}} = 1005 \text{ g}$$

$$\text{Percent by mass} = \frac{\text{g solute}}{\text{g solution}} \times 100\% = \frac{54.50 \text{ g}}{1005 \text{ g}} \times 100\% = 5.423\%$$

11.55 We often express the concentrations of solutions composed of liquid solutes as percent by volume. We do this primarily for convenience, because we usually measure liquids by volume.

11.57 Percent by volume is:

$$\text{Percent by volume} = \frac{\text{volume solute}}{\text{volume solution}} \times 100\%$$

$$\text{Percent by volume} = \frac{35.0 \text{ mL}}{115.0 \text{ mL}} \times 100\% = 30.4\%$$

11.59 (a) Since the EPA has established the safe drinking level for TCE to be 5 ppb, we need to calculate the concentration of TCE in the sample and see if it falls below this level. If we find that it is below, the water is considered safe to drink. The concentration, in ppb, is calculated using the formula:

$$\text{Parts Per Billion (ppb)} = \frac{\text{mass solute}}{\text{mass solution}} \times 10^9$$

Technically, the mass of solution is the mass of the solvent plus the mass of the solute. However, since the mass of the solute is so small, the solute does not have a significant effect on the total mass:

$m_{\text{solution}} = 200.0 \text{ kg} + 0.43 \text{ mg} = 200.0 \text{ kg} + 0.00000043 \text{ kg}$ (not enough to make a difference!)

$m_{\text{solution}} = 200.0 \text{ kg}$

The mass of TCE needs to be expressed in kg so that the units will cancel properly:

$$m_{\text{TCE}} = 0.43 \text{ mg} \times \frac{10^{-3} \text{ g}}{1 \text{ mg}} \times \frac{1 \text{ kg}}{10^3 \text{ g}} = 4.3 \times 10^{-7} \text{ kg}$$

$$\text{Parts Per Billion (ppb)} = \frac{4.3 \times 10^{-7} \text{ kg}}{200.0 \text{ kg}} \times 10^9 = 2.2 \text{ ppb}$$

(b) The concentration is less than 5 ppb, so the water is safe to drink.

11.61   The molarity of a solution is the number of moles of solvent dissolved in 1 L of the solution. Molarity is a useful conversion between a number of moles of solute and the volume of solution in liters. For a 0.15 M HCl solution, we can write:

$$\frac{0.15 \text{ mol HCl}}{1 \text{ L}} \quad \text{and} \quad \frac{1 \text{ L}}{0.15 \text{ mol HCl}}$$

Because the solution volume in this case is given in milliliters (100.0 mL), we must express it in liters before we can use it in our calculation.

$$\text{Volume in liters} = 100.0 \text{ mL} \times \frac{1 \text{ L}}{1000 \text{ mL}} = 0.100 \text{ L}$$

$$\text{Moles of HCl} = 0.100 \text{ L} \times \frac{0.15 \text{ mol HCl}}{1 \text{ L}} = 0.015 \text{ mol HCl}$$

11.63   To calculate molality we need to know the number of moles of solute and the mass of the solvent (not the solution) in kilograms. We determine the number of moles of solute from the mass (20.5 g NaCl) and molar mass of NaCl (58.44 g/mol).

$$\text{Moles of NaCl} = 20.5 \text{ g NaCl} \times \frac{1 \text{ mol NaCl}}{58.44 \text{ g NaCl}} = 0.351 \text{ mol NaCl}$$

Because we know the mass of the solution, we need to calculate the mass of solvent in kilograms:

Mass of solvent = 166.2 g solution − 20.5 g solute = 145.7 g solvent

$$\text{Mass in kilograms} = 145.7 \text{ g solvent} \times \frac{1 \text{ kg solvent}}{1000 \text{ g solvent}} = 0.1457 \text{ kg solvent}$$

$$\text{Molality of solution} = \frac{\text{moles solute}}{\text{mass of solvent (kg)}} = \frac{0.351 \text{ mol NaCl}}{0.1457 \text{ kg}} = 2.41 \ m$$

11.65   When potassium nitrate dissolves, it produces two ions:

$$KNO_3(s) \rightarrow K^+(aq) + NO_3^-(aq)$$

Because there are twice as many ions as there are KNO_3 formula units, the ion concentration is two times the molarity, or 3.0 M. Going about this another way, we can also calculate the concentration of ions as shown below:

$$\text{Moles of ions per liter} = \frac{1.5 \text{ mol } KNO_3}{1 \text{ L}} \times \frac{2 \text{ mol ions}}{1 \text{ mol } KNO_3} = \frac{3.0 \text{ mol ions}}{1 \text{ L}} = 3.0 \ M \text{ ions}$$

Similarly, the molal concentration of ions is 3.0 m.

$$\text{Moles of ions per kilogram solvent} = \frac{1.5 \text{ mol } KNO_3}{1 \text{ kg}} \times \frac{2 \text{ mol ions}}{1 \text{ mol } KNO_3} = \frac{3.0 \text{ mol ions}}{1 \text{ kg}} = 3.0 \ m \text{ ions}$$

11.67 We determine the percent (or parts per hundred), parts per million (ppm), and parts per billion (ppb) concentrations of a solution as shown:

$$\text{Percent by mass} = \frac{\text{mass solute}}{\text{mass solution}} \times 100\%$$

$$\text{Parts per million} = \frac{\text{mass solute}}{\text{mass solution}} \times 10^6 \text{ ppm} \qquad \text{Parts per billion} = \frac{\text{mass solute}}{\text{mass solution}} \times 10^9 \text{ ppb}$$

The mass units in the equation must be the same. Because the mass of solute (arsenic) is given as 2.4 mg, we convert milligrams to grams before we calculate the concentrations.

$$\text{Mass in grams} = 2.4 \text{ mg} \times \frac{1 \text{ g}}{1000 \text{ mg}} = 2.4 \times 10^{-3} \text{ g}$$

$$\text{Percent by mass} = \frac{2.4 \times 10^{-3} \text{ g}}{250.0 \text{ g}} \times 100\% = 9.6 \times 10^{-4}\%$$

$$\text{Parts per million} = \frac{2.4 \times 10^{-3} \text{ g}}{250.0 \text{ g}} \times 10^6 \text{ ppm} = 9.6 \text{ ppm}$$

$$\text{Parts per billion b} = \frac{2.4 \times 10^{-3} \text{ g}}{250.0 \text{ g}} \times 10^9 \text{ ppb} = 9600 \text{ ppb}$$

11.69 (a) A solution that is 0.0090 ppm in lead contains 0.0090 g of lead in every one million (1×10^6) grams of solution. To express this concentration in ppb, we assume that we have 1×10^6 g of solution:

$$\text{Parts per billion} = \frac{\text{mass solute}}{\text{mass solution}} \times 10^9 \text{ ppb} = \frac{0.0090 \text{ g}}{10^6 \text{ g}} \times 10^9 \text{ ppb} = 9.0 \text{ ppb}$$

The water is safe to drink because its lead concentration, 9.0 ppb, is lower than the EPA established limit of 15.0 ppb.

(b) From part (a) we know that 1×10^6 g of water contains 0.0090 grams of lead. The density of water is 1.0 g/mL, so we can say that this mass of water represents 1×10^6 mL of water. To calculate the number of milligrams of lead per milliliter of water, we first convert the mass of lead to milligrams

$$\text{mg lead} = 0.0090 \text{ g Pb}^{2+} \times \frac{1000 \text{ mg Pb}^{2+}}{1 \text{ g Pb}^{2+}} = 9.0 \text{ mg}$$

$$\text{Concentration in mg/mL} = \frac{9.0 \text{ mg Pb}^{2+}}{10^6 \text{ mL}} = 9.0 \times 10^{-6} \text{ mg/mL}$$

(c) $\text{Mass of lead in 100.0 mL} = 100.0 \text{ mL} \times \frac{9.0 \times 10^{-6} \text{ mg}}{\text{mL}} = 9.0 \times 10^{-4} \text{ mg}$

(d) To calculate the number of moles of lead ion in 100.0 mL of water, we convert the number of milligrams from part (c) to grams, and use the molar mass of lead (207.2 g/mol) to calculate the number of moles of lead in 100.0 mL of water.

$$\text{Moles lead} = 9.0 \times 10^{-4} \text{ mg Pb}^{2+} \times \frac{1 \text{ g Pb}^{2+}}{1000 \text{ mg Pb}^{2+}} \times \frac{1 \text{ mol Pb}^{2+}}{207.2 \text{ g Pb}^{2+}} = 4.3 \times 10^{-9} \text{ mol Pb}^{2+}$$

11.71 (a) For problems involving solution concentrations, we begin with the solution for which we have both a volume and concentration (the Pb^{2+} solution), and use the following problem solving map:

$$mL\ Pb^{2+} \xrightarrow{\ \ 1000\ mL = 1\ L\ \ } L\ Pb^{2+} \xrightarrow{\ \ M\ Pb^{2+}\ \ } mol\ Pb^{2+} \xrightarrow{\ \ mole\ ratio\ \ } mol\ I^- \xrightarrow{\ \ M\ I^-\ \ } L\ I^-$$

$$Volume\ I^- = 100.0\ mL\ Pb^{2+} \times \frac{1\ L\ Pb^{2+}}{1000\ mL\ Pb^{2+}} \times \frac{0.15\ mol\ Pb^{2+}}{1\ L\ Pb^{2+}} \times \frac{2\ mol\ I^-}{1\ mol\ Pb^{2+}} \times \frac{1\ L\ I^-}{1.0\ mol\ I^-} = 0.030\ L\ I^-$$

Because there is one mole of I^- per mole of KI, we need 0.030 L of 1.0 M KI.

(b) We can use a problem solving map similar to the one we used in (a) to determine the mass of PbI_2 that forms.

$$mL\ Pb^{2+} \xrightarrow{\ \ 1000\ mL = 1\ L\ \ } L\ Pb^{2+} \xrightarrow{\ \ M\ Pb^{2+}\ \ } mol\ Pb^{2+} \xrightarrow{\ \ mole\ ratio\ \ } mol\ PbI_2 \xrightarrow{\ \ MM\ PbI_2\ \ } g\ PbI_2$$

$$Mass\ PbI_2 = 100.0\ mL\ Pb^{2+} \times \frac{1\ L\ Pb^{2+}}{1000\ mL\ Pb^{2+}} \frac{0.15\ mol\ Pb^{2+}}{1\ L\ Pb^{2+}} \times \frac{1\ mol\ PbI_2}{1\ mol\ Pb^{2+}} \times \frac{461.0\ g\ PbI_2}{1\ mol\ PbI_2} = 6.9\ g\ PbI_2$$

11.73 The balanced chemical equation is:

$$Na_2CO_3(aq) + CaCl_2(aq) \rightarrow CaCO_3(s) + 2NaCl(aq)$$

We begin by determining the number of moles of $CaCl_2$ present, and then determine the number of moles $NaCO_3$ necessary to react with the $CaCl_2$:

$$mL\ CaCl_2 \xrightarrow{\ \ 1000\ mL = 1\ L\ \ } L\ CaCl_2 \xrightarrow{\ \ M\ CaCl_2\ \ } mol\ CaCl_2 \xrightarrow{\ \ mole\ ratio\ \ } mol\ Na_2CO_3$$

$$Moles\ Na_2CO_3 = 850.0\ mL\ CaCl_2 \times \frac{1\ L\ CaCl_2}{1000\ mL\ CaCl_2} \times \frac{0.35\ mol\ CaCl_2}{1\ L\ CaCl_2} \times \frac{1\ mol\ Na_2CO_3}{1\ mol\ CaCl_2} = 0.30\ mol\ Na_2CO_3$$

11.75 The balanced equation for the complete neutralization of H_2SO_4 is:

$$H_2SO_4(aq) + 2NaOH(aq) \rightarrow Na_2SO_4(aq) + 2H_2O(l)$$

We use the following problem solving map to calculate the volume of NaOH required for the neutralization:

$$L\ H_2SO_4 \xrightarrow{\ \ M\ H_2SO_4\ \ } mol\ H_2SO_4 \xrightarrow{\ \ mole\ ratio\ \ } mol\ NaOH \xrightarrow{\ \ M\ NaOH\ \ } L\ NaOH$$

First, we must express the volume of H_2SO_4 solution in liters:

$$Volume\ in\ Liters = 20.00\ mL\ H_2SO_4 \times \frac{1\ L\ H_2SO_4}{1000\ mL\ H_2SO_4} = 0.02000\ L\ H_2SO_4$$

$$Volume\ NaOH = 0.02000\ L\ H_2SO_4 \times \frac{0.1500\ mol\ H_2SO_4}{1\ L\ H_2SO_4} \times \frac{2\ mol\ NaOH}{1\ mol\ H_2SO_4} \times \frac{1\ L\ NaOH}{0.1050\ mol\ NaOH} = 0.05714\ L$$

We need 0.05714 L or 57.14 mL of 0.1050 M NaOH for the complete neutralization of the acid.

11.77 The balanced equation for the complete neutralization of H_2SO_4 is:

$$H_2SO_4(aq) + 2NaOH(aq) \rightarrow Na_2SO_4(aq) + 2H_2O(l)$$

To calculate the molarity of the H_2SO_4 solution, we first need to determine the number of moles of H_2SO_4 (the volume is given as 20.00 mL) present. Then we can calculate the molarity of the H_2SO_4 solution.

$$\text{L NaOH} \xrightarrow{\;M\ \text{NaOH}\;} \text{mol NaOH} \xrightarrow{\;\text{mole ratio}\;} \text{mol } H_2SO_4$$

The volume of NaOH is 35.77 mL or 0.03577 L.

$$\text{Moles } H_2SO_4 = 0.03577 \text{ L NaOH} \times \frac{0.9854 \text{ mol NaOH}}{1 \text{ L NaOH}} \times \frac{1 \text{ mol } H_2SO_4}{2 \text{ mol NaOH}} = 0.01762 \text{ mol } H_2SO_4$$

To calculate the molarity of H_2SO_4 we divide the number of moles by the volume of the solution, in liters (20.00 mL or 0.02000 L).

$$\text{Molarity} = \frac{0.01762 \text{ mol } H_2SO_4}{0.02000 \text{ L}} = 0.8812 \, M \, H_2SO_4$$

11.79 We base the calculations for each part of this problem on the problem solving map below:

$$\text{L acid} \xrightarrow{\;M\ \text{acid}\;} \text{mol acid} \xrightarrow{\;\text{mole ratio with hydroxide}\;} \text{mol NaOH} \xrightarrow{\;M\ \text{NaOH}\;} \text{L NaOH}$$

We determine the mole ratio from the chemical equation. We can take a shortcut if we realize that each mole of hydrogen ion we neutralize consumes one mole of NaOH. This means that 1 mol of HCl or of HNO_3 reacts with 1 mol of NaOH; and 1 mol of H_3PO_4 reacts with 3 mol of NaOH. We also need to remember to convert volumes to liters at the beginning of the each calculation.

(a) $\text{Volume NaOH} = 0.01000 \text{ L HCl} \times \dfrac{0.1000 \text{ mol HCl}}{1 \text{ L HCl}} \times \dfrac{1 \text{ mol NaOH}}{1 \text{ mol HCl}} \times \dfrac{1 \text{ L NaOH}}{0.1000 \text{ mol NaOH}}$

$= 0.01000 \text{ L NaOH (or 10.00 mL)}$

(b) $\text{Volume NaOH} = 0.01500 \text{ L } HNO_3 \times \dfrac{0.3500 \text{ mol } HNO_3}{1 \text{ L } HNO_3} \times \dfrac{1 \text{ mol NaOH}}{1 \text{ mol } HNO_3} \times \dfrac{1 \text{ L NaOH}}{0.1000 \text{ mol NaOH}}$

$= 0.05250 \text{ L NaOH (or 52.50 mL)}$

(c) $\text{Volume NaOH} = 0.02500 \text{ L } H_3PO_4 \times \dfrac{0.0500 \text{ mol } H_3PO_4}{1 \text{ L } H_3PO_4} \times \dfrac{3 \text{ mol NaOH}}{1 \text{ mol } H_3PO_4} \times \dfrac{1 \text{ L NaOH}}{0.1000 \text{ mol NaOH}}$

$= 0.0375 \text{ L NaOH (or 37.5 mL)}$

11.81 A colligative property is a physical property of a solution that varies with the number of solute particles dissolved in the solvent, and does not depend on the identity of those solute particles.

11.83 Osmosis occurs when two solutions with different solute particle concentrations are separated by a membrane that allows solvent, but not solute, particles to pass through (a semipermeable membrane). The volume of the more concentrated solution increases (often seen as increasing height of solution). Solvent particles pass through the membrane to equalize the concentrations of the solutions on both sides.

11.85 Reverse osmosis occurs when solvent is forced to travel in a direction opposite the direction it would naturally travel during osmosis. For example, if we separate a salt solution and pure water with a semipermeable membrane, osmosis would cause water to flow through the membrane into the salt solution. By applying pressure (osmotic pressure) to the salt solution, we can cause water to pass from the salt solution into the pure water, increasing the concentration of the salt solution.

11.87 The red blood cell will shrink as water moves from the inside of the cell into the aqueous solution (Figure 11.30).

11.89 (a) vapor pressure decreases; (b) boiling point increases; (c) freezing point decreases

11.91 Addition of sugar, a nonvolatile solute, decreases the vapor pressure of the solution. This vapor pressure decrease results in an increase in the boiling point of the solution (the sweet tea).

11.93 We calculate boiling point elevation using the equation: $\Delta T_b = K_b m$ where the boiling point elevation constant, K_b, depends on the solvent. The boiling point elevation constant for water is $0.52°C/m$.
(a) We calculate the difference in boiling points using the boiling point elevation constant for water and the molality of the solute particles. Because $C_6H_{12}O_6$ is a molecular compound, it produces only one solute particle per molecule.

$$\Delta T_b = \frac{0.52\,°C}{m} \times 2.5\,m = 1.3°C$$

The difference in boiling points is 1.3°C.
(b) We calculate the boiling point of the solution by adding the boiling point elevation to the boiling point of the pure solvent (100.0°C for water).

$$T_b = 100.0°C + 1.3°C = 101.3°C$$

11.95 The sodium chloride solution should have the higher boiling point because one formula unit of NaCl produces two solute particles when it dissolves in water ($Na^+(aq)$ and $Cl^-(aq)$). Glucose is a molecular compound, so it produces only one solute particle per molecule.

11.97 Solution B has the highest concentration of dissolved particles, and therefore it has the lowest freezing point. Glucose is a molecular compound, and images B and C show molecular compounds. Sodium chloride produces two ions of different size, as shown in image A.

11.99 Non-acidic molecular compounds produce 1 solute particle per molecule. Ionic compounds produce 1 particle per ion in the formula unit. Sodium chloride, for example, has two ions (Na+ and Cl−) in its formula unit so it produces two solute particles (ions) per formula unit. (a) CH3OH is a molecular compound. The solution contains 2.0 mol of particles per kilogram of solvent. (b) NaCl is an ionic compound that produces 2 ions per formula unit. The solution contains 5.0 mol of ions (2.5 mol of Na+ ions and 2.5 mol of Cl− ions). (c) Al(NO3)3 is an ionic compound that produces 4 mol of ions per formula unit. The solution contains 4.0 mol of ions (1 mol of Al3+ ions and 3 mol of NO3− ions).

11.101 One formula unit of magnesium chloride produces 3 particles when dissolved (one Mg^{2+} ion and two Cl^- ions), whereas one formula unit of $NaNO_3$ produces 2 ions (one Na^+ ion and one NO_3^- ion). The magnitude of colligative properties depends on the number of solute particles in the solution. Therefore, the osmotic pressure of the $MgCl_2$ solution will be greater than that of the $NaNO_3$ solution.

11.103 Vitamin D is nonpolar. We make that assumption based on the "like dissolves like" rule. Substances that are soluble in each other have similar intermolecular forces.

11.105 Energy is required to separate molecules or oppositely charged ions. This is because disruption of a crystal lattice and disruption of intermolecular attractions between water molecules both require energy. The formation of ion-dipole interactions between water and the ions releases energy. In general, energy is released any time two objects that are attracted to one another come together.

11.107 We can express the solubility of NaCl in water (36.3 g per 100 g water) in two ways:

$$\frac{36.3 \text{ g NaCl}}{100 \text{ g H}_2\text{O}} \quad \text{and} \quad \frac{100 \text{ g H}_2\text{O}}{36.3 \text{ g NaCl}}$$

We know the mass of water, and can use the solubility information to cancel 'g H_2O' and give us the desired units, 'g NaCl.'

$$\text{Mass NaCl} = 500.0 \text{ g H}_2\text{O} \times \frac{36.3 \text{ g NaCl}}{100 \text{ g H}_2\text{O}} = 182 \text{ g NaCl}$$

11.109 Molarity expresses the number of moles of solute dissolved in 1 L of solution. To calculate the molarity of a solution, we need to find the number of moles of solute and the solution volume. Assume we have exactly 100 g of the 48.0% HBr solution. This means that we have 48.0 g of HBr. Using the molar mass of HBr (80.91 g/mol), we can calculate the equivalent number of moles of HBr represented by 48.0 g. Also, because we know the solution density, we can determine the volume occupied by 100 g of the solution.

$$\text{Moles HBr} = 48.0 \text{ g HBr} \times \frac{1 \text{ mol HBr}}{80.91 \text{ g HBr}} = 0.593 \text{ mol HBr}$$

$$\text{Volume in mL} = 100 \text{ g HBr} \times \frac{1 \text{ mL}}{1.50 \text{ g HBr}} = 66.7 \text{ mL}$$

$$\text{Volume in L} = 66.7 \text{ mL} \times \frac{1 \text{ L}}{1000 \text{ mL}} = 6.67 \times 10^{-2} \text{ L}$$

$$\text{Molarity} = \frac{0.593 \text{ mol}}{6.67 \times 10^{-2} \text{ L}} = 8.90 \text{ } M$$

11.111 The volume of a solution changes with temperature. Concentration units based only on mass or number of moles will not change with temperature changes. Percent by mass (a) and molality (d) do not change with temperature. This is one of the reasons we use molality in colligative property calculations.

11.113 The densities of dilute aqueous solutions are approximately equal to the density of water. The mass of one liter of water is one kilogram. Therefore, the molarity (number of moles of solute/liter of solution) and molality (number of moles of solute/kilogram of solvent) are very similar.

11.115 The balanced equation for each of these situations is the same:

$$\text{BaCl}_2(aq) + \text{K}_2\text{SO}_4(aq) \rightarrow 2\text{KCl}(aq) + \text{BaSO}_4(s)$$

We are given the volumes and concentrations of two solutions, so we must determine which, if either, of the two solutions is the limiting reactant. Because $BaCl_2$ and K_2SO_4 react in a 1:1 molar ratio, the solution that provides the smaller number of moles of reactant is limiting.
(a) 0.5000 L × 0.100 M $BaCl_2$ = 0.0500 mol; 0.0900 L × 0.500 M K_2SO_4 = 0.0450 mol; K_2SO_4 is limiting.

$$\text{Mass of BaSO}_4 \text{ produced} = 0.0450 \text{ mol K}_2\text{SO}_4 \times \frac{1 \text{ mol BaSO}_4}{1 \text{ mol K}_2\text{SO}_4} \times \frac{233.36 \text{ g BaSO}_4}{1 \text{ mol BaSO}_4} = 10.5 \text{ g BaSO}_4$$

(b) $0.1000 \text{ L} \times 0.100 \, M \, BaCl_2 = 0.0100 \text{ mol}$; $0.1000 \text{ L} \times 0.500 \, M \, K_2SO_4 = 0.0500 \text{ mol}$; $BaCl_2$ is limiting.

$$\text{Mass of } BaSO_4 \text{ produced} = 0.0100 \text{ mol } BaCl_2 \times \frac{1 \text{ mol } BaSO_4}{1 \text{ mol } BaCl_2} \times \frac{233.36 \text{ g } BaSO_4}{1 \text{ mol } BaSO_4} = 2.33 \text{ g } BaSO_4$$

(c) $0.1000 \text{ L} \times 0.100 \, M \, BaCl_2 = 0.0100 \text{ mol}$; $0.5000 \text{ L} \times 0.500 \, M \, K_2SO_4 = 0.250 \text{ mol}$; $BaCl_2$ is limiting

$$\text{Mass of } BaSO_4 \text{ produced} = 0.0100 \text{ mol } BaCl_2 \times \frac{1 \text{ mol } BaSO_4}{1 \text{ mol } BaCl_2} \times \frac{233.36 \text{ g } BaSO_4}{1 \text{ mol } BaSO_4} = 2.33 \text{ g } BaSO_4$$

11.117 The balanced chemical equation is:

$$H_2SO_4(aq) + 2NaHCO_3(s) \rightarrow Na_2SO_4(aq) + 2H_2O(l) + 2CO_2(g)$$

To answer this question, we use the following problems solving map:

$$\text{L } H_2SO_4 \xrightarrow{M \, H_2SO_4} \text{mol } H_2SO_4 \xrightarrow{\text{mole ratio}} \text{mol } NaHCO_3 \xrightarrow{MM \, NaHCO_3} \text{g } NaHCO_3$$

The molar mass of $NaHCO_3$ is 84.01 g/mol.

$$\text{Mass } NaHCO_3 = 2.0 \text{ L } H_2SO_4 \times \frac{6.0 \text{ mol } H_2SO_4}{1 \text{ L } H_2SO_4} \times \frac{2 \text{ mol } NaHCO_3}{1 \text{ mol } H_2SO_4} \times \frac{84.01 \text{ g } NaHCO_3}{1 \text{ mol } NaHCO_3} = 2.0 \times 10^3 \text{ g}$$

11.119 Freezing point depression depends on the number of moles of solute particles in the solution. Solution B has the lowest freezing point (the largest freezing point depression) so it is the solution with the highest NaCl concentration.

11.121 Olive oil is nonpolar and vinegar is a solution of polar acetic acid in polar water. Olive oil molecules are attracted more strongly to one another than to acetic acid or to water, so these molecules will separate out of the mixture. Water and acetic acid molecules are attracted more strongly to one another than to olive oil molecules, so they will also separate out of the mixture.

11.123 Freezing point decreases as the concentration of particles in solution increases. All the solutions listed are strong electrolytes, so the concentration of particles will be the concentration of ions. NaCl has two ions per formula unit, so the concentration of ions is 0.40 m. Na_2SO_4 has three ions per formula unit, so the concentration of ions is 0.30 m. KBr has two ions per formula unit, so the concentration of ions is 0.2 m. VCl_3 has four ions per formula unit, so the concentration of ions is 0.60 m. The order of decreasing freezing point is : 0.10 m KBr, 0.10 m Na_2SO_4, 0.20 m NaCl, 0.15 m VCl_3.

11.125 Lettuce becomes limp when its cells lose water. Placing the lettuce in water causes water molecules to cross the cell membrane into the cells by osmosis since there is a higher concentration of ions inside the cells than outside.

11.127 Barium sulfate is a very insoluble ionic salt (Table 11.1). Since very little of the barium sulfate dissolves, the digestive system contains very few barium ions. In addition, barium sulfate has a high density, so it passes through the digestive system rather quickly.

11.129 Blood cells are found in polar water in the bloodstream. Fatty tissues are largely nonpolar. Following the rule "like-dissolves-like", we can predict that nonpolar vitamin A is more soluble in fatty tissues and therefore more likely to be found there.

Chapter 12 – Reaction Rates and Chemical Equilibrium

12.1   (a) activation energy, E_a; (b) intermediate; (c) homogeneous equilibrium; (d) Le Chatelier's principle; (e) collision theory; (f) equilibrium constant expression

12.3   Some common methods for increasing reaction rates are increasing reactant concentrations, increasing temperature, and providing a catalyst. Increasing concentration increases the collision frequencies. Increasing the reaction temperature increases the energy of the collisions. Adding a catalyst provides an alternate, lower energy pathway for the reactants to take as they form reaction products.

12.5   Increasing the temperature increases the reaction rate by increasing the energy that reactant molecules have when they collide. Increasing the surface area of solid reactants will often increase the reaction rate. This can be done by grinding the solid to a fine powder. Increasing the HCl concentration will also increase the reaction rate.

12.7   Catalysts increase the rates of reactions.

12.9   No, not all reactant collisions are effective. In order for a reactant collision to be effective, the reactants must collide with enough energy and in the proper orientation to form the required activated complex. Because these activated complexes are higher in energy than the reactants, they are not stable. If the activated complex cannot form, the reaction cannot proceed to form products.

12.11   We can get some clue to the proper orientation from the balanced chemical equation for the reaction. Two HI molecules likely collide to produce one H_2 molecule and one I_2 molecule. For the collision to be effective, the iodine atoms and hydrogen atoms must align with one another. (a) This image represents an improper orientation; the H atom strikes the I atom. (b) This image represents the proper orientation; the I atoms and the H atoms are aligned. (c) This image represents an improper orientation; the H atoms are not going to collide with one another.

12.13   An activated complex forms as the result of an effective collision of reactant species.

12.15   Activation energy is the minimum amount of energy required to disrupt critical chemical bonds in the reactants so that the reactants can be converted into reaction products.

12.17   Generally, reactions with higher activation energies proceed more slowly than reactions with lower activation energies. If you've ever run the hurdles, you will know this well. Athletes run more slowly when they cross higher hurdles than when they cross lower hurdles. Similarly, reactions proceed more slowly when the activation energy barrier is higher than when it is lower.

12.19   The energy diagram should look something like the one below. Although the structure of the activated complex can vary, for the reaction to produce products in a single collision, the activated complex should look something like the one in the diagram.

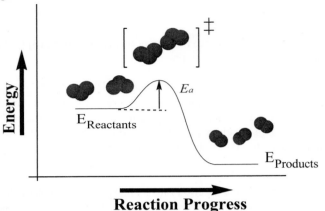

12.21   Increasing reaction temperature increases the kinetic energy and velocity of the reacting particles. Two related factors cause the resulting reaction rate to increase. The most significant factor is the increase in collision energy that results in a larger fraction of productive reactant collisions. A less significant factor is the increase in collision frequency that occurs because the molecules are moving faster.

12.23   Of the three factors, only increasing temperature increases the average kinetic energy of the reactants. Increasing concentration only increases the frequency of collisions and adding a catalyst provides a reaction pathway with a lower activation energy.

12.25   At room temperature, propane and oxygen molecules do not collide with sufficient energy to cause a reaction. However, by providing a spark (heat energy), the kinetic energy of some of the reactant molecules increases enough for their collisions to be productive. Combustion reactions are exothermic; once they begin, they are sustained by the heat they produce.

12.27   Roasts are generally thicker and more massive than steaks. In a sense, for a given mass, steaks have greater surface areas than roasts. Therefore, a greater percentage of the protein in a steak is exposed to the heat than is the case for a roast, so the steak cooks faster.

12.29   A catalyst provides a lower energy route (pathway) to the same products produced by the uncatalyzed reaction. Imagine you are hiking over a high hill with a friend. Your friend decides to go over the top, but you know a shortcut through a tunnel.   You will likely arrive on the other side before your friend does, because you do not have to overcome the same energy barrier. Most catalysts increase the number of effective reactant collisions by holding the reactants in the proper orientation so the reaction can take place more easily.

12.31   Catalytic converters contain palladium and platinum catalysts (Figure 12.11).

12.33   The region of the enzyme where all the "action" takes place is called the active site (Figure 12.13).

12.35   Catalysts are only temporarily changed during the course of a reaction (e.g. when an enzyme binds with its substrate). Once the reaction takes place, the catalyst usually reverts to its original form (e.g. the reaction products leave the active site of the enzyme, leaving the enzyme in its original configuration). If you were to write a balance equation for an enzyme-catalyzed reaction, it might look like:

enzyme + reactant(s) → product(s) + enzyme

As is the case for net-ionic reactions where we cancel the spectator ions, the enzyme does not belong in the net equation above. Typically, chemists write:

$$reactant(s) \xrightarrow{\text{Enzyme}} product(s)$$

12.37   Chlorine atoms in the upper atmosphere act as catalysts for the decomposition of ozone. Because they act catalytically, they remain effective until they diffuse into outer space (which is a slow process).

12.39   We call the temporary substances that form during reactions intermediates.

12.41   (a) In this reaction, NO_3 is an intermediate and there are no catalysts present. Intermediates are produced during the reaction, but consumed later. Catalysts are consumed during the reaction, and then produced later. An additional requirement for catalysts is that they increase the reaction rate. From the reaction mechanism we can see that NO_3 is formed during the first step and consumed during the second step:

$$N_2O_5 \rightleftharpoons \cancel{NO_3} + NO_2$$
$$\underline{\cancel{NO_3} + NO \rightarrow 2NO_2}$$
$$N_2O_5 + NO \rightarrow 3NO_2 \qquad \textit{net reaction}$$

(b) By canceling the intermediates (and catalysts, if present) and adding the reaction steps, we arrive at the equation for the net reaction:

$$N_2O_5 + NO \rightarrow 3NO_2 \qquad \textit{net reaction}$$

12.43   (a) H^+ and Br^- are catalysts, and Br_2 is an intermediate. Intermediates are produced during the reaction, but consumed later. Catalysts are consumed during the reaction and then produced later. An additional requirement for catalysts is that they increase the reaction rate.

$$H_2O_2 + \cancel{2Br^-} + \cancel{2H^+} \rightarrow 2H_2O + \cancel{Br_2}$$
$$\underline{H_2O_2 + \cancel{Br_2} \rightarrow \cancel{2H^+} + O_2 + \cancel{2Br^-}}$$
$$2H_2O_2 \rightarrow 2H_2O + O_2 \qquad \textit{net reaction}$$

(b) $2H_2O_2 \rightarrow 2H_2O + O_2 \qquad \textit{net reaction}$

12.45   When a reaction reaches equilibrium, the reactant and product concentrations stop changing.

12.47   A chemical equilibrium is a state reached by a reaction when the rates of the forward and reverse reactions are equal.

12.49   We can determine that the system has reached equilibrium in the last two panels. In these panels the concentrations of reactants and products are the same, which means that these concentrations are not changing with time.

12.51   For equilibrium to be established all products and reactants must be present. The reaction does not need to have both initially. They can be formed by the reaction in either the forward or reverse direction (a) Yes. Sulfur trioxide will produce SO_2 and O_2 in the forward reaction. (b) No. In order for the reverse reaction to occur, both SO_2 and O_2 must be present. (c) Yes. The reverse reaction can take place and produce SO_3. (d) Yes. The SO_3 will produce O_2 and SO_2, so the reverse reaction is possible.

12.53   We could measure the mass of the bromine liquid in the container, but doing so would be tricky because $Br_2(l)$ is very reactive and volatile. A better option is to measure the $Br_2(g)$ pressure in the space above the liquid in the container. When the $Br_2(g)$ pressure stabilizes, we know that the system is at equilibrium. Also, bromine gas is dark brown. By measuring the amount of light the gas absorbs, we can determine when its concentration stops changing and the system is at equilibrium.

12.55   The position of equilibrium tells us whether a reaction is product-favored or reactant-favored. When an equilibrium mixture contains higher concentrations of products than of reactants, we say that the equilibrium is product-favored. Conversely, if an equilibrium mixture contains higher concentrations of reactants than of products, we say that the equilibrium is reactant-favored.

12.57   We use the coefficients of the reactants and products from the balanced equation as exponents in the equilibrium constant expression.

12.59   The brackets represent the molar concentration of the substance in the brackets. For example, [HCl] represents the molar concentration of HCl.

12.61   For each equilibrium expression, we write the product concentrations in the numerator, and the reactant concentrations in the denominator of the expression. Then, we raise each concentration to the power equal to its stoichiometric coefficient in the balanced chemical equation.

(a) $K_{eq} = \dfrac{[HF]^2}{[H_2][F_2]}$

(b) $K_{eq} = \dfrac{[CS_2][H_2]^4}{[CH_4][H_2S]^2}$

(c) $K_{eq} = \dfrac{[NO_2]^2}{[N_2O_4]}$

12.63   To write a balanced chemical equation from an equilibrium constant expression, we recognize that the reaction products appear in the numerator, and the reactants appear in the denominator of the expression. Each term in the expression is raised to the power of its stoichiometric coefficient in the balanced chemical equation.
(a) $C(g) \rightleftharpoons A(g) + B(g)$
(b) $2A(g) \rightleftharpoons 4B(g) + C(g) + D(g)$

12.65   (a) Mathematically, one equilibrium constant expression is the reciprocal of the other.
(b) $2NO_2(g) \rightleftharpoons 2NO(g) + O_2(g)$
    $2NO(g) + O_2(g) \rightleftharpoons 2NO_2(g)$

12.67   When we reverse the direction of a reaction, the equilibrium constant of the new reaction is the reciprocal of the equilibrium constant of the initial reaction:

$K_{eq} = \dfrac{1}{4.0} = 0.25$

12.69   An equilibrium constant changes only if the reaction temperature changes.

12.71 (a) Before we calculate the equilibrium constant, we write the equilibrium constant expression from the balanced chemical equation. Then, we substitute the molar concentrations of the reactants and products into the expression, as shown below.

$$K_{eq} = \frac{[PCl_3][Cl_2]}{[PCl_5]}$$

$$K_{eq} = \frac{[0.025][0.025]}{[0.20]} = 3.1 \times 10^{-3}$$

Note: Normally, we do not assign units to equilibrium constants.

(b) Because K_{eq} is small (much less than one), we describe the equilibrium position as "reactant favored." This is also apparent when we look at the relative concentrations of the reactants and products in the equilibrium mixture.

12.73 Before we calculate an equilibrium constant, we write the equilibrium expression from the balanced chemical equation. Then, we substitute the molar concentrations of the reactants and products into the expression, as shown below.

(a) $$K_{eq} = \frac{[CO_2][H_2]^4}{[CH_4][H_2O]^2}$$

$$K_{eq} = \frac{[0.00090][0.0036]^4}{[0.049][0.048]^2} = 1.3 \times 10^{-9}$$

(b) The reaction is reactant-favored because K_{eq} is extremely small ($K_{eq} \lll 1$).

12.75 If we substitute the initial reactant and product concentrations into the equilibrium constant expression and the quotient is equal to the equilibrium constant, we can say that the reaction is at equilibrium. Once the reaction reaches equilibrium, the ratio of reactant and product concentrations will not shift unless the system is disturbed (by changing temperature or concentrations).

12.77 By substituting the actual concentrations into the equilibrium expression, we determine whether the reaction will proceed in the forward or reverse direction (to the right or to the left) to reach an equilibrium position.

$$K_{eq} = \frac{[NH_3]^2}{[N_2][H_2]^3} = \frac{[0.050]^2}{[0.050][0.050]^3} = 400$$

Because the quotient (400) is larger than the equilibrium constant (224), we conclude that the reaction is not at equilibrium and will proceed in the reverse direction (to the left) to reach an equilibrium position. By shifting left, the concentration of ammonia will decrease and the N_2 and H_2 concentrations will increase. This shift will continue until the quotient and the equilibrium constant are equal.

12.79 When we take a careful inventory of the four types of molecules in the drawing, we find six of each kind. When we substitute these numbers into the equilibrium constant expression, the quotient is 1:

$$K_{eq} = \frac{[O_2][NO_2]}{[O_3][NO]} = \frac{[6][6]}{[6][6]} = 1$$

Because 1 is smaller than the equilibrium constant (25), we know that the reaction is not at equilibrium and will proceed in the forward direction (to the right) to reach an equilibrium state. As more products form, the numerator will increase and the denominator will decrease, bringing the ratio closer to 25.

12.81   In heterogeneous equilibria, the products and reactants are not all in the same physical state.

12.83   The concentrations of pure liquids and solids do not change during chemical reactions; only the relative amounts of these substances change. The magnitude of the equilibrium constant takes into account the unchanging concentrations of these substances. Gases can also be pure substances, but when gases confined in closed systems react, their concentrations change.

12.85   The equilibrium constant expressions for heterogeneous equilibria do not include terms for solids or liquids (the "pure substances"). Otherwise, the guidelines for writing the equilibrium constant expressions are the same as those for writing homogenous equilibrium constant expressions.

(a) $K_{eq} = \left[NH_3 \right]\left[HCl \right]$

(b) $K_{eq} = \left[Ca^{2+} \right]\left[CO_3^{2-} \right]$

(c) $K_{eq} = \dfrac{\left[H_3O^+ \right]\left[F^- \right]}{\left[HF \right]}$

12.87   The equilibrium constant expression for this reaction is: $K_{eq} = \dfrac{\left[CO_2 \right]}{\left[CO \right]} = 0.67$

We know that $[CO] = 0.40\ M$, so we can calculate the concentration of CO_2:
$[CO_2] = 0.67 \times 0.40\ M = 0.27\ M$

12.89   Le Chatelier's principle describes the way in which reactions that are at equilibrium respond to changes in reaction conditions. If we put stress on a reaction (for example, by changing the temperature, or the concentration of one of the reaction components), the equilibrium responds to counteract that stress. For example, if a concentration of a reactant is increased (the stress), the equilibrium shifts to the right (toward the products) to counteract that stress.

12.91   (a) When we remove CH_4 from the equilibrium system, we disturb the equilibrium. The equilibrium shifts to the left, producing some of the CH_4 as a means of reestablishing equilibrium. (b) As the equilibrium shifts to the left, products are consumed and more reactants form. This means that the CO_2 and H_2 concentrations decrease and the H_2O concentration increases.

12.93   In the following table, the larger arrows (⇑) indicate the stress on the equilibrium mixture, and the smaller arrows indicate the resultant concentration changes as the reaction reaches a new equilibrium position. The central arrow indicates the direction in which the equilibrium shifts:

	NO(g)	SO₃(g)		NO₂	SO₂
(a)	⇑	↓	→	↑	↑
(b)	↓	⇑	→	↑	↑
(c)	↓	↓	→	⇓	↑
(d)	↑	↑	←	⇑	↓

Notice that when a reactant concentration changes, we observe a change of the same type in the product concentrations. For example, in (a) the concentration of a reactant is increased. The equilibrium shifts right, and the product concentrations increase.

12.95   The concentration of lead in drinking water decreases when the equilibrium shifts to the left. (a) Adding PbI_2 has no effect on the position of the chemical equilibrium because the solid concentration does not change when we add more to the reaction mixture. Only substances that appear in the equilibrium constant expression have an effect on the position of the equilibrium. (b) Adding KI decreases the concentration of Pb^{2+} by causing the equilibrium to shift left. (c) Adding $Pb(NO_3)_2$ causes an increase in Pb^{2+} concentration. The Pb^{2+} concentration increase causes the equilibrium to shift left, but it does not shift far enough to completely remove all the added Pb^{2+}.

12.97   We can manipulate the molar concentration of a gas by changing the volume of its container. For a given number of moles of a gas, increasing the container size results in a decrease in molarity; conversely, decreasing the container size results in an increase in molarity. For chemical equilibria involving at least one substance in the gas phase, changes in container volumes produce shifts in the equilibrium positions of the reactions. If the container volume increases, the equilibrium shifts to the side with more moles of gas. Solids, liquids, and aqueous species are not affected by such volume changes. (a) There are more moles of gas on the left (5) than on the right (3). The equilibrium shifts to the left. (b) There is one mole of gas on the right and there are none on the left. The equilibrium will shift to the right. (c) There are the same number of moles of gas on the left and the right (2). The equilibrium does not respond to a volume change.

12.99   (a) No. Volume changes affect only those equilibria that include substances in the gas phase. (b) Yes. Adding water increases the volume of the solution. The concentrations of products decrease, and the equilibrium shifts right to produce more moles of products. The solid is not affected by the addition of the solvent unless all of it dissolves. If the solid all dissolves, the system cannot reach a new equilibrium position.

12.101   Increasing the temperature causes the equilibrium position of an exothermic reaction to shift to the left, and the equilibrium position of an endothermic reaction to shift to the right. (a) left; (b) right

12.103   When an equilibrium shifts to the right, the product concentrations increase and the reactant concentrations decrease. Because the equilibrium constant expression represents "products over reactants," the equilibrium constant increases as a result of these concentration changes.

12.105   (a) Increasing the temperature of an exothermic forward reaction causes the equilibrium to shift to the left. This shift decreases the product/reactant ratio, decreasing the equilibrium constant. (b) Increasing the temperature of an endothermic forward reaction causes the equilibrium to shift to the right. This shift increases the product/reactant ratio, increasing the equilibrium constant.

12.107   The equilibrium constant decreases at higher temperatures, indicating that the equilibrium shifts to the left as temperature increases. This indicates that heat is a product of the forward reaction, so the reaction is exothermic.

12.109   The number of moles of NO increases when the equilibrium shifts to the right. (a) Because water is in the liquid state; removing it does not affect the equilibrium position or the number of moles of NO. (b) There are fewer moles of gas on the right than on the left side of the equation. Decreasing the container volume causes the equilibrium to shift to the right, increasing the number of moles of NO. (c) The reaction is exothermic, so decreasing the temperature (removing heat) causes the equilibrium to shift to the right, increasing the number of moles of NO. (d) Adding oxygen causes the equilibrium to shift to the right, increasing the number of moles of NO. (e) Adding a catalyst has no effect on the equilibrium concentrations of any of the reaction components.

12.111   The light-producing reaction slows down so the reactants are consumed more slowly than is the case when the stick is left out of the freezer.

12.113   The catalyst does not actually lower the activation energy of the pathway taken by the uncatalyzed reaction. Instead, the catalyst facilitates a different reaction mechanism with a lower activation energy. In general, it splits one high activation energy step into two or more lower activation energy steps.

12.115   (a) $K_{eq} = \dfrac{[H_2][I_2]}{[HI]^2}$

(b) $K_{eq} = \dfrac{[0.0275][0.0275]}{[0.195]^2} = 0.0199$

(c) Because the equilibrium constant is small (less than one), the equilibrium position favors the reactants.

12.117 The effect of temperature depends on whether the forward reaction is endothermic or exothermic. If the forward reaction is exothermic, heat behaves as a product; if the forward reaction is endothermic, heat behaves as a reactant.

12.119 The forward reaction must be exothermic because removal of heat (the cold winter temperatures) causes the reaction to shift to the right.

12.121 The simplest treatment is to increase the O_2 concentration in the air the patient breathes. Increasing the O_2 concentration shifts the second equation to the left, displacing CO from HbCO.

12.123 (a) When a reaction system is heated to a higher temperature, average kinetic energy increases, so the energy of each collision should be greater on average.
(b) Increased kinetic energy means increased molecular velocities. When the molecules are moving faster, there are more collisions per unit time, so collision frequency increases.
(c) While the total number of collisions per unit time increases, the fraction of collisions with proper molecular orientation remains the same.
(d) Since the average kinetic energy increases, more molecules will have the minimum energy required for effective collision.
(e) Reaction rate increases because collision frequency increases and a greater fraction of the collisions have sufficient energy for reaction.

12.125 As reactants convert to products, the concentration of reactants decreases. As concentration decreases, there are fewer collisions per unit time and therefore fewer effective collisions, resulting in a lower reaction rate. The rate of the forward reaction decreases until equilibrium is reached, where the rates of the forward and reverse reactions are the same.

12.126 Enzymes are protein molecules that act as catalysts to specific reactions. The function of the enzyme lipase is similar to that of the enzyme sucrase in that it catalyzes the breakdown of larger molecules into smaller molecules during digestion. Sucrase provides a molecular cavity for the sucrose molecule to bind to during reaction, weakening the bonds and lowering the activation energy. The lipase enzyme likely lowers the activation energy in a similar way.

12.127 (a) False. The activation energy for a reaction does not change with temperature. Increasing the temperature increases reaction rate by increasing collision frequency and the number of reactants with sufficient energy (the activation energy) to convert to products.
(b) False. One characteristic of a catalyst is that it is recovered unchanged at the end of a reaction. Reactants in a net balanced equation are the substances that undergo a permanent chemical change.
(c) True. This supports the definition given in the explanation to part (b). When the equations in a mechanism are added together, the catalyst occurs on the reactant and on the product side, canceling out, and not appearing in the net balanced equation.
(d) False. An intermediate is different from a catalyst. An intermediate is generated in an early step of the mechanism and exists temporarily, reacting in a later step. Like a catalyst, the intermediate does not appear in the net balanced equation.

12.129 (a) False. While equal concentrations of reactants and products at equilibrium is theoretically possible, it is not common and is not a definition of equilibrium. When a reaction system reaches a state of equilibrium, the concentrations of reactants and products no longer change, and they are usually very different from one another.

(b) True. The equilibrium constant value is a ratio with product concentrations in the numerator and reactant concentrations in the denominator. When there are much more reactants than products, the ratio will be less than 1.

(c) False. This statement would only be true in a rare circumstance where the equilibrium lies exactly half-way between reactants and products. A system is at equilibrium when the ratio of products to reactants defined by the equilibrium constant expression is equal to the equilibrium constant value for that reaction.

(d) True. This is why we sometimes call chemical equilibrium a *dynamic* equilibrium. Both the forward and reverse processes occur, but at the same rate.

(e) False. The addition of a catalyst does not affect the position of equilibrium. It only changes the rate at which equilibrium is reached.

12.131 (a) The reaction $AgCl(s) \rightleftharpoons Ag^+(aq) + Cl^-(aq)$ is classified as a heterogeneous equilibrium because it involves more than one physical state, solid and aqueous.

(b) By convention, we exclude the concentration of silver chloride from the equilibrium constant expression because it is a solid. The equilibrium constant expression therefore includes only the aqueous ions:
$K_{eq} = [Ag^+][Cl^-]$

(c) Use the equilibrium constant value and your equilibrium constant expression to calculate the concentration of each ion in a saturated solution of silver chloride. The value of K_{eq} given is 1.70×10^{-10}, so we can substitute that into our expression:

$$1.70 \times 10^{-10} = [Ag^+][Cl^-]$$

To determine the concentrations of Ag^+ and Cl^-, we look to see their mole ratio in the balanced equation. They both have a coefficient of 1, so they are produced in a one to one mole ratio and their concentrations must be equal. Setting $x = [Ag^+] = [Cl^-]$ gives:

$$1.70 \times 10^{-10} = (x)^2$$

We solve for x by taking the square root of both sides of the equation:

$$x = 1.30 \times 10^{-5}$$

$$[Ag^+] = [Cl^-] = 1.30 \times 10^{-5} \, M$$

We expect the concentrations of silver and chloride ions to be small because the compound is relatively insoluble.

Chapter 13 – Acids and Bases

13.1   (a) strong base; (b) Brønsted-Lowry theory; (c) conjugate acid; (d) polyprotic acid; (e) acidic solution; (f) weak base; (g) amphoteric substance; (h) ion-product constant of water, K_w; (i) self-ionization

13.3   The significant figures of the log of a number are in the decimal portion of the number. The log of 7.4×10^3 is 3.87 because 3.87 has two decimal places. (a) $\log [10^{-9}] = -9$; (b) $\log [1 \times 10^{-11}] = -11.0$; (c) $\log [7.4 \times 10^3] = 3.87$; (d) $\log [10^5] = 5$; (e) $\log[1] = 0.0$

13.5   The antilog button is often a second function of your log button. The antilog is calculated by raising 10 to the power of the logarithm (i.e. 10^x). The significant figures are found in the decimal places. If a number has two decimal places, the answer should have two significant figures.
        (a) $10^{1.2} = 15.8$ (this should be written as 20 since the answer should only have 1 significant figure).
        (b) $10^{-6.2} = 6.3 \times 10^{-7}$ (6×10^{-7} with significant figures); (c) $10^0 = 1$

13.7   Acids taste sour (but you should never taste anything to see if it's an acid!), are corrosive to many metals, turn blue litmus red, and neutralize bases.

13.9   The list below includes some common "household" bases and their uses:

Formula	Name	Use
$Mg(OH)_2$	magnesium hydroxide	Antacid
$Al(OH)_3$	aluminum hydroxide	Antacid
NH_3	ammonia	Cleaning solutions Smelling Salts
$Ca(OH)_2$	calcium hydroxide	Concrete

13.11   An Arrhenius acid produces hydrogen ions (H^+) when it dissolves in water. An Arrhenius base produces hydroxide ion (OH^-) when it dissolves in water.

13.13   The Brønsted-Lowry theory of acids and bases focuses on the transfer of hydrogen ions from an acid to a base. While Arrhenius acids release hydrogen ions into a solution, Arrhenius bases are metal hydroxides (such as NaOH and KOH) that release OH^- ions into a solution. According to the Brønsted-Lowry theory, a base is any substance (including, but not limited to, OH^-) that can accept a hydrogen ion.

13.15   Acids donate hydrogen ions, and bases accept hydrogen ions. We can recognize an acid because, as a reactant, its formula will contain at least one H atom. As a product, its formula will be missing at least one H atom. Bases accept hydrogen ions, so their formulas gain at least one hydrogen ion during the reaction. a) HCN is an acid. After HCN donates its hydrogen ion, it becomes CN^-. (b) Sulfate, SO_4^{2-}, is a base. After it accepts a hydrogen ion, it becomes HSO_4^-. (c) C_6H_5OH is an acid. After it donates a hydrogen ion, it becomes $C_6H_5O^-$.

13.17   After an acid donates a hydrogen ion, its charge becomes more negative. For example, HNO_2 becomes NO_2^-. The species that remains after an acid donates a hydrogen ion (H^+) is the conjugate base of the acid.
        (a) NO_2^-; (b) F^-; (c) $H_2BO_3^-$

13.19   When a base gains a hydrogen ion, its charge becomes more positive. For example, OH^- accepts H^+ and becomes H_2O. The species that forms after the base gains a hydrogen ion (H^+) is the conjugate acid of the base.
        (a) H_2O; (b) $C_6H_5NH_3^+$; (c) H_2CO_3

13.21   For convenience, when we write the chemical formulas of acids we indicate the acidic hydrogen atom first in the formula. Bases are substances with amine groups ($-NH_2$ in chemical formulas), hydroxide ions, or selected anions in their formulas.
(a) HCl is an acid that donates a hydrogen ion to H_2O, forming H_3O^+.

$$HCl(g) + H_2O(l) \rightarrow H_3O^+(aq) + Cl^-(aq)$$

(b) $HClO_4$ is an acid that donates a hydrogen ion to H_2O, forming H_3O^+.

$$HClO_4(l) + H_2O(l) \rightarrow H_3O^+(aq) + ClO_4^-(aq)$$

(c) $CH_3CO_2^-$ is a base that accepts a hydrogen ion from water to form CH_3COOH.

$$CH_3CO_2^-(aq) + H_2O(l) \rightarrow HCH_3CO_2(aq) + OH^-(aq)$$

13.23   An amphoteric substance can both donate and accept hydrogen ions. Both water (a) and hydrogen sulfite (b) are amphoteric. Sulfate (c) cannot be amphoteric because it lacks a hydrogen ion to donate.

13.25   Carbonic acid has two acidic hydrogen atoms. In polyatomic oxoanions (anions containing oxygen) acidic hydrogen atoms are always connected to oxygen atoms.

13.27   When strong acids dissolve in water, 100% of the dissolved molecules ionize. When weak acids dissolve in water, the percentage of molecules that ionize is much lower (typically, 5% or less). This means that when we dissolve 100 strong acid molecules in water, 100 H_3O^+ ions form, leaving none of the unionized acid molecules. When we dissolve 100 weak acid molecules in water, only a few of them ionize and produce H_3O^+ ions. As a result, most of the weak acid molecules remain unionized in solution. When strong bases dissolve in water, 100% of the dissolved molecules ionize to produce OH^- ions. In contrast, molecules of weak bases remain largely unionized.

13.29   Refer to Table 13.1 & 13.2 for lists of strong acids and bases. Consult Tables 13.3 and 13.4 for lists of commonly-encountered weak acids and weak bases. Practically, it is helpful to learn a short list of strong acids and bases, and then to consider any acids or bases that are not on the list as if they were weak (see (d)). Most commonly-encountered weak bases contain nitrogen, in the form of an amine ($-NH_2$), in their formulas or are conjugate bases of weak acids.

(a) strong acid (Table 13.1); $H_2SO_4(aq) + H_2O(l) \rightarrow HSO_4^-(aq) + H_3O^+(aq)$

(b) strong base (Table 13.2); $Ca(OH)_2(aq) \rightarrow Ca^{2+}(aq) + 2OH^-(aq)$

(c) weak base (conjugate base of a weak acid); $Na_2CO_3(aq) + H_2O(l) \rightleftharpoons HCO_3^-(aq) + 2Na^+(aq) + OH^-(aq)$

(d) weak acid (not on the strong acid list); $H_3C_6H_5O_7(aq) + H_2O(l) \rightleftharpoons H_3O^+(aq) + H_2C_6H_5O_7^-(aq)$

(e) weak base (amine); $C_6H_5NH_2 + H_2O(l) \rightleftharpoons C_6H_5NH_3^+(aq) + OH^-(aq)$

13.31   Image A represents NH_3. NH_3 is a weak base, so it only partially ionizes in water. Therefore, we see only a few hydroxide and ammonium ions in relation to the number of unionized NH_3 molecules present. Image B represents HF. HF is a weak acid, so it only partially ionizes in water. Therefore, we see only a few hydronium ions and fluoride ions in relation to the number of HF molecules in the image. Image C represents HCl. HCl is a strong acid, so it ionizes completely in solution. Therefore, we see none of the original HCl molecules, but only hydronium ions and chloride ions.

13.33   The conjugate base of a strong acid will not act as a base in solution (it is too weak to act as a base). Only the conjugate base of a weak acid will act as a weak base. KBr comes from HBr (strong acid) so it will not act as a base. KF is derived from the weak acid HF so will act as a base in solution.

13.35   Acid A is a stronger acid than acid B. The stronger acid is the one that ionizes to the greater extent.

13.37   Formic acid has a larger acid ionization constant, K_a, than acetic acid $(1.8 \times 10^{-4} > 1.8 \times 10^{-5})$. Therefore, a higher percentage of formic acid molecules than of acetic acid molecules ionize in water. Because the concentrations of the two acid solutions are the same and formic acid experiences a higher degree of ionization, the hydronium ion concentration will be higher in the formic acid solution than in the acetic acid solution.

$$HCO_2H(aq) + H_2O(l) \rightleftharpoons H_3O^+(aq) + CO_2H^-(aq)$$

13.39   A simple rule to remember is that the stronger acid produces the weaker conjugate base. The opposite is also true; a weaker acid makes a stronger conjugate base. Table 13.5 lists the acids HF and HCH_3CO_2 with their acid ionization constants. HCH_3CO_2 is the weaker acid, so its conjugate base will be stronger. A solution of $NaCH_3CO_2$ would be more basic and have a higher OH^- concentration.

13.41   A strong acid solution contains acid molecules that fully ionize in water. Regardless of whether the solution is concentrated or dilute, the acid molecules all ionize. A concentrated acid solution contains many acid molecules in a given volume but the acid molecules are not necessarily ionized to any great extent.

13.43   A polyprotic acid is an acid with more than one acidic hydrogen atom. Diprotic acids have two acidic hydrogen atoms; triprotic acids have three acidic hydrogen atoms. We often refer to acids such as HCl as monoprotic acids.

13.45   Hydrosulfuric acid is a weak acid with two acidic hydrogen atoms. The equations that correspond to the two K_as are:

$$H_2S(aq) + H_2O(l) \rightleftharpoons HS^-(aq) + H_3O^+(aq) \quad K_{a1} = 8.9 \times 10^{-8}$$

$$HS^-(aq) + H_2O(l) \rightleftharpoons S^{2-}(aq) + H_3O^+(aq) \quad K_{a2} = 1.0 \times 10^{-19}$$

13.47   We begin by writing the ionization equations for oxalic acid:

$$H_2C_2O_4(aq) + H_2O(l) \rightleftharpoons HC_2O_4^-(aq) + H_3O^+(aq) \quad K_{a1} = 5.6 \times 10^{-2}$$

$$HC_2O_4^-(aq) + H_2O(l) \rightleftharpoons C_2O_4^{2-}(aq) + H_3O^+(aq) \quad K_{a2} = 1.5 \times 10^{-4}$$

We can find all of the following substances in an oxalic acid solution: $H_2C_2O_4$, $HC_2O_4^-$, $C_2O_4^{2-}$, H_3O^+, and H_2O. Oxalate ions, $C_2O_4^{2-}$, are present in the lowest concentration. Oxalate ions form from hydrogen oxalate, $HC_2O_4^-$. Only a small amount of hydrogen oxalate is formed in the first reaction, so an even smaller amount of oxalate ion, $C_2O_4^{2-}$, is produced by the second reaction.

13.49   Like other equilibrium constants, K_w remains constant so long as the temperature does not change.

13.51   In pure water, $[OH^-] = [H_3O^+] = 1.0 \times 10^{-7} \, M$.

13.53   Refer to 13.11. (a) The solution is basic because the hydronium ion concentration is lower than the hydroxide ion concentration. (b) The solution is acidic because the H_3O^+ concentration is greater than $1.0 \times 10^{-7} \, M$. (c) The solution is basic because the OH^- concentration is greater than $1.0 \times 10^{-7} \, M$.

13.55   To calculate $[H_3O^+]$ from $[OH^-]$, we use the ion-product constant for water $K_w = [H_3O^+][OH^-]$ where $K_w = 1.0 \times 10^{-14}$.

$$[H_3O^+] = \frac{K_w}{[OH^-]}$$

(a) $[H_3O^+] = \dfrac{1.0 \times 10^{-14}}{1.0 \times 10^{-3}\,M} = 1.0 \times 10^{-11}\,M$

Because $[H_3O^+] < 1.0 \times 10^{-7}\,M$, the solution is basic.

(b) $[H_3O^+] = \dfrac{1.0 \times 10^{-14}}{1.0 \times 10^{-11}\,M} = 1.0 \times 10^{-3}\,M$

Because $[H_3O^+] > 1.0 \times 10^{-7}\,M$, the solution is acidic.

(c) $[H_3O^+] = \dfrac{1.0 \times 10^{-14}}{3.2 \times 10^{-8}\,M} = 3.1 \times 10^{-7}\,M$

Because $[H_3O^+] > 1.0 \times 10^{-7}\,M$, the solution is acidic (but only very slightly acidic).

13.57   For solutions of strong, monoprotic acids, the hydronium ion concentration is equal to the acid concentration. To calculate the hydronium ion concentration in a solution of a strong base, we use the ion-product constant of water.
(a) $[H_3O^+] = 0.010\,M$
(b) $[H_3O^+] = 0.020\,M$
(c) $[H_3O^+] = \dfrac{K_w}{[OH^-]} = \dfrac{1.0 \times 10^{-14}}{0.015\,M} = 6.7 \times 10^{-13}\,M$

13.59   Olivia's garden became more acidic. A decrease in pH represents an increase in acidity.

13.61   The pH scale is a logarithmic scale. This means that every 10-fold increase in hydrogen ion concentration is represented by a 1 pH-unit change (i.e. log 10 = 1). If the hydronium ion concentration increases by a factor of 10, the pH drops one unit.

13.63   A 1 M HCl solution has a pH of about 0.0. However to get a pH of −1.0, you would have to have a 10 M solution. Similarly, to get a pH of 15 you would have to have a 10 M NaOH solution. Practically speaking, most solutions of acids and bases are lower in concentration than 1 M so the pH scale typically ranges between 0 and 14. However, the pH values can exceed this range.

13.65   No. A solution is acidic if its pH is less than 7.00, so long as the temperature is 25°C.

13.67   pH is defined as $-\log [H_3O^+]$. A solution with pH less than 7.0 is acidic. A solution with pH greater than 7.0 is basic.
(a) pH = $-\log(1.0 \times 10^{-3}) = 3.00$, acidic
(b) pH = $-\log(1.0 \times 10^{-13}) = 13.00$, basic
(c) pH = $-\log(3.4 \times 10^{-10}) = 9.47$, basic

13.69   There are two ways to calculate pH from $[OH^-]$. Which problem solving map you choose to follow is really a matter of preference. We will use map 1.

Problem solving map 1:

$[OH^-] \xrightarrow{\ pOH = -\log[OH^-]\ } pOH \xrightarrow{\ pH + pOH = 14\ } pH$

Problem solving map 2:

$[OH^-] \xrightarrow{\ K_w = [H_3O^+][OH^-]\ } [H_3O^+] \xrightarrow{\ pH = -\log[H_3O^+]\ } pH$

(a) pH = 10.00, basic

$$pOH = -\log[OH^-] = -\log(1.0 \times 10^{-4}) = 4.00$$

$$pH = 14.00 - 4.00 = 10.00$$

(b) pH = 7.00, neutral

$$pOH = -\log[OH^-] = -\log(1.0 \times 10^{-7}) = 7.00$$

$$pH = 14.00 - 7.00 = 7.00$$

(c) pH = 4.91, acidic

$$pOH = -\log[OH^-] = -\log(8.2 \times 10^{-10}) = 9.09$$

$$pH = 14.00 - 9.09 = 4.91$$

13.71   (a) Because HNO_3 is a strong acid, we assume that the hydronium ion concentration is the same as the acid concentration.
pH = −log(0.010) = 2.00, acidic
(b) $HClO_4$ is a strong acid.
pH = −log(0.020) = 1.70, acidic
(c) NaOH is a strong base, so we assume that the hydroxide ion concentration is equal to the sodium hydroxide concentration.

$$pOH = -\log[OH^-] = -\log(0.015) = 1.82$$
$$pH = 14.00 - 1.82 = 12.18, \text{ basic}$$

13.73   The underlined values are given in the problem. The equations that relate the variables are given below.

$pH = -\log[H_3O^+]$	$[H_3O^+] = 10^{-pH}$	$pH + pOH = 14$
$pOH = -\log[OH^-]$	$[OH^-] = 10^{-pOH}$	$K_w = [H_3O^+][OH^-] = 1.0 \times 10^{-14}$

	$[H_3O^+]$	$[OH^-]$	pH	pOH	Acidic or Basic?
a	$\underline{1.0 \times 10^{-5}}$	1.0×10^{-9}	5.00	9.00	Acidic
b	1.0×10^{-10}	$\underline{1.0 \times 10^{-4}}$	10.00	4.00	Basic
c	1.0×10^{-6}	1.0×10^{-8}	6.00	$\underline{8.00}$	Acidic
d	2.9×10^{-9}	3.5×10^{-6}	$\underline{8.54}$	5.46	Basic
e	1.1×10^{-5}	$\underline{9.0 \times 10^{-10}}$	4.95	9.05	Acidic

13.75   pH = −log(0.0050) = 2.30, pOH = 14 − 2.30 = 11.70. pH plus pOH equals 14.00 at 25°C.

13.77   We determine the H_3O^+ concentration using the relationship: $[H_3O^+] = 10^{-pH}$.
(a) pH = 5.00, acidic (Table 13.7)
$[H_3O^+] = 10^{-5.00} = 1.0 \times 10^{-5}\ M$

(b) pH = 12.00, basic
$[H_3O^+] = 10^{-12.0} = 1.0 \times 10^{-12}\ M$

(c) pH = 5.90, acidic
$[H_3O^+] = 10^{-5.90} = 1.3 \times 10^{-6}\ M$

13.79   We determine the H_3O^+ concentration using the relationship: $[H_3O^+] = 10^{-pH}$.

  (a)   household ammonia, pH = 11.00, basic (Table 13.7)
        $[H_3O^+] = 10^{-11.00} = 1.0 \times 10^{-11}$ M

  (b)   blood, pH = 7.40, basic
        $[H_3O^+] = 10^{-7.40} = 4.0 \times 10^{-8}$ M

  (c)   lime juice, pH = 1.90, acidic
        $[H_3O^+] = 10^{-1.90} = 1.3 \times 10^{-2}$ M

13.81   We determine the OH^- concentration using the relationships: $[OH^-] = 10^{-pOH}$ and pH + pOH = 14.00.

  (a)   gastric juice, pH = 1.00, acidic (Table 13.7)
        pOH = 14.00 − pH = 14.00 − 1.00 = 13.00
        $[OH^-] = 10^{-13.00} = 1.0 \times 10^{-13}$ M

  (b)   magnesium hydroxide, pH = 10.50, basic
        pOH = 14.00 − pH = 14.00 − 10.50 = 3.50
        $[OH^-] = 10^{-3.50} = 3.2 \times 10^{-4}$ M

  (c)   soft drink, pH = 3.60, acidic
        pOH = 14.00 − pH = 14.00 − 3.60 = 10.40
        $[OH^-] = 10^{-10.40} = 4.0 \times 10^{-11}$ M

13.83   Higher. The pH of a 0.010 M solution of a strong acid, which completely ionizes, is 2.0. Acetic acid is a weak acid, so the hydronium ion concentration in a 0.010 M acetic acid solution is less than 0.010 M, causing the pH of that solution to be higher than 2.0 (more basic).

13.85   Sodium hydroxide is a strong base. When NaOH dissolves in water, it dissociates into its ions:

$$NaOH(s) \rightarrow Na^+(aq) + OH^-(aq)$$

This means that the concentrations of NaOH and OH^- are the same. If pH = 13.0, then pOH = 14.0 − 13.0 = 1.0. From the pOH we calculate the hydroxide ion concentration: $[OH^-] = 10^{-1.0} = 0.1$ M
The NaOH concentration is 0.1 M.

13.87   A pH meter typically gives measurements accurate to two decimal places (Figure 13.17). Other methods (pH paper or indicator solutions) generally give lower precision.

13.89   Acid-base indicators are typically colorful compounds that are also weak acids or bases. The colors of the acid and base forms of these compounds are different. Many natural food colors are acid-base indicators (i.e. colors from blueberry, purple cabbage, curry powder, etc.). Litmus is actually an extract from lichens.

13.91   The color range of an indicator indicates the pH range over which the indicator changes color. We can use the indicator hue to estimate the pH of a solution. The range over which an indicator changes color is usually about 1.5 pH units.

13.93   Thymolphthalein is a good choice. In a solution with pH less than about 9.2, thymolphthalein is colorless. Above that pH, thymolphthalein gradually shifts from colorless to blue.

13.95   At pH 3.5, methyl orange should be red, with a small amount of yellow-orange.

13.97   Thymol blue is red in solutions with pH less than 2.5, yellow in solutions with pH between about 2.5 and 8.0, and blue (or with a blue hue) in solutions with pH greater than 8.0. Thymol blue is blue in solutions with pH greater than 9.5.

13.99   A universal indicator is a mixture of several different indicators. The mixture is selected so the indicator undergoes many distinct color changes throughout a wide pH range. Many newer pH papers contain mixtures of several different indicators and we use a color chart to relate the indicator color to a solution pH.

13.101   The color of indicators at different pH ranges is shown in Figure 13.18. Bromocresol green is yellow at low pH (i.e. pH = 1.0) and blue at high pH (i.e. greater than about 4.5). This means the solution would be yellow at the start of the titration and turn blue just past or very near the endpoint. If you are careful, you will observe the pale green color at the endpoint (a combination of yellow and blue).

13.103   A buffered solution contains both a weak acid and its conjugate base (or a weak base and its conjugate acid). As a result, when small amounts of acid or base are added to the solution, the buffer can absorb (react with) the added hydronium ions or hydroxide ions with only a small change in pH. Weak acid solutions cannot absorb any added acid, and weak base solutions cannot absorb any added base.

13.105   The buffer system in blood helps to maintain a pH between 7.35 and 7.45.

13.107   Acetic acid (CH_3CO_2H) is a weak acid. To make a buffered solution, we should add any soluble ionic compound containing $CH_3CO_2^-$, the conjugate base of CH_3CO_2H (e.g., $NaCH_3CO_2$).

13.109   (a) Na_2HPO_4 dissolves in water and produces sodium ions and hydrogen phosphate ions (HPO_4^{2-}). Na_3PO_4 dissolves in water and produces Na^+ and phosphate (PO_4^{3-}) ions. Hydrogen phosphate, HPO_4^{2-}, is the acid, and the conjugate base is phosphate, PO_4^{3-}. The solution pH reflects the free hydronium ion concentration in the solution ("free" means "H_3O^+ that is unreacted").

$$HPO_4^{2-}(aq) + H_2O(l) \rightleftharpoons PO_4^{3-}(aq) + H_3O^+(aq)$$

(b) When we add acid to this buffered solution, hydronium ions (H_3O^+) are produced and the pH momentarily drops (more acidic). The added hydronium quickly reacts with the PO_4^{3-}, shifting the equilibrium to the left to produce more HPO_4^{2-}. Because most of the added hydronium ions are effectively consumed by the PO_4^{3-}, the pH change is not as large as would be the case if the buffer were not present.

13.111   The solutions described in (b) and (c) are buffered. To prepare a buffered solution we need both the *weak* acid and its conjugate base (or *weak* base and its conjugate acid) in approximately equal concentrations. The solution in (b) will contain $\mathbf{HNO_2}$ (weak acid) and $\mathbf{NO_2^-}$, the conjugate base of HNO_2 (potassium ions are spectators):

$$\mathbf{HNO_2}(aq) + H_2O(l) \rightleftharpoons \mathbf{NO_2^-}(aq) + H_3O^+(aq)$$

The solution in (c) contains $\mathbf{CH_3NH_2}$ (weak base) and its conjugate acid, $\mathbf{CH_3NH_3^+}$, (chloride ions are spectators) in the reaction:

$$\mathbf{CH_3NH_2}(aq) + H_2O(l) \rightleftharpoons \mathbf{CH_3NH_3^+}(aq) + OH^-(aq)$$

In (a) NaCl is not the conjugate base of HOCl.

13.113   Carbon dioxide is a reactant in the reaction given:

$$\mathbf{CO_2}(g) + H_2O(l) \rightleftharpoons H_2CO_3(aq) \rightleftharpoons HCO_3^-(aq) + H^+(aq)$$

As the concentration of carbon dioxide increases, the equilibrium of the reaction would shift to the right. This leads to the formation of carbonic acid, $H_2CO_3(aq)$. The ionization of carbonic acid would cause a decrease in the pH of the extracellular fluid.

13.115 There is an unshared electron pair on the nitrogen atom that gives ammonia, NH_3, its basic properties. The carbon atom in methane does not have an unshared electron pair, so it cannot accept a hydrogen ion (proton) in an acid-base reaction.

ammonia   ammonium ion   methane

13.117 When NaF dissolves in water, sodium and fluoride ions are produced. The fluoride ions react with water to produce a weak acid, HF:

$$F^-(aq) + H_2O(l) \rightleftharpoons HF(aq) + OH^-(aq)$$

In general, the conjugate base of a weak acid (F^- in this case) reacts to some extent with water, producing the weak acid and hydroxide ions. The resulting solution is slightly basic.

13.119 No. At 25°C, a neutral solution (e.g. pure water) has pH = pOH = 7.00. If we add any sodium hydroxide to the solution, the hydroxide ion concentration will increase, giving the solution a pH greater than 7.00.

13.121 Although both H_2SO_4 and HNO_3 are strong acids, the pH of the H_2SO_4 solution is lower. Both acids dissociate completely *for the first ionization*. However, H_2SO_4 is a polyprotic acid. The additional hydronium ion produced from the second ionization will make the pH of the sulfuric acid solution lower.

$$H_2SO_4(aq) + H_2O(l) \rightarrow HSO_4^-(aq) + H_3O^+(aq) \qquad \textit{completely ionized}$$

$$HSO_4^-(aq) + H_2O(l) \rightleftharpoons SO_4^{2-}(aq) + H_3O^+(aq) \qquad \textit{partially ionized}$$

13.123 (a) $HOCl(aq) + H_2O(l) \rightleftharpoons OCl^-(aq) + H_3O^+(aq)$
(b) The added HCl reacts with the pool water, producing hydronium ions and chloride ions
($HCl(aq) + H_2O(l) \rightleftharpoons Cl^-(aq) + H_3O^+(aq)$). The added H_3O^+ causes the HOCl/OCl$^-$ equilibrium to shift to the left, decreasing the impact of the added hydronium ion. The HOCl concentration increases and the OCl$^-$ concentration decreases. While the pH of the pool water may not drop below 7.00, *anytime* acid is added to a solution, the solution becomes at least slightly more acidic (pH is lowered).

13.125 Sulfuric acid is a strong polyprotic acid and is completely ionized. This can be illustrated in the series of reactions:

$$H_2SO_4(aq) \rightarrow H^+(aq) + HSO_4^-(aq)$$
$$HSO_4^-(aq) \rightleftharpoons H^+(aq) + SO_4^{2-}(aq)$$

A strong acid completely dissociates in the first reaction. In this case HSO_4^- is formed and all of the H_2SO_4 is consumed. As a result, H_2SO_4 is absent from the image. HSO_4^- is a much weaker acid than sulfuric acid, so very little of it dissociates.

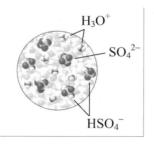

13.127 To prepare a buffer you must have both the acid and its conjugate base present in approximately equal concentrations. The simplest way to do this is to mix equal molar amounts of the acid and the salt of its conjugate base. In this example you might choose to mix formic acid and sodium formate (HCO_2H and $NaCO_2H$). You can also partially neutralize a solution of the acid with sodium hydroxide to get the same result. This works because the product of the reaction is sodium formate:

$$HCO_2H(aq) + NaOH(aq) \rightarrow NaCO_2H(aq) + H_2O(l)$$

13.129 The acidic hydrogens are attached to oxygen atoms:

(a)

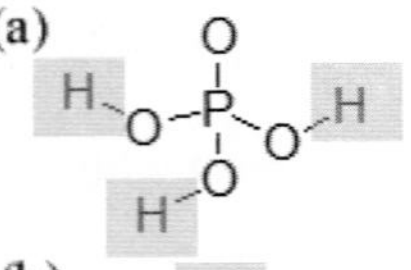

3 acidic hydrogens

(b)

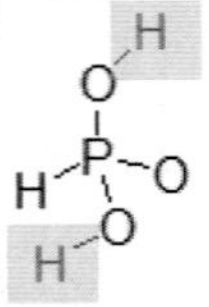

2 acidic hydrogens

(c)

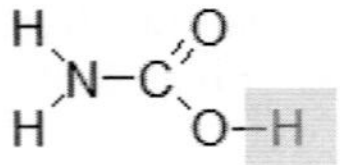

1 acidic hydrogen

(d)

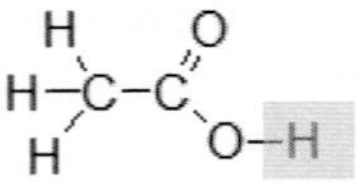

1 acidic hydrogen

13.131 The following statements about acids are incorrect. Change each statement to make it correct.
   (a) **Some** strong acids have H atoms bonded to electronegative oxygen atoms. (Binary acids generally do not contain oxygen atoms.)
   (b) The conjugate base of a strong acid is itself a **weak** base.
   (c) Strong acids **may be** very concentrated acids. (Acid strength is a measure of the extent to which the acid dissociates in water, not a measure of its concentration.)
   (d) Strong acids produce solutions with a **low** pH. (Acidic solutions have a low pH, and basic solutions have a high pH.)

13.133 Under these conditions, $[H^+][OH^-] = 1.0 \times 10^{-13}$. Taking logarithms of both sides give us:
$$\log [H^+] + \log [OH^-] = \log (1.0 \times 10^{-13}) = -13.0$$
If we change the sign of each side of the equation, we get:
$$-\log [H^+] + -\log [OH^-] = -13.0$$
The left-hand terms are just the definitions of pH and pOH:
$$pH + pOH = 13.0$$
For pure water, pH = pOH, so we have:
$$pH + pH = 13.0 \text{ or } pH = 6.5$$

13.135   Arranging the solutions in order of increasing hydroxide ion concentration is the same as arranging them in order of decreasing hydrogen ion concentration or increasing pH. We must first determine the pH of the solutions for which this information was not given.

pH of 0.1 M HNO$_3$ = $-$ log (0.1) = 1.0
pH of 0.5 M HCl = $-$ log (0.5) = 0.3
Pure water has a pH of 7.0

We were given the pH of the other two solutions:
a solution of HNO$_2$ with a pH of 4.0
a buffer solution with a pOH of 11.0

We get the following order of increasing pH and increasing [OH$^-$]:
0.5 M HCl < 0.1 M HNO$_3$ < a solution of HNO$_2$ with a pH of 4.0 < water < a buffer solution with a pOH of 11.0

13.137   For a solution of sodium acetate to become a buffer, some acetic acid (its conjugate acid) must be present. This condition can be accomplished by adding acetic acid. It can also be accomplished by adding hydrochloric acid, which will react with acetate ion to form acetic acid.

Chapter 14 – Oxidation-Reduction Reactions

14.1   (a) electrolytic cell; (b) cathode; (c) oxidation; (d) half-reaction; (e) electrochemistry; (f) electrode; (g) oxidizing agent; (h) oxidation-reduction reaction; (i) corrosion; (j) oxidation number

14.3   An oxidation-reduction reaction (redox reaction) is a reaction where electrons are transferred from one reactant to another.

14.5   A substance is oxidized when its oxidation number increases. For monatomic ions, the charge becomes more positive as a result of losing electrons (i.e. $Zn \rightarrow Zn^{2+} + 2e^-$; the charge on Zn changes from zero to 2+).

14.7   (a) $Mg(s) + Cu(NO_3)_2(aq) \rightarrow Cu(s) + Mg(NO_3)_2(aq)$

         (b)   Magnesium is in its elemental form as a reactant, so we know that its charge is zero. However, as a product, the charge on magnesium is 2+ (because the charge of each nitrate ion is 1−). The charge on magnesium increases by two when it transfers two electrons to copper ($Mg \rightarrow Mg^{2+} + 2e^-$).

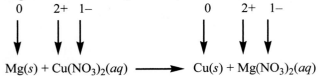

         (c)   As a reactant, the charge on copper is 2+ (because the charge of each nitrate ion is 1−). As a product, copper is in its elemental form so its charge is zero. The charge on copper decreases by 2 when it gains 2 electrons from magnesium ($Cu^{2+} + 2e^- \rightarrow Cu$).

14.9   To begin, we determine the charge on each reactant and product. The species whose charge increases is oxidized, and the species whose charge decreases is reduced. The species that contains the element being reduced is the oxidizing agent, and the species that contains the element being oxidized is the reducing agent.

         (a)   The oxidation number of magnesium increases ($0 \rightarrow 2+$), so Mg is oxidized in this reaction.
         (b)   The oxidation number of tin decreases ($2+ \rightarrow 0$), so Sn^{2+} is reduced in this reaction.
         (c) SnO_4 is the oxidizing agent because it accepts electrons from Mg.
         (d)   Mg is the reducing agent because it gives two electrons to Sn^{2+}.

(e) Sulfate is a spectator ion. Tin ions (red) make contact with the surface of the magnesium metal. Magnesium atoms (gray) transfer two electrons to the tin ions. The tin ions deposit as tin metal.

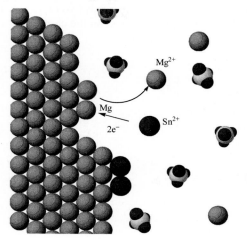

14.11   An oxidation number is part of a bookkeeping method for tracking where electrons go during a chemical reaction. We assign oxidation numbers to atoms in covalent compounds or polyatomic ions as if they were ions, giving them charges established by a set of rules. Then we can determine which substance contains an atom(s) that loses electrons, and which contains an atom(s) that gains electrons. Substances with atoms that lose electrons are oxidized, and the oxidation number of the oxidized atom(s) increases. Substances with atoms that gain electrons are reduced, and the oxidation number of the oxidized atom(s) decreases.

14.13   The oxidation number of atoms in their elemental form is zero. The charges of monatomic ions are their oxidation numbers.
(a) Cr, 0; (b) Br_2, 0; (c) Cr^{3+}, 3+; (d) Br^-, 1−

14.15   In compounds and ions, we generally assign to the most electronegative element in a substance the oxidation number we expect from its position in the periodic table. For example, oxygen is more electronegative than sulfur, so the oxidation number of oxygen is 2− (because an oxygen atom needs two more electrons to fill its valence shell, Rule 6). We determine the oxidation number of sulfur from the chemical formula and net charge (if any) on the species.
For SO_3 we can write:

$$\text{net charge} = \begin{pmatrix} \text{total positive} \\ \text{oxidation numbers} \\ 1 \times S \end{pmatrix} + \begin{pmatrix} \text{total negative} \\ \text{oxidation numbers} \\ 3 \times O \end{pmatrix}$$

SO_3 does not have a charge, so its net charge is zero. From table 14.1 (Rule 6), assign oxygen an oxidation number of oxygen of 2−. We assign an oxidation number to sulfur that gives SO_3 a net charge of zero.
Net charge = 0

Total negative oxidation numbers = $3 \times (-2) = -6$.
This gives us:

$$0 = \begin{pmatrix} \text{total positive} \\ \text{oxidation numbers} \\ S \end{pmatrix} + (-6)$$

The oxidation number of sulfur in SO_3 is 6+. S = 6+, O = 2−

14.17   In compounds and ions, we assign to the most electronegative element in a substance the oxidation number we expect from its position in the periodic table. For example, oxygen is more electronegative than phosphorus, so the oxidation number we assign to oxygen is 2– (because an oxygen atom needs two more electrons to fill its valence shell, Rule 6). We assign an oxidation number to phosphorus based on the chemical formula and net charge (if any) of the substance.

(a) P_4O_{10}

$$\text{net charge} \atop P_4O_{10} = \left(\begin{array}{c} \text{total positive} \\ \text{oxidation numbers} \\ 4 \times N \end{array} \right) + \left(\begin{array}{c} \text{total negative} \\ \text{oxidation numbers} \\ 10 \times O \end{array} \right)$$

Net charge = 0

Total negative oxidation numbers = $10 \times (-2) = -20$

This gives us:

$$0 = \left(\begin{array}{c} \text{total positive} \\ \text{oxidation numbers} \\ 4 \times P \end{array} \right) + (-20)$$

$$P = \frac{20}{4} = 5$$

Each phosphorus atom in P_4O_{10} has an oxidation number of 5+.

(b) P_4O_6

$$\text{net charge} \atop P_4O_6 = \left(\begin{array}{c} \text{total positive} \\ \text{oxidation numbers} \\ 4 \times N \end{array} \right) + \left(\begin{array}{c} \text{total negative} \\ \text{oxidation numbers} \\ 6 \times O \end{array} \right)$$

Net charge = 0

Total negative oxidation numbers = $6 \times (-2) = -12$

This gives us:

$$0 = \left(\begin{array}{c} \text{total positive} \\ \text{oxidation numbers} \\ 4 \times P \end{array} \right) + (-12)$$

$$P = \frac{12}{4} = 3$$

Each phosphorus atom in P_4O_6 has an oxidation number of 3+.

(c) P_4O_8

$$\text{net charge} \atop P_4O_8 = \left(\begin{array}{c} \text{total positive} \\ \text{oxidation numbers} \\ 4 \times N \end{array} \right) + \left(\begin{array}{c} \text{total negative} \\ \text{oxidation numbers} \\ 8 \times O \end{array} \right)$$

Net charge = 0

Total negative oxidation numbers = $8 \times (-2) = -16$

This gives us:

$$0 = \begin{pmatrix} \text{total positive} \\ \text{oxidation numbers} \\ 4 \times P \end{pmatrix} + (-16)$$

$$P = \frac{16}{4} = 4$$

Each phosphorus atom in P_4O_8 has an oxidation number of 4+.

14.19   In compounds and ions, the most electronegative element in a substance generally takes the oxidation number we expect from its position in the periodic table. When a substance contains hydrogen, we must remember (Rule 4) that the oxidation number of hydrogen is 1+, unless it is combined with a metal (where it is 1−).
(a) $AlPO_4$ – Aluminum's oxidation number is 3+ (predicted from the periodic table).

$$\begin{matrix} \text{net charge} = \\ AlPO_4 \end{matrix} \begin{pmatrix} \text{total positive} \\ \text{oxidation numbers} \\ 1 \times Al + 1 \times P \end{pmatrix} + \begin{pmatrix} \text{total negative} \\ \text{oxidation numbers} \\ 4 \times O \end{pmatrix}$$

Net charge = 0

Total negative oxidation numbers = $4 \times (-2) = -8$

This gives us:

$$0 = \begin{pmatrix} \text{total positive} \\ \text{oxidation numbers} \\ 3 + P \end{pmatrix} + (-8)$$

$$0 = 3 + P - 8$$

$$P = 5$$

The oxidation number of P in $AlPO_4$ is 5+.
(b) PF_5 – Fluorine's oxidation number is 1−(predicted from Rule 3 and the periodic table).

$$\begin{matrix} \text{net charge} = \\ PF_5 \end{matrix} \begin{pmatrix} \text{total positive} \\ \text{oxidation numbers} \\ P \end{pmatrix} + \begin{pmatrix} \text{total negative} \\ \text{oxidation numbers} \\ 5 \times F \end{pmatrix}$$

Net charge = 0

Total negative oxidation numbers = $5 \times (-1) = -5$

This gives:

$$0 = \begin{pmatrix} \text{total positive} \\ \text{oxidation numbers} \\ \text{P} \end{pmatrix} + (-5)$$

P = 5

The oxidation number of P in PF_5 is 5+.

(c) H_3PO_4 – The oxidation number of hydrogen is 1+ (predicted from Rule 4 and the periodic table). Because we analyzed the phosphate ion in part (a), we can predict that the oxidation number of phosphorus is 5+.

$$\begin{array}{c} \text{net charge} = \\ H_3PO_4 \end{array} \begin{pmatrix} \text{total positive} \\ \text{oxidation numbers} \\ 3 \times H + 1 \times P \end{pmatrix} + \begin{pmatrix} \text{total negative} \\ \text{oxidation numbers} \\ 4 \times O \end{pmatrix}$$

Net charge = 0

Total negative oxidation numbers = $4 \times (-2) = -8$

This gives us:

$$0 = \begin{pmatrix} \text{total positive} \\ \text{oxidation numbers} \\ 3 + P \end{pmatrix} + (-8)$$

0 = 3 + P – 8

P = 5

The oxidation number of P in H_3PO_4 is 5+.

(d) H_3PO_2 – The oxidation number of hydrogen is 1+ (predicted from Rule 4 and the periodic table).

$$\begin{array}{c} \text{net charge} = \\ H_3PO_2 \end{array} \begin{pmatrix} \text{total positive} \\ \text{oxidation numbers} \\ 3 \times H + 1 \times P \end{pmatrix} + \begin{pmatrix} \text{total negative} \\ \text{oxidation numbers} \\ 2 \times O \end{pmatrix}$$

Net charge = 0

Total negative oxidation numbers = $2 \times (-2) = -4$

This gives:

$$0 = \begin{pmatrix} \text{total positive} \\ \text{oxidation numbers} \\ 3 + P \end{pmatrix} + (-4)$$

0 = 3 + P – 4

P = 1

The oxidation number of P in H_3PO_2 is 1+.

(e) PH_3 – The oxidation number of hydrogen is 1+ (predicted from Rule 4 and the periodic table). In this example, hydrogen has a positive oxidation number, so phosphorus has a negative oxidation number.

$$\text{net charge} \atop PH_3 = \left(\begin{array}{c} \text{total positive} \\ \text{oxidation numbers} \\ 3 \times H \end{array} \right) + \left(\begin{array}{c} \text{total negative} \\ \text{oxidation numbers} \\ P \end{array} \right)$$

Net charge = 0

Total positive oxidation numbers = $3 \times (+1) = -3$

This gives us:

$$0 = \left(\begin{array}{c} \text{total positive} \\ \text{oxidation numbers} \\ 3 \end{array} \right) + (P)$$

$0 = 3 + P$

$P = 3-$

The oxidation number of P in PH_3 is 3–.

(f) H_3PO_3 – The oxidation number of hydrogen is 1+ (predicted from Rule 4 and the periodic table).

$$\text{net charge} \atop H_3PO_3 = \left(\begin{array}{c} \text{total positive} \\ \text{oxidation numbers} \\ 3 \times H + 1 \times P \end{array} \right) + \left(\begin{array}{c} \text{total negative} \\ \text{oxidation numbers} \\ 3 \times O \end{array} \right)$$

Net charge = 0

Total negative oxidation numbers = $3 \times (-2) = -6$

This gives us:

$$0 = \left(\begin{array}{c} \text{total positive} \\ \text{oxidation numbers} \\ 3 + P \end{array} \right) + (-6)$$

$0 = 3 + P - 6$

$P = 3$

The oxidation number of P in H_3PO_3 is 3+.

14.21 SO_4^{2-}

$$\text{net charge} \atop SO_4^{2-} = \left(\begin{array}{c} \text{total positive} \\ \text{oxidation numbers} \\ S \end{array} \right) + \left(\begin{array}{c} \text{total negative} \\ \text{oxidation numbers} \\ 4 \times O \end{array} \right)$$

Net charge = –2

Total negative oxidation numbers = $4 \times (-2) = -8$

This gives us:

$$-2 = \left(\begin{array}{c} \text{total positive} \\ \text{oxidation numbers} \\ \text{S} \end{array} \right) + (-8)$$

$S = 6$

The oxidation numbers of sulfur and oxygen in SO_4^{2-} are 6+ and 2–, respectively.

14.23   (a) BrF_7 – Fluorine's oxidation number is 1– (predicted from Rule 3 and the periodic table).

$$\begin{array}{c} \text{net charge} = \\ BrF_7 \end{array} \left(\begin{array}{c} \text{total positive} \\ \text{oxidation numbers} \\ 1 \times \text{Br} \end{array} \right) + \left(\begin{array}{c} \text{total negative} \\ \text{oxidation numbers} \\ 7 \times \text{F} \end{array} \right)$$

Net charge = 0

Total negative oxidation numbers = $7 \times (-1) = -7$

This gives us:

$$0 = \left(\begin{array}{c} \text{total positive} \\ \text{oxidation numbers} \\ \text{Br} \end{array} \right) + (-7)$$

$Br = 7$

The oxidation number of Br in BrF_7 is 7+.

(b) BrO_3^-

$$\begin{array}{c} \text{net charge} = \\ BrO_3^- \end{array} \left(\begin{array}{c} \text{total positive} \\ \text{oxidation numbers} \\ 1 \times \text{Br} \end{array} \right) + \left(\begin{array}{c} \text{total negative} \\ \text{oxidation numbers} \\ 3 \times \text{O} \end{array} \right)$$

Net charge = -1

Total negative oxidation numbers = $3 \times (-2) = -6$

This gives us:

$$-1 = \left(\begin{array}{c} \text{total positive} \\ \text{oxidation numbers} \\ \text{Br} \end{array} \right) + (-6)$$

$Br = 5$

The oxidation number of Br in BrO_3^- is 5+.

(c) For monatomic ions, the oxidation number and charge are the same.
Br$^-$ has an oxidation number of 1−.

(d) BrCl$_3$ − Chloride has an oxidation number of 1− (predicted from Rule 5c and the periodic table).

$$\text{net charge} \atop \text{BrCl}_3 = \left({\text{total positive} \atop \text{oxidation numbers} \atop 1 \times \text{Br}} \right) + \left({\text{total negative} \atop \text{oxidation numbers} \atop 3 \times \text{Cl}} \right)$$

Net charge = 0

Total negative oxidation numbers = $3 \times (-2) = -3$

This gives us:

$$0 = \left({\text{total positive} \atop \text{oxidation numbers} \atop \text{I}} \right) + (-3)$$

Br = 3

The oxidation number of Br in BrCl$_3$ is 3+.

(e) BrOCl$_3$ − We predict the oxidation numbers of oxygen (2−) and chlorine (1−) from Rules 4 and 6, and the periodic table.

$$\text{net charge} \atop \text{BrOCl}_3 = \left({\text{total positive} \atop \text{oxidation numbers} \atop 1 \times \text{Br}} \right) + \left({\text{total negative} \atop \text{oxidation numbers} \atop 1 \times \text{O} + 3 \times \text{Cl}} \right)$$

Net charge = 0

Total negative oxidation numbers = $1 \times (2-) + 3 \times (-1) = -5$

This gives us:

$$0 = \left({\text{total positive} \atop \text{oxidation numbers} \atop \text{Br}} \right) + (-5)$$

Br = 5

The oxidation number of Br in BrOCl$_3$ is 5+.

14.25   (a) ClO$_2$

$$\text{net charge} \atop \text{ClO}_2 = \left({\text{total positive} \atop \text{oxidation numbers} \atop 1 \times \text{Cl}} \right) + \left({\text{total negative} \atop \text{oxidation numbers} \atop 2 \times \text{O}} \right)$$

Net charge = 0

Total negative oxidation numbers = $2 \times (-2) = -4$

This gives us:

$$0 = \begin{pmatrix} \text{total positive} \\ \text{oxidation numbers} \\ \text{Cl} \end{pmatrix} + (-4)$$

$Cl = 4$

The oxidation numbers of oxygen and chlorine in ClO_2 are 2– and 4+, respectively.

(b) CaF_2 – Because fluorine is the most electronegative element, it's oxidation number is 1–. Because there are two fluoride ions and only one calcium ion, calcium's oxidation number is 2+.

(c) H_2TeO_3 – We predict the oxidation numbers of hydrogen (1+) and oxygen (2–) from Rules 4 and 6, and the periodic table.

$$\begin{matrix} \text{net charge} = \\ H_2TeO_3 \end{matrix} \begin{pmatrix} \text{total positive} \\ \text{oxidation numbers} \\ 2 \times H + 1 \times Te \end{pmatrix} + \begin{pmatrix} \text{total negative} \\ \text{oxidation numbers} \\ 3 \times O \end{pmatrix}$$

Net charge = 0

Total negative oxidation numbers = $3 \times (-2) = -6$

This gives us:

$$0 = \begin{pmatrix} \text{total positive} \\ \text{oxidation numbers} \\ 2 + Te \end{pmatrix} + (-6)$$

$Te = 4$

The oxidation numbers of H, O, and Te in H_2TeO_3 are 1+, 2–, and 4+, respectively.

(d) NaH – In compounds with metals, the oxidation number of hydrogen is 1– (Rule 4). The oxidation number of sodium is 1+.

14.27 (a) NO_2^-

$$\begin{matrix} \text{net charge} = \\ NO_2^- \end{matrix} \begin{pmatrix} \text{total positive} \\ \text{oxidation numbers} \\ N \end{pmatrix} + \begin{pmatrix} \text{total negative} \\ \text{oxidation numbers} \\ 2 \times O \end{pmatrix}$$

Net charge = –1

Total negative oxidation numbers = $2 \times (-2) = -4$

This gives us:

$$-1 = \begin{pmatrix} \text{total positive} \\ \text{oxidation numbers} \\ N \end{pmatrix} + (-4)$$

$N = 3$

The oxidation numbers of N and O in NO_2^- are 3+ and 2–, respectively.

(b) $Cr_2O_7^{2-}$

$$\text{net charge} \atop Cr_2O_7^{2-} = \left(\begin{array}{c} \text{total positive} \\ \text{oxidation numbers} \\ 2 \times Cr \end{array} \right) + \left(\begin{array}{c} \text{total negative} \\ \text{oxidation numbers} \\ 7 \times O \end{array} \right)$$

Net charge = −2

Total negative oxidation numbers = $7 \times (-2) = -14$

This gives us:

$$-2 = \left(\begin{array}{c} \text{total positive} \\ \text{oxidation numbers} \\ 2 \times Cr \end{array} \right) + (-14)$$

Cr = 6

The oxidation numbers of Cr and O in $Cr_2O_7^{2-}$ are 6+ and 2−, respectively.

(c) $AgCl_2^-$ – From the periodic table we can predict that chlorine has an oxidation number of 1−. Because there are two chloride ions in the species, and the overall charge of the species is 1−, we can predict that silver has an oxidation number of 1+. To verify this, we can calculate the overall charge and see if it matches the net charge on the ion.

$$\text{net charge} \atop AgCl_2^- = \left(\begin{array}{c} \text{total positive} \\ \text{oxidation numbers} \\ Ag \end{array} \right) + \left(\begin{array}{c} \text{total negative} \\ \text{oxidation numbers} \\ 2 \times Cl \end{array} \right)$$

Net charge = −1

Total negative oxidation numbers = $2 \times (1-) = -2$

Total positive oxidation numbers = +1

Overall charge = $1 - 2 = -1$

The overall charge and the charge we calculated from the predicted oxidation numbers are the same. The oxidation numbers of Ag and Cl in $AgCl_2^-$ are 1+ and 1−, respectively.

(d) SO_3^{2-}

$$\text{net charge} \atop SO_3^{2-} = \left(\begin{array}{c} \text{total positive} \\ \text{oxidation numbers} \\ 1 \times S \end{array} \right) + \left(\begin{array}{c} \text{total negative} \\ \text{oxidation numbers} \\ 3 \times O \end{array} \right)$$

Net charge = −2

Total negative oxidation numbers = $3 \times (-2) = -6$.

This gives us:

$$-2 = \left(\begin{array}{c} \text{total positive} \\ \text{oxidation numbers} \\ S \end{array} \right) + (-6)$$

S = 4

The oxidation numbers of S and O in SO_3^{2-} are 4+ and 2−, respectively.

(e) CO_3^{2-}

$$\text{net charge} = \begin{pmatrix} \text{total positive} \\ \text{oxidation numbers} \\ 1 \times C \end{pmatrix} + \begin{pmatrix} \text{total negative} \\ \text{oxidation numbers} \\ 3 \times O \end{pmatrix}$$
CO_3^{2-}

Net charge = –2
Total negative oxidation numbers = $3 \times (-2) = -6$.
This gives us:

$$-2 = \begin{pmatrix} \text{total positive} \\ \text{oxidation numbers} \\ C \end{pmatrix} + (-6)$$

$C = 4$
The oxidation numbers of C and O in CO_3^{2-} are 4+ and 2–, respectively.

14.29 To be classified as a redox reaction, the oxidation number of at least one element must change. For each reaction, we calculate, or predict, the oxidation numbers of each element in each species and determine if a change occurs. If an element's oxidation number becomes more positive, that element loses electrons by reducing another substance and, the substance containing that element is the reducing agent. If an element's oxidation number becomes more negative, that element gains electrons by oxidizing another substance, and the substance containing that element is the oxidizing agent.

(a) This is not a redox reaction because none of the oxidation numbers change.

$$\underset{BaCl_2(aq)}{\overset{2+\ \ 1-}{\downarrow\ \downarrow}} + \underset{H_2SO_4(aq)}{\overset{1+\ 6+\ 2-}{\downarrow\ \downarrow\ \downarrow}} \longrightarrow \underset{BaSO_4(s)}{\overset{2+\ 6+\ 2-}{\downarrow\ \downarrow\ \downarrow}} + \underset{2HCl(aq)}{\overset{1+\ \ 1-}{\downarrow\ \downarrow}}$$

(b) The oxidation number of N decreases because N atoms gain electrons from H atoms. N_2 is the oxidizing agent. H_2 is the reducing agent because the oxidation number of H increases as it supplies electrons to nitrogen.

$$\underset{3H_2(g)}{\overset{0}{\downarrow}} + \underset{N_2(g)}{\overset{0}{\downarrow}} \longrightarrow \underset{2NH_3(g)}{\overset{3-\ \ 1+}{\downarrow\ \downarrow}}$$

(c) This is not a redox reaction because none of the oxidation numbers change.

$$\underset{H_2CO_3(aq)}{\overset{1+\ \ 4+\ \ 2-}{\downarrow\ \downarrow\ \downarrow}} \longrightarrow \underset{H_2O(l)}{\overset{1+\ \ 2-}{\downarrow\ \downarrow}} + \underset{CO_2(g)}{\overset{4+\ 2-}{\downarrow\ \downarrow}}$$

(d) This is not a redox reaction because none of the oxidation numbers change.

$$\underset{AgNO_3(aq)}{\overset{1+\ 5+\ 2-}{\downarrow\ \downarrow\ \downarrow}} + \underset{NaCl(aq)}{\overset{1+\ \ 1-}{\downarrow\ \downarrow}} \longrightarrow \underset{AgCl(s)}{\overset{1+\ \ 1-}{\downarrow\ \downarrow}} + \underset{NaNO_3(aq)}{\overset{1+\ 5+\ 2-}{\downarrow\ \downarrow\ \downarrow}}$$

(e) The oxidation number of O decreases because it gains electrons by oxidizing C. O_2 is the oxidizing agent. C_2H_6 is the reducing agent because the oxidation number of C increases when it supplies electrons to O.

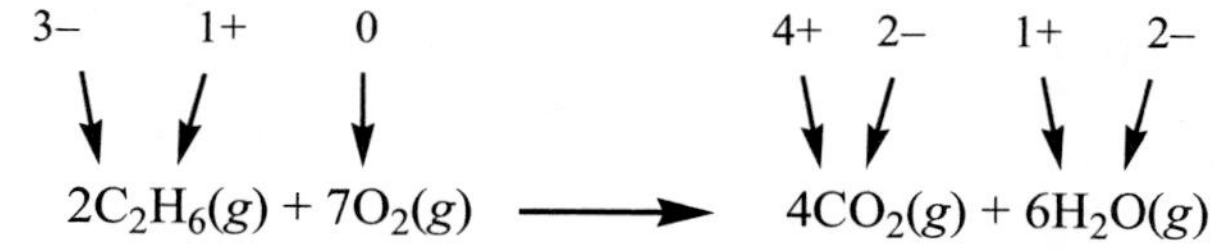

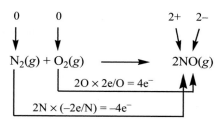

$$2C_2H_6(g) + 7O_2(g) \longrightarrow 4CO_2(g) + 6H_2O(g)$$

14.31   (a)

$$N_2(g) + O_2(g) \longrightarrow 2NO(g)$$

$2O \times 2e/O = 4e^-$

$2N \times (-2e/N) = -4e^-$

(b) When N_2 is transformed into NO, the oxidation number of nitrogen increases by 2. Because there are two nitrogen atoms in N_2, a total of four electrons are transferred. Each N atom loses two electrons.

(c) In any redox reaction, the numbers of electrons gained and lost must be equal. When O_2 reacts with N_2 to produce NO, the oxidation number of oxygen decreases by 2. Because there are two oxygen atoms in O_2, a total of four electrons are transferred from N to O atoms. Each O atom gains two electrons.

14.33   The oxidation numbers of both vanadium and chromium change in this reaction (see below).

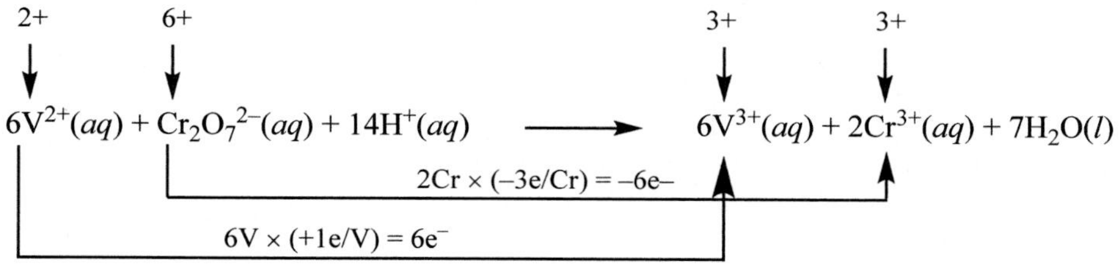

$$6V^{2+}(aq) + Cr_2O_7^{2-}(aq) + 14H^+(aq) \longrightarrow 6V^{3+}(aq) + 2Cr^{3+}(aq) + 7H_2O(l)$$

$2Cr \times (-3e/Cr) = -6e-$

$6V \times (+1e/V) = 6e^-$

(a) V^{2+} is oxidized from 2+ to 3+.
(b) $Cr_2O_7^{2-}$ is reduced (Cr changes from 6+ to 3+).
(c) Because Cr (in $Cr_2O_7^{2-}$) gains electrons, $Cr_2O_7^{2-}$ is the oxidizing agent.
(d) The reducing agent transfers electrons to reduce another element. Because vanadium (in V^{2+}) loses electrons, V^{2+} is the reducing agent.

14.35   (a) In this reaction, I_2 is both the oxidizing agent and the reducing agent. The oxidation number of I atoms in molecular iodine (I_2) is zero. The oxidation number of iodine in I^- is 1−, indicating a gain of one electron. In IO^-, iodine has an oxidation number of 1+, indicating a loss of one electron. In this reaction, 1 electron is transferred between I atoms.

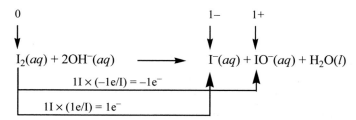

$$I_2(aq) + 2OH^-(aq) \longrightarrow I^-(aq) + IO^-(aq) + H_2O(l)$$

$1I \times (-1e/I) = -1e^-$

$1I \times (1e/I) = 1e^-$

(b) Cr = reducing agent, H$^+$ = oxidizing agent; 2 electrons are transferred.

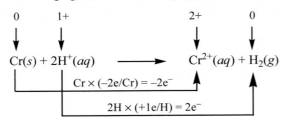

(c) Cr$_2$O$_7^{2-}$ is both the oxidizing and reducing agent, and 12 electrons are transferred. Note that this reaction is different than the one in 14.35(a) where the same element (iodine) is both oxidized and reduced. In the case of Cr$_2$O$_7^{2-}$, all of the chromium atoms are reduced (oxidation number changes from 6+ to 3+), while only 6 oxygen atoms are oxidized (oxidation number changes from 2− to 0). The remaining oxygen atoms contribute to the production of water.

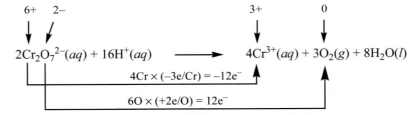

(d) Fe^{3+} = oxidizing agent, Al = reducing agent; 3 electrons are transferred.

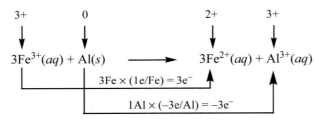

14.37   In an electrochemical cell, oxidation takes place at the anode and reduction takes place at the cathode. Iron is being oxidized, so the iron electrode in the Fe(NO$_3$)$_2$ solution is the anode half-cell. Nickel is being reduced, so the nickel electrode in the Ni(NO$_3$)$_2$ solution is the cathode half-cell.

The half-reaction in the anode half-cell is: Fe(s) → Fe^{2+}(aq) + 2e$^-$

The half-reaction in the cathode half-cell is: Ni^{2+}(aq) + 2e$^-$ → Ni(s)

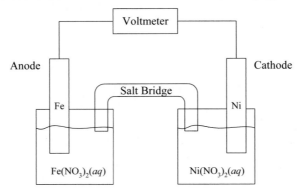

14.39 As the cell runs, iron oxidizes causing the reduction of Ni^{2+} ions. The iron electrode loses mass as Fe atoms on its surface oxidize and become $Fe^{2+}(aq)$. The nickel electrode gains mass as Ni^{2+} ions from the solution are reduced to Ni(*s*) which deposits on the electrode. The nitrate ions are not shown. The concentration of Fe^{2+} in the solution in the anode half-cell increases and the concentration of Ni^{2+} in the solution in the cathode half-cell decreases.

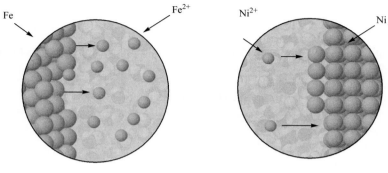

Iron Electrode | Nickel Electrode

14.41 To answer this question, we determine the oxidation numbers of the elements in each substance. The substances whose oxidation numbers change are shown below:

$$\overset{0}{Pb}(s) + \overset{4+}{Pb}O_2(s) + 2H_2SO_4(aq) \longrightarrow 2\overset{2+}{Pb}SO_4(s) + 2H_2O(l)$$

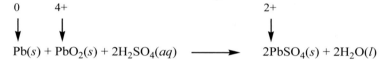

Pb(*s*) is oxidized, and the Pb in $PbO_2(s)$ is reduced. The oxidizing agent is $PbO_2(s)$, and the reducing agent is Pb(*s*).

14.43 Note that in the cases of both half-reactions, there is a material balance (the atoms are balanced) but not a charge balance. To determine which half-reaction represents oxidation and which represents reduction, we begin by assigning oxidation numbers. In the Cd/$Cd(OH)_2$ half-reaction, the oxidation number of cadmium changes from 0 to 2+. This is the oxidation half-reaction because cadmium is oxidized.

$$\overset{0}{Cd}(s) + 2OH^-(aq) \longrightarrow \overset{2+}{Cd}(OH)_2(s)$$

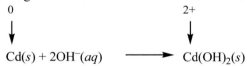

By adding two electrons as reaction products, we balance the overall charge in the half-reaction. Both the product and reactant sides of the equation have a net 2– charge. When a half-reaction is properly balanced, both the atoms and charge balance.

$$\underbrace{Cd(s) + 2OH^-(aq)}_{2-} \longrightarrow \underbrace{Cd(OH)_2(s) + 2e^-}_{2-}$$

In the NiO_2/$Ni(OH)_2$ half-reaction, the oxidation number of nickel changes from 4+ to 2+. This is the reduction half-reaction because nickel is reduced.

$$\overset{4+}{Ni}O_2(s) + 2H_2O(l) \longrightarrow \overset{2+}{Ni}(OH)_2(s) + 2OH^-(aq)$$

By adding two electrons as reactants, we balance the overall charge in the half-reaction. Both the product and reactant sides of the equation have a net 2– charge. When a half-reaction is properly balanced, both the atoms and charge balance.

$$2e^- + NiO_2(s) + 2H_2O(l) \longrightarrow Ni(OH)_2(s) + 2OH^-(aq)$$

$$\underbrace{}_{2-} \qquad \underbrace{}_{2-}$$

14.45   To balance a simple half-reaction, we first balance the atoms and then add electrons to the side with more positive charge to balance the charge.

(a) $Fe^{3+}(aq) \rightarrow Fe(s)$

The atoms are balanced. We balance the charge by adding three electrons to the reactant side of the equation.

$$3e^- + Fe^{3+}(aq) \longrightarrow Fe(s)$$

$$\underbrace{\phantom{3e^- + Fe^{3+}(aq)}}_{0} \qquad \underbrace{}_{0}$$

(b) $Zn(s) \rightarrow Zn^{2+}(aq)$

The atoms are balanced. We balance the charge by adding two electrons to the product side of the equation.

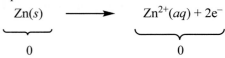

$$Zn(s) \longrightarrow Zn^{2+}(aq) + 2e^-$$

$$\underbrace{}_{0} \qquad \underbrace{\phantom{Zn^{2+}(aq) + 2e^-}}_{0}$$

(c) $Cl^-(aq) \rightarrow Cl_2(g)$

The atoms are not balanced, so we add a coefficient of 2 in front of Cl^-.

$2Cl^-(aq) \rightarrow Cl_2(g)$

Then we balance the charge by adding two electrons to the product side of the equation.

$$2Cl^-(aq) \longrightarrow Cl_2(g) + 2e^-$$

$$\underbrace{}_{2-} \qquad \underbrace{}_{2-}$$

(d) $Fe^{2+}(aq) \rightarrow Fe^{3+}(aq)$

The atoms are balanced. We balance the charge by adding one electron to the product side of the equation.

$$Fe^{2+}(aq) \longrightarrow Fe^{3+}(aq) + e^-$$

$$\underbrace{\phantom{Fe^{2+}(aq)}}_{2+} \qquad \underbrace{\phantom{Fe^{3+}(aq) + e^-}}_{2+}$$

14.47   (a) $Zn(s) + Fe(NO_3)_3(aq) \rightarrow Zn(NO_3)_2(aq) + Fe(s)$

Nitrate ions, NO_3^-, are spectator ions in this reaction so we can eliminate them in the first part of the balancing process, and add them back into the equation after we have balanced the half-reactions. Eliminating the nitrate ions gives us the ionic equation:

$Zn(s) + Fe^{3+}(aq) \rightarrow Zn^{2+}(aq) + Fe(s)$ *skeletal ionic equation*

We can write the oxidation half-reaction as:

$Zn(s) \rightarrow Zn^{2+}(aq)$ *oxidation half-reaction*

The atoms are balanced. We can balance the charge by adding two electrons to the right side of the equation:

$Zn(s) \rightarrow Zn^{2+}(aq) + 2e^-$ *balanced oxidation half-reaction*

The reduction half-reaction is:

$Fe^{3+}(aq) \rightarrow Fe(s)$ *reduction half-reaction*

We can balance the charge by adding three electrons to the reactant side of the equation:

$3e^- + Fe^{3+}(aq) \rightarrow Fe(s)$ *balanced reduction half-reaction*

Then we equalize the number of electrons lost and gained in the two half-reactions by multiplying each by an appropriate coefficient:

$3 \times [Zn(s) \rightarrow Zn^{2+}(aq) + 2e^-]$

$2 \times [3e^- + Fe^{3+}(aq) \rightarrow Fe(s)]$

$3Zn(s) \rightarrow 3Zn^{2+}(aq) + 6e^-$

$6e^- + 2Fe^{3+}(aq) \rightarrow 2Fe(s)$

Next we add the two half-reactions and cancel the six electrons that appear on both sides of the equation.

$$3Zn(s) \rightarrow 3Zn^{2+}(aq) + \cancel{6e^-}$$
$$\underline{\cancel{6e^-} + 2Fe^{3+}(aq) \rightarrow 2Fe(s)}$$
$$3Zn(s) + 2Fe^{3+}(aq) \rightarrow 3Zn^{2+}(aq) + 2Fe(s) \quad \textit{balanced skeletal equation}$$

Finally, we replace the nitrate ions to complete the balanced equation.

$3Zn(s) + 2Fe(NO_3)_3(aq) \rightarrow 3Zn(NO_3)_2(aq) + 2Fe(s)$ *balanced*

(b) $Mn(s) + HCl(aq) \rightarrow MnCl_2(aq) + H_2(g)$

Chloride ions, Cl^-, are spectator ions in this reaction so we can eliminate them in the first part of the balancing process, and add them back into the equation after we have balanced the half-reactions. Eliminating the chloride ions gives us the ionic equation:

$Mn(s) + H^+(aq) \rightarrow Mn^{2+}(aq) + H_2(g)$ *skeletal ionic equation*

The oxidation half-reaction is:

$Mn(s) \rightarrow Mn^{2+}(aq)$ *oxidation half-reaction*

The atoms are balanced, and we can balance the charge by adding two electrons to the right side of the equation:

$Mn(s) \rightarrow Mn^{2+}(aq) + 2e^-$ *balanced oxidation half-reaction*

The reduction half-reaction is:

$H^+(aq) \rightarrow H_2(g)$ *reduction half-reaction*

First, we balance the atoms by adding a coefficient of 2 in front of H^+. Next we balance the charge by adding two electrons to the reactant side of the equation.

$2e^- + 2H^+(aq) \rightarrow H_2(g)$ *balanced reduction half-reaction*

The electrons exchanged between the two half-reactions are already balanced, so we add the two half-reactions and cancel the two electrons that appear on both sides of the equation.

$$Mn(s) \rightarrow Mn^{2+}(aq) + \cancel{2e^-}$$
$$\underline{\cancel{2e^-} + 2H^+(aq) \rightarrow H_2(g)}$$
$$Mn(s) + 2H^+(aq) \rightarrow Mn^{2+}(aq) + H_2(g) \quad \textit{balanced skeletal equation}$$

Finally, we replace the chloride ions to complete the balanced equation.

$$Mn(s) + 2HCl(aq) \rightarrow MnCl_2\,(aq) + H_2(g) \qquad \textit{balanced}$$

14.49   From the problem we can write:

$$Fe_2(SO_4)_3(aq) + KI(aq) \rightarrow FeSO_4(aq) + K_2SO_4(aq) + I_2(aq)$$

Sulfate ions and potassium ions are spectator ions in this reaction, so we can eliminate them when we balance the half-reactions. The oxidation number of Fe in $Fe_2(SO_4)_3$ is 3+.

$$Fe^{3+}(aq) + I^-(aq) \rightarrow Fe^{2+}(aq) + I_2(aq) \quad \textit{skeletal ionic equation}$$

The oxidation half-reaction is:

$$I^-(aq) \rightarrow I_2(aq) \qquad\qquad\qquad \textit{oxidation half-reaction}$$

First, we balance the atoms by adding a coefficient of 2 in front of I^-. Next, we balance the charge by adding two electrons to the right side of the equation.

$$2I^-(aq) \rightarrow I_2(aq) + 2e^- \qquad\qquad \textit{balanced oxidation half-reaction}$$

The reduction half-reaction is:

$$Fe^{3+}(aq) \rightarrow Fe^{2+}(aq) \qquad\qquad \textit{reduction half-reaction}$$

The atoms are balanced, so we can balance the charge by adding one electron to the reactant side of the equation.

$$e^- + Fe^{3+}(aq) \rightarrow Fe^{2+}(aq) \qquad\qquad \textit{balanced reduction half-reaction}$$

To balance the electrons between the two half-reactions we multiply each half-reaction by the appropriate coefficient:

$$1 \times [2I^-(aq) \rightarrow I_2(aq) + 2e^-]$$

$$2 \times [e^- + Fe^{3+}(aq) \rightarrow Fe^{2+}(aq)]$$

$$2I^-(aq) \rightarrow I_2(aq) + 2e^-$$

$$2e^- + 2Fe^{3+}(aq) \rightarrow 2Fe^{2+}(aq)$$

The electrons between the two half-reactions are already balanced, so we add the two half-reactions and cancel the two electrons that appear on both sides of the equation.

$$2I^-(aq) \qquad\qquad \rightarrow I_2(aq) + \cancel{2e^-}$$
$$\underline{\cancel{2e^-} + 2Fe^{3+}(aq) \quad \rightarrow 2Fe^{2+}(aq)}$$
$$2Fe^{3+}(aq) + 2I^-(aq) \quad \rightarrow 2Fe^{2+}(aq) + I_2(aq) \qquad \textit{balanced skeletal equation}$$

Finally, we replace the sulfate and potassium ions to complete the balanced equation. Because the formula for $Fe_2(SO_4)_3$ includes two iron ions, we incorporate both Fe^{3+} from the balanced skeletal equation into that chemical formula.

$$Fe_2(SO_4)_3(aq) + 2KI(aq) \rightarrow 2FeSO_4(aq) + K_2SO_4(aq) + I_2(aq) \qquad\qquad \textit{balanced}$$

14.51   For simple half-reactions (see part (a) below) we determine the oxidation numbers of the atoms and add the number of electrons needed to balance the charge. To balance more complicated half-reactions (in acidic conditions) we follow the steps outlined below. Because these reactions occur in acidic solution, we add H^+ to balance the charge.

(a) $Ba(s) \rightarrow Ba^{2+}(aq)$

We balance the charge by adding two electrons to the product side:

$$Ba(s) \quad \rightarrow \quad Ba^{2+}(aq) + 2e^-$$

$$\underbrace{\qquad}_{0} \qquad \underbrace{\qquad}_{0}$$

(b) $HNO_2(aq) \rightarrow NO(g)$

$$\underset{HNO_2(aq)}{\overset{3+}{\downarrow}} \quad \rightarrow \quad \underset{NO(g)}{\overset{2+}{\downarrow}}$$ *determine oxidation numbers*

$e^- + HNO_2(aq) \rightarrow NO(g)$ *add e^- to balance oxidation numbers*

$e^- + H^+(aq) + HNO_2(aq) \rightarrow NO(g)$ *add H^+ to balance charge*

$e^- + H^+(aq) + HNO_2(aq) \rightarrow NO(g) + H_2O$ *add H_2O to balance atoms*

(c) $H_2O_2(aq) \rightarrow H_2O(l)$

$$\underset{H_2O_2(aq)}{\overset{1-}{\downarrow}} \quad \rightarrow \quad \underset{H_2O(l)}{\overset{2-}{\downarrow}}$$ *determine oxidation numbers*

$H_2O_2(aq) \rightarrow 2H_2O(l)$ *balance atoms being oxidized or reduced*

$2e^- + H_2O_2(aq) \rightarrow 2H_2O(l)$ *add e^- to balance oxidation numbers*

$2e^- + 2H^+(aq) + H_2O_2(aq) \rightarrow 2H_2O(l)$ *add H^+ to balance charge, equation balanced*

(d) $Cr^{3+}(aq) \rightarrow Cr_2O_7^{2-}(aq)$

$$\underset{Cr^{3+}}{\overset{3+}{\downarrow}} \quad \rightarrow \quad \underset{Cr_2O_7^{2-}}{\overset{6+}{\downarrow}}$$ *determine oxidation numbers*

$2Cr^{3+}(aq) \rightarrow Cr_2O_7^{2-}(aq)$ *balance atoms being oxidized or reduced*

$2Cr^{3+}(aq) \rightarrow Cr_2O_7^{2-}(aq) + 6e^-$ *add e^- to balance oxidation numbers*

$2Cr^{3+}(aq) \rightarrow Cr_2O_7^{2-}(aq) + 14H^+ + 6e^-$ *add H^+ to balance charge*

$2Cr^{3+}(aq) + 7H_2O(l) \rightarrow Cr_2O_7^{2-}(aq) + 14H^+ + 6e^-$ *add H_2O to balance atoms*

14.53 To balance more complicated half-reactions that occur in basic solutions, we follow the steps outlined below. Because these reactions occur in basic solutions, we add OH⁻ to balance the charge.

(a) $La(s) \rightarrow La(OH)_3(s)$

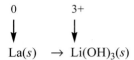

$La(s) \rightarrow Li(OH)_3(s)$	*determine oxidation numbers*
$La(s) \rightarrow La(OH)_3(s) + 3e^-$	*add e^- to balance oxidation numbers*
$3OH^-(aq) + La(s) \rightarrow La(OH)_3(s) + 3e^-$	*add OH^- to balance charge*

(b) $NO_3^-(aq) \rightarrow NO_2^-(aq)$

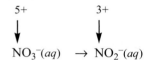

$NO_3^-(aq) \rightarrow NO_2^-(aq)$	*determine oxidation numbers*
$2e^- + NO_3^-(aq) \rightarrow NO_2^-(aq)$	*add e^- to balance oxidation numbers*
$2e^- + NO_3^-(aq) \rightarrow NO_2^-(aq) + 2OH^-(aq)$	*add OH^- to balance charge*
$2e^- + H_2O(l) + NO_3^-(aq) \rightarrow NO_2^-(aq) + 2OH^-(aq)$	*add H_2O to balance atoms*

(c) $H_2O_2(aq) \rightarrow O_2(g)$

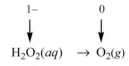

$H_2O_2(aq) \rightarrow O_2(g)$	*determine oxidation numbers*
$H_2O_2(aq) \rightarrow O_2(g) + 2e^-$	*add e^- to balance oxidation numbers*
$H_2O_2(aq) + 2OH^-(aq) \rightarrow O_2(g) + 2e^-$	*add OH^- to balance charge*
$H_2O_2(aq) + 2OH^-(aq) \rightarrow O_2(g) + 2e^- + 2H_2O(l)$	*add H_2O to balance atoms*

(d) $Cl_2O_7(aq) \rightarrow 2ClO_2^-(aq)$

$$7+ \qquad\qquad 3+$$

$$\downarrow \qquad\qquad\quad \downarrow$$

$Cl_2O_7(aq) \rightarrow ClO_2^-(aq)$	*determine oxidation numbers*
$Cl_2O_7(aq) \rightarrow 2ClO_2^-(aq)$	*balance atoms being oxidized or reduced*
$8e^- + Cl_2O_7(aq) \rightarrow 2ClO_2^-(aq)$	*add e^- to balance oxidation numbers*
$8e^- + Cl_2O_7(aq) \rightarrow 2ClO_2^-(aq) + 6OH^-(aq)$	*add OH^- to balance charge*
$8e^- + 3H_2O(l) + Cl_2O_7(aq) \rightarrow 2ClO_2^-(aq) + 6OH^-(aq)$	*add H_2O to balance atoms*

14.55 (a) $H_2S(aq) + Cr_2O_7^{2-}(aq) \rightarrow S(s) + Cr^{3+}(aq)$

$$\underset{\downarrow}{\overset{2-}{}} \qquad \underset{\downarrow}{\overset{6+}{}} \qquad\qquad \underset{\downarrow}{\overset{0}{}} \qquad \underset{\downarrow}{\overset{3+}{}}$$

$H_2S(aq) + Cr_2O_7^{2-}(aq) \quad \rightarrow \quad S(s) + Cr^{3+}(aq)$ *determine oxidation numbers*

Balance the oxidation half-reaction:

$H_2S(aq) \rightarrow S(s) + 2e^-$ *add e^- to balance oxidation numbers*

$H_2S(aq) \rightarrow S(s) + 2H^+(aq) + 2e^-$ *add H^+ to balance charge*

Balance the reduction half-reaction:

$Cr_2O_7^{2-}(aq) \rightarrow 2Cr^{3+}(aq)$ *balance atoms being oxidized or reduced*

$6e^- + Cr_2O_7^{2-}(aq) \rightarrow 2Cr^{3+}(aq)$ *add e^- to balance oxidation numbers*

$6e^- + 14H^+(aq) + Cr_2O_7^{2-}(aq) \rightarrow 2Cr^{3+}(aq)$ *add H^+ to balance charge*

$6e^- + 14H^+(aq) + Cr_2O_7^{2-}(aq) \rightarrow 2Cr^{3+}(aq) + 7H_2O(l)$ *add H_2O to balance atoms*

Equalize the number of electrons exchanged between the two half-reactions:

$3 \times [H_2S(aq) \rightarrow S(s) + 2H^+(aq) + 2e^-]$

$1 \times [6e^- + 14H^+(aq) + Cr_2O_7^{2-}(aq) \rightarrow 2Cr^{3+}(aq) + 7H_2O(l)]$

$3H_2S(aq) \rightarrow 3S(s) + 6H^+(aq) + 6e^-$

$6e^- + 14H^+(aq) + Cr_2O_7^{2-}(aq) \rightarrow 2Cr^{3+}(aq) + 7H_2O(l)$

$3H_2S(aq) \rightarrow 3S(s) + 6H^+(aq) + \cancel{6e^-}$

$\underline{\cancel{6e^-} + 14H^+(aq) + Cr_2O_7^{2-}(aq) \rightarrow 2Cr^{3+}(aq) + 7H_2O(l)}$

$\mathbf{14H^+}(aq) + 3H_2S(aq) + Cr_2O_7^{2-}(aq) \rightarrow 3S(s) + 2Cr^{3+}(aq) + 7H_2O(l) + \mathbf{6H^+}(aq)$

Eliminate duplicate substances (bold type):

$8H^+(aq) + 3H_2S(aq) + Cr_2O_7^{2-}(aq) \rightarrow 3S(s) + 2Cr^{3+}(aq) + 7H_2O(l)$ *balanced equation*

(b) $V^{2+}(aq) + MnO_4^-(aq) \rightarrow VO_2^+(aq) + Mn^{2+}(aq)$

$$\underset{\downarrow}{\overset{2+}{}} \qquad \underset{\downarrow}{\overset{7+}{}} \qquad\qquad \underset{\downarrow}{\overset{5+}{}} \qquad \underset{\downarrow}{\overset{2+}{}}$$

$V^{2+}(aq) + MnO_4^-(aq) \quad \rightarrow \quad VO_2^+(aq) + Mn^{2+}(aq)$ *determine oxidation numbers*

Balance the oxidation half-reaction:

$V^{2+}(aq) \rightarrow VO_2^+(aq) + 3e^-$ *add e^- to balance oxidation numbers*

$V^{2+}(aq) \rightarrow VO_2^+(aq) + 4H^+ + 3e^-$ *add H^+ to balance charge*

$2H_2O(l) + V^{2+}(aq) \rightarrow VO_2^+(aq) + 4H^+ + 3e^-$ *add H_2O to balance atoms*

Balance the reduction half-reaction:

$5e^- + MnO_4^-(aq) \rightarrow Mn^{2+}(aq)$ *add e^- to balance oxidation numbers*

$5e^- + 8H^+(aq) + MnO_4^-(aq) \rightarrow Mn^{2+}(aq)$ *add H^+ to balance charge*

$5e^- + 8H^+(aq) + MnO_4^-(aq) \rightarrow Mn^{2+}(aq) + 4H_2O(l)$ *add H_2O to balance atoms*

Equalize the number of electrons exchanged between the two half-reactions:

$5 \times [2H_2O(l) + V^{2+}(aq) \rightarrow VO_2^+(aq) + 4H^+ + 3e^-]$

$3 \times [5e^- + 8H^+(aq) + MnO_4^-(aq) \rightarrow Mn^{2+}(aq) + 4H_2O(l)]$

$10H_2O(l) + 5V^{2+}(aq) \rightarrow 5VO_2^+(aq) + 20H^+ + 15e^-$

$15e^- + 24H^+(aq) + 3MnO_4^-(aq) \rightarrow 3Mn^{2+}(aq) + 12H_2O(l)$

$10H_2O(l) + 5V^{2+}(aq) \rightarrow 5VO_2^+(aq) + 20H^+(aq) + \cancel{15e^-}$

$\cancel{15e^-} + 24H^+(aq) + 3MnO_4^-(aq) \rightarrow 3Mn^{2+}(aq) + 12H_2O(l)$

_____

$\mathbf{24H^+}(aq) + \mathbf{10H_2O}(l) + 5V^{2+}(aq) + 3MnO_4^-(aq) \rightarrow 5VO_2^+(aq) + 3Mn^{2+}(aq) + \mathbf{20H^+}(aq) + \mathbf{12H_2O}(l)$

Eliminate duplicate substances (bold type):

$4H^+(aq) + 5V^{2+}(aq) + 3MnO_4^-(aq) \rightarrow 5VO_2^+(aq) + 3Mn^{2+}(aq) + 2H_2O(l)$ *balanced equation*

(c) $Fe^{2+}(aq) + ClO_3^-(aq) \rightarrow Fe^{3+}(aq) + Cl^-(aq)$

$\begin{array}{cccc} 2+ & 5+ & 3+ & 1- \\ \downarrow & \downarrow & \downarrow & \downarrow \end{array}$

$Fe^{2+}(aq) + ClO_3^-(aq) \quad \rightarrow \quad Fe^{3+}(aq) + Cl^-(aq)$ *determine oxidation numbers*

Balance the oxidation half-reaction:

$Fe^{2+}(aq) \rightarrow Fe^{3+}(aq) + e^-$ *add e^- to balance oxidation numbers*

Balance the reduction half-reaction:

$6e^- + ClO_3^-(aq) \rightarrow Cl^-(aq)$ *add e^- to balance oxidation numbers*

$6e^- + 6H^+(aq) + ClO_3^-(aq) \rightarrow Cl^-(aq)$ *add H^+ to balance charge*

$6e^- + 6H^+(aq) + ClO_3^-(aq) \rightarrow Cl^-(aq) + 3H_2O(l)$ *add H_2O to balance atoms*

Equalize the number of electrons exchanged between the two half-reactions:

$6 \times [Fe^{2+}(aq) \rightarrow Fe^{3+}(aq) + e^-]$

$1 \times [6e^- + 6H^+(aq) + ClO_3^-(aq) \rightarrow Cl^-(aq) + 3H_2O(l)]$

$6Fe^{2+}(aq) \rightarrow 6Fe^{3+}(aq) + 6e^-$

$6e^- + 6H^+(aq) + ClO_3^-(aq) \rightarrow Cl^-(aq) + 3H_2O(l)$

$6Fe^{2+}(aq) \rightarrow 6Fe^{3+}(aq) + \cancel{6e^-}$

$\cancel{6e^-} + 6H^+(aq) + ClO_3^-(aq) \rightarrow Cl^-(aq) + 3H_2O(l)$

_____

$6H^+(aq) + 6Fe^{2+}(aq) + ClO_3^-(aq) \rightarrow 6Fe^{3+}(aq) + Cl^-(aq) + 3H_2O(l)$ *balanced equation*

14.57 (a) $NH_3(aq) + ClO^-(aq) \rightarrow N_2H_4(aq) + Cl^-(aq)$

$\begin{array}{cccc} 3- & 1+ & 2- & 1- \\ \downarrow & \downarrow & \downarrow & \downarrow \end{array}$

$NH_3(aq) + ClO^-(aq) \quad \rightarrow \quad N_2H_4(aq) + Cl^-(g)$ *determine oxidation numbers*

Balance oxidation half-reaction:

$2NH_3(aq) \rightarrow N_2H_4(aq)$ *balance atoms being oxidized or reduced*

$2NH_3(aq) \rightarrow N_2H_4(aq) + 2e^-$ *add e^- to balance oxidation numbers*

14–21

$2OH^-(aq) + 2NH_3(aq) \rightarrow N_2H_4(aq) + 2e^-$ *add OH$^-$ to balance charge*

$2OH^-(aq) + 2NH_3(aq) \rightarrow N_2H_4(aq) + 2H_2O(l) + 2e^-$ *add H$_2$O to balance atoms*

Balance reduction half-reaction:

$2e^- + ClO^-(aq) \rightarrow Cl^-(aq)$ *add e$^-$ to balance oxidation numbers*

$2e^- + ClO^-(aq) \rightarrow Cl^-(aq) + 2OH^-(aq)$ *add OH$^-$ to balance charge*

$2e^- + H_2O(l) + ClO^-(aq) \rightarrow Cl^-(aq) + 2OH^-(aq)$ *add H$_2$O to balance atoms*

Equalize the number of electrons exchanged between the two half-reactions:

$$2OH^-(aq) + 2NH_3(aq) \rightarrow N_2H_4(aq) + 2H_2O(l) + \cancel{2e^-}$$

$$\cancel{2e^-} + H_2O(l) + ClO^-(aq) \rightarrow Cl^-(aq) + 2OH^-(aq)$$

$$\overline{\mathbf{2OH^-}(aq) + \mathbf{H_2O}(l) + 2NH_3(aq) + ClO^-(aq) \rightarrow N_2H_4(aq) + Cl^-(aq) + \mathbf{2OH^-}(aq) + \mathbf{2H_2O}(l)}$$

Eliminate duplicate substances (bold type):

$2NH_3(aq) + ClO^-(aq) \rightarrow N_2H_4(aq) + Cl^-(aq) + H_2O(l)$ *balanced equation*

(b) $Cr(OH)_4^-(aq) + HO_2^-(aq) \rightarrow CrO_4^{2-}(aq) + H_2O(l)$

$$
\begin{array}{cccc}
3+ & 1- & 6+ & 2- \\
\downarrow & \downarrow & \downarrow & \downarrow \\
\end{array}
$$

$Cr(OH)_4^-(aq) + HO_2^-(aq) \quad \rightarrow \quad CrO_4^{2-}(aq) + H_2O(g)$ *determine oxidation numbers*

Balance oxidation half-reaction:

$Cr(OH)_4^-(aq) \rightarrow CrO_4^{2-}(aq) + 3e^-$ *add e$^-$ to balance oxidation numbers*

$4OH^-(aq) + Cr(OH)_4^-(aq) \rightarrow CrO_4^{2-}(aq) + 3e^-$ *add OH$^-$ to balance charge*

$4OH^-(aq) + Cr(OH)_4^-(aq) \rightarrow CrO_4^{2-}(aq) + 4H_2O(l) + 3e^-$ *add H$_2$O to balance atoms*

Balance reduction half-reaction:

$HO_2^-(aq) \rightarrow 2H_2O(l)$ *balance atoms being oxidized or reduced*

$2e^- + HO_2^-(aq) \rightarrow 2H_2O(l)$ *add e$^-$ to balance oxidation numbers*

$2e^- + HO_2^-(aq) \rightarrow 2H_2O(l) + 3OH^-(aq)$ *add OH$^-$ to balance charge*

$2e^- + 3H_2O(l) + HO_2^-(aq) \rightarrow 2H_2O(l) + 3OH^-(aq)$ *add H$_2$O to balance atoms*

Equalize the number of electrons exchanged between the two half-reactions:

$2 \times [4OH^-(aq) + Cr(OH)_4^-(aq) \rightarrow CrO_4^{2-}(aq) + 4H_2O(l) + 3e^-]$

$3 \times [2e^- + 3H_2O(l) + HO_2^-(aq) \rightarrow 2H_2O(l) + 3OH^-(aq)]$

$8OH^-(aq) + 2Cr(OH)_4^-(aq) \rightarrow 2CrO_4^{2-}(aq) + 8H_2O(l) + 6e^-$

$$6e^- + 9H_2O(l) + 3HO_2^-(aq) \rightarrow 6H_2O(l) + 9OH^-(aq)$$

$$8OH^-(aq) + 2Cr(OH)_4^-(aq) \rightarrow 2CrO_4^{2-}(aq) + 8H_2O(l) + \cancel{6e^-}$$

$$\cancel{6e^-} + 9H_2O(l) + 3HO_2^-(aq) \rightarrow 6H_2O(l) + 9OH^-(aq)$$

$$\mathbf{8OH^-}(aq) + \mathbf{9H_2O}(l) + 2Cr(OH)_4^-(aq) + 3HO_2^-(aq) \rightarrow 2CrO_4^{2-}(aq) + \mathbf{14H_2O}(l) + \mathbf{9OH^-}(aq)$$

Eliminate duplicate substances (bold type):

$$2Cr(OH)_4^-(aq) + 3HO_2^-(aq) \rightarrow 2CrO_4^{2-}(aq) + 5H_2O(l) + OH^-(aq) \quad \textit{balanced equation}$$

(c) $Br_2(aq) \rightarrow Br^-(aq) + BrO^-(aq)$

  0   1−   1+

$$Br_2(aq) \rightarrow Br^-(aq) + BrO^-(aq) \qquad \textit{determine oxidation numbers}$$

Balance oxidation half-reaction:

$Br_2(aq) \rightarrow 2BrO^-(aq)$ *balance atoms being oxidized or reduced*

$Br_2(aq) \rightarrow 2BrO^-(aq) + 2e^-$ *add e^- to balance oxidation numbers*

$4OH^-(aq) + Br_2(aq) \rightarrow 2BrO^-(aq) + 2e^-$ *add OH^- to balance charge*

$4OH^-(aq) + Br_2(aq) \rightarrow 2BrO^-(aq) + 2H_2O(l) + 2e^-$ *add H_2O to balance atoms*

Balance reduction half-reaction:

$Br_2(aq) \rightarrow 2Br^-(aq)$ *balance atoms being oxidized or reduced*

$2e^- + Br_2(aq) \rightarrow 2Br^-(aq)$ *add e^- to balance oxidation numbers*

Equalize the number of electrons exchanged between the two half-reactions:

$$4OH^-(aq) + Br_2(aq) \rightarrow 2BrO^-(aq) + 2H_2O(l) + \cancel{2e^-}$$

$$\cancel{2e^-} + Br_2(aq) \rightarrow 2Br^-(aq)$$

$$4OH^-(aq) + \mathbf{2}Br_2(aq) \rightarrow \mathbf{2}Br^-(aq) + \mathbf{2}BrO^-(aq) + \mathbf{2}H_2O(l)$$

Divide coefficients by 2 (bold type):

$$2OH^-(aq) + Br_2(aq) \rightarrow Br^-(aq) + BrO^-(aq) + H_2O(l) \qquad \textit{balanced equation}$$

14.59 We begin by calculating the oxidation numbers of each atom in the chemical equation:

  0   5+   4+   0

$$C_6H_{12}O_6(aq) + NO_3^-(aq) \rightarrow CO_2(g) + N_2(g) \qquad \textit{determine oxidation numbers}$$

Balance oxidation half-reaction:

$C_6H_{12}O_6(aq) \rightarrow 6CO_2(g)$ *balance atoms being oxidized or reduced*

$C_6H_{12}O_6(aq) \rightarrow 6CO_2(g) + 24e^-$ *add e^- to balance oxidation numbers*

$C_6H_{12}O_6(aq) \rightarrow 6CO_2(g) + 24H^+(aq) + 24e^-$ *add H^+ to balance charge*

$6H_2O(l) + C_6H_{12}O_6(aq) \rightarrow 6CO_2(g) + 24H^+(aq) + 24e^-$ *add H_2O to balance atoms*

Balance reduction half-reaction:

$2NO_3^-(aq) \rightarrow N_2(g)$ *balance atoms being oxidized or reduced*

$10e^- + 2NO_3^-(aq) \rightarrow N_2(g)$ *add e^- to balance oxidation numbers*

$10e^- + 12H^+(aq) + 2NO_3^-(aq) \rightarrow N_2(g)$ *add H^+ to balance charge*

$10e^- + 12H^+(aq) + NO_3^-(aq) \rightarrow N_2(g) + 6H_2O(l)$ *add H_2O to balance atoms*

Equalize the number of electrons exchanged between the two half-reactions:
Rather than multiplying by 10 and 24 respectively, we can use the factors 5 and 12 to produce smaller coefficients:

$5 \times [6H_2O(l) + C_6H_{12}O_6(aq) \rightarrow 6CO_2(g) + 24H^+(aq) + 24e^-]$

$12 \times [10e^- + 12H^+(aq) + 2NO_3^-(aq) \rightarrow N_2(g) + 6H_2O(l)]$

$30H_2O(l) + 5C_6H_{12}O_6(aq) \rightarrow 30CO_2(g) + 120H^+(aq) + 120e^-$

$120e^- + 144H^+(aq) + 24NO_3^-(aq) \rightarrow 12N_2(g) + 72H_2O(l)$

$30H_2O(l) + 5C_6H_{12}O_6(aq) \rightarrow 30CO_2(g) + 120H^+(aq) + \cancel{120e^-}$

$\cancel{120e^-} + 144H^+(aq) + 24NO_3^-(aq) \rightarrow 12N_2(g) + 72H_2O(l)$

$\overline{\textbf{144H}^+(aq) + \textbf{30H}_2\textbf{O}(l) + 5C_6H_{12}O_6(aq) + 24NO_3^-(aq) \rightarrow 30CO_2(g) + 12N_2(g) + \textbf{72H}_2\textbf{O}(l) + \textbf{120H}^+(aq)}$

Eliminate duplicate substances (bold type):

$24H^+(aq) + 5C_6H_{12}O_6(aq) + 24NO_3^-(aq) \rightarrow 30CO_2(g) + 12N_2(g) + 42H_2O(l)$ *balanced equation*

14.61   The anode can be any metal that is more active than iron (e.g. aluminum or zinc; see Figure 14.22). The supporting electrolyte can be any soluble salt that does not precipitate with Fe^{3+} (e.g. potassium nitrate, because all nitrates are soluble). The anodic half-cell must also contain a salt that contains the metal used for the anode (e.g. aluminum nitrate for an aluminum anode, or zinc nitrate for a zinc anode).

14.63   Au < Bi < Ni < Zn < Al < Ca (see Figure 14.22). The activity series orders metals according to reducing power. The most active reducing metals are listed at the top of the activity series, with potassium being the most active metal of those listed. Of the metals listed, gold (Au) is the least active.

14.65   Electrolysis is the process of using electricity (electrical energy) to cause a nonspontaneous redox reaction to occur.

14.67   When the temperature of sodium iodide is above its melting point, NaI is a molten salt. Two types of ions are present, Na^+ and I^-. Both are in the molten state (which is designated as (l)). The molten salt is similar to a liquid, except that the ions are not bound together (like H_2O in liquid water) and can move separately from one another. The sodium ions cannot lose any additional electrons because they already have a noble gas configuration; they can only be reduced. This means that iodide ions are oxidized (to I_2). At the high temperatures employed with molten salts, molecular iodine is in the gas state.
(a) Reduction takes place at the cathode: sodium metal forms from sodium ions in the molten salt.
(b) Oxidation takes place at the anode: molecular iodine forms from iodide ions in the molten salt.
(c) $2I^-(l) \rightarrow I_2(g) + 2e^-$ (anode, oxidation)
    $Na^+(l) + e^- \rightarrow Na(l)$ (cathode, reduction)
(d) To properly balance the overall equation, we double the reduction half-reaction and add it to the oxidation half-reaction.
    $2Na^+ + 2I^-(l) \rightarrow 2Na(l) + I_2(g)$

14.69   At the anode, the iodine atoms decrease in size as iodide ions in solution lose electrons and combine to form neutral I$_2$ molecules. At the catholde, the sodium atoms increase in size as the netural metal atoms gain electrons to become ions in solution.

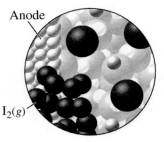 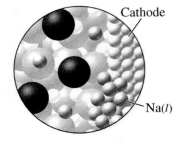

Anode

Cathode

I$_2$(g)

Na(l)

Anode reaction   Cathode reaction

14.71   Steel is an alloy, or mixture, of metals. The principle component is usually iron. Steel corrodes when the iron actively reduces oxygen in the presence of water. The metal is oxidized and the oxygen is reduced. For example, pure iron, Fe, corrodes eventually forming iron(III) oxide (Fe$_2$O$_3$, which is Fe^{3+} and O^{2-}). Magnesium reduces oxygen more actively than iron does. When the two metals are connected, the magnesium will reduce the O$_2$(g), preventing the Fe from being oxidized. When all the Mg(s) is oxidized, the iron will begin to corrode.

14.73   Tin is less active than iron, so it does not prevent corrosion in the same way as a coating with a more active metal would. An alternative method of preventing corrosion is to prevent contact of an easily corroded metal with oxygen and moisture. The tin coating on a tin can serves as a barrier to moisture and oxygen. Paints serve the same purpose.

14.75   When iron is exposed to oxygen and moisture, it will corrode (rust). Chrome plating protects iron in two ways. By isolating the iron from oxygen and moisture, it can not corrode. Chromium is also a more active metal than iron. If the chromium coating is scratched, the chromium oxidizes preferentially because of its higher activity. As a result, the iron will not rust.

14.77   See Table 14.1
         (a) NH$_3$

$$\text{net charge} \atop \text{NH}_3 = \left(\begin{array}{c} \text{total positive} \\ \text{oxidation numbers} \\ 3 \times \text{H} \end{array} \right) + \left(\begin{array}{c} \text{total negative} \\ \text{oxidation numbers} \\ 1 \times \text{N} \end{array} \right)$$

Net charge = 0

Total positive oxidation numbers = 3 × (+1) = 3

This gives us:

$$0 = (3) + \left(\begin{array}{c} \text{total negative} \\ \text{oxidation numbers} \\ \text{N} \end{array} \right)$$

The nitrogen in NH$_3$ has an oxidation number of 3−. This is also the oxidation number you expect for nitrogen based on its position in the periodic table. The most electronegative element in a compound usually takes the oxidation number predicted from the periodic table. We base our assignment of the oxidation number of H on Rule 4 (see Table 14.1).

(b) N_2H_4

$$\text{net charge} \atop N_2H_4 = \begin{pmatrix} \text{total positive} \\ \text{oxidation numbers} \\ 4 \times H \end{pmatrix} + \begin{pmatrix} \text{total negative} \\ \text{oxidation numbers} \\ 2 \times N \end{pmatrix}$$

Net charge = 0

Total positive oxidation numbers = $4 \times (+1) = 4$

This gives us:

$$0 = (4) + \begin{pmatrix} \text{total negative} \\ \text{oxidation numbers} \\ 2 \times N \end{pmatrix}$$

$$N = \frac{-4}{2} = -2$$

The nitrogen in N_2H_4 has an oxidation number of 2−.

(c) NF_3

$$\text{net charge} \atop NF_3 = \begin{pmatrix} \text{total positive} \\ \text{oxidation numbers} \\ N \end{pmatrix} + \begin{pmatrix} \text{total negative} \\ \text{oxidation numbers} \\ 3 \times F \end{pmatrix}$$

Net charge = 0

Total negative oxidation numbers = $3 \times (-1) = -3$

This gives us:

$$0 = \begin{pmatrix} \text{total positive} \\ \text{oxidation numbers} \\ N \end{pmatrix} + (-3)$$

The nitrogen in NF_3 has an oxidation number of 3+.

(d) NH_2OH

$$\text{net charge} \atop NH_2OH = \begin{pmatrix} \text{total positive} \\ \text{oxidation numbers} \\ N + 3 \times H \end{pmatrix} + \begin{pmatrix} \text{total negative} \\ \text{oxidation numbers} \\ 1 \times O \end{pmatrix}$$

Net charge = 0

Total negative oxidation numbers = $1 \times (-2) = -2$

Total positive oxidation numbers = $N + 3 \times (+1) = N + 3$

This gives us:

$$0 = \left(\begin{array}{c} \text{total positive} \\ \text{oxidation numbers} \\ N + 3 \end{array} \right) + (-2)$$

$0 = N + 1$

The nitrogen in NH_2OH has an oxidation number of 1−.

Note that even though we assumed at first that the nitrogen had a positive oxidation number, this assumption did affect the fact that the oxidation number turned out to be 1−.

(e) $Fe(NO_3)_3$ – The easiest way to calculate the oxidation number N in iron(III) nitrate is to recognize that nitrate ion has a chemical formula of NO_3^-. The oxidation number of nitrogen will be the same in nitrate as it is in compounds containing nitrate.

$$\begin{array}{c} \text{net charge} = \\ NO_3^- \end{array} \left(\begin{array}{c} \text{total positive} \\ \text{oxidation numbers} \\ N \end{array} \right) + \left(\begin{array}{c} \text{total negative} \\ \text{oxidation numbers} \\ 3 \times O \end{array} \right)$$

Net charge = −1

Total negative oxidation numbers = $3 \times (-2) = -6$

This gives us:

$$-1 = \left(\begin{array}{c} \text{total positive} \\ \text{oxidation numbers} \\ N \end{array} \right) + (-6)$$

$N = 5$
The oxidation number of nitrogen in NO_3^- is 5+.

(f) HNO_2

$$\begin{array}{c} \text{net charge} = \\ HNO_2 \end{array} \left(\begin{array}{c} \text{total positive} \\ \text{oxidation numbers} \\ 1 \times N + 1 \times H \end{array} \right) + \left(\begin{array}{c} \text{total negative} \\ \text{oxidation numbers} \\ 2 \times O \end{array} \right)$$

Net charge = 0

Total negative oxidation numbers = $2 \times (-2) = -4$

Total positive oxidation numbers = N + 1

This gives us:

$$0 = \left(\begin{array}{c} \text{total positive} \\ \text{oxidation numbers} \\ N + 1 \end{array} \right) + (-4)$$

$N = 3$
The oxidation number of nitrogen in HNO_2 is 3+.

14.79 (a) *Answer:* $2H_2O(l) + 2OH^-(aq) + 2CoCl_2(s) + Na_2O_2(aq) \rightarrow 2Co(OH)_3(s) + 4Cl^-(aq) + 2Na^+(aq)$

The oxidizing agent is Na_2O_2 (oxygen is being reduced from 1− to 2−).

The reducing agent is $CoCl_2$ (cobalt is being oxidized from 2+ to 3+).

Solution: First, we determine the oxidation numbers of each atom and separate the half-reactions. Because chloride is part of $CoCl_2$, we include chloride in that half-reaction. Sodium is part of the peroxide half-reaction.

$$\overset{2+}{\downarrow} \qquad \overset{1-}{\downarrow} \qquad \overset{3+}{\downarrow}\overset{2-}{\downarrow}$$

$CoCl_2(s) + Na_2O_2(aq) \rightarrow Co(OH)_3(s) + Cl^-(aq) + Na^+(aq)$ *determine oxidation numbers*

The oxidizing agent is Na_2O_2 (oxygen is being reduced from 1− to 2−).

The reducing agent is $CoCl_2$ (cobalt is being oxidized from 2+ to 3+).

The peroxide ion is reduced to hydroxide under basic conditions, so the reduction half-reaction is:

$Na_2O_2(aq) \rightarrow OH^-(aq) + Na^+(aq)$

For the oxidation reaction, the reaction is:

$CoCl_2(s) \rightarrow Co(OH)_3(s) + Cl^-(aq)$

Balance oxidation half-reaction:

$CoCl_2(s) \rightarrow Co(OH)_3(s) + 2Cl^-(aq)$	*balance atoms*
$CoCl_2(s) \rightarrow Co(OH)_3(s) + 2Cl^-(aq) + e^-$	*add e^- to balance oxidation numbers*
$3OH^- + CoCl_2(s) \rightarrow Co(OH)_3(s) + 2Cl^-(aq) + e^-$	*add OH^- to balance charge*

Balance reduction half-reaction:

$Na_2O_2(aq) \rightarrow 2OH^-(aq) + 2Na^+(aq)$	*balance atoms*
$2e^- + Na_2O_2(aq) \rightarrow 2OH^-(aq) + 2Na^+(aq)$	*add e^- to balance oxidation numbers*
$2e^- + Na_2O_2(aq) \rightarrow 2OH^-(aq) + 2Na^+(aq) + 2OH^-(aq)$	*add OH^- to balance charge*
$2e^- + 2H_2O(l) + Na_2O_2(aq) \rightarrow 4OH^-(aq) + 2Na^+(aq)$	*add H_2O to balance atoms*

Balance the electrons between the two half-reactions:

$2 \times [3OH^- + CoCl_2(s) \rightarrow Co(OH)_3(s) + 2Cl^-(aq) + e^-]$

$1 \times [2e^- + 2H_2O(l) + Na_2O_2(aq) \rightarrow 4OH^-(aq) + 2Na^+(aq)]$

$6OH^- + 2CoCl_2(aq) \rightarrow 2Co(OH)_3(s) + 4Cl^-(aq) + 2e^-$

$2e^- + 2H_2O(l) + Na_2O_2(aq) \rightarrow 4OH^-(aq) + 2Na^+(aq)$

$6OH^- + 2CoCl_2(aq) \rightarrow 2Co(OH)_3(s) + 4Cl^-(aq) + \cancel{2e^-}$

$\cancel{2e^-} + 2H_2O(l) + Na_2O_2(aq) \rightarrow 4OH^-(aq) + 2Na^+(aq)$

6OH$^-(aq) + 2H_2O(l) + 2CoCl_2(s) + Na_2O_2(aq) \rightarrow$ $2Co(OH)_3(s) +$
$2Na^+(aq) + 4Cl^-(aq) +$ **4OH**$^-(aq)$

Eliminate duplicate substances (bold type):

$2OH^-(aq) + 2H_2O(l) + 2CoCl_2(s) + Na_2O_2(aq) \rightarrow 2Co(OH)_3(s) + 2Na^+(aq) + 4Cl^-(aq)$

(b) *Answer*: $2OH^-(aq) + Bi_2O_3(s) + 2ClO^-(aq) \rightarrow 2BiO_3^-(aq) + 2Cl^-(aq) + H_2O(l)$

The oxidizing agent is ClO^- (chlorine is reduced from 1+ to 1−).

The reducing agent is Bi_2O_3 (bismuth is oxidized from 3+ to 5+).

Solution:

3+   1+   5+   1−

↓   ↓   ↓   ↓

$Bi_2O_3(s) + ClO^-(aq) \rightarrow BiO_3^-(aq) + Cl^-(aq)$ *determine oxidation numbers*

Balance oxidation half-reaction:

$Bi_2O_3(s) \rightarrow 2BiO_3^-(aq)$ *balance atoms*

$Bi_2O_3(s) \rightarrow 2BiO_3^-(aq) + 4e^-$ *add e^- to balance oxidation numbers*

$6OH^- + Bi_2O_3(s) \rightarrow 2BiO_3^-(aq) + 4e^-$ *add OH^- to balance charge*

$6OH^- + Bi_2O_3(s) \rightarrow 2BiO_3^-(aq) + 3H_2O(l) + 4e^-$ *add H_2O to balance atoms*

Balance reduction half-reaction:

$2e^- + ClO^-(aq) \rightarrow Cl^-(aq)$ *add e^- to balance oxidation numbers*

$2e^- + ClO^-(aq) \rightarrow Cl^-(aq) + 2OH^-(aq)$ *add OH^- to balance charge*

$2e^- + H_2O(l) + ClO^-(aq) \rightarrow Cl^-(aq) + 2OH^-(aq)$ *add H_2O to balance atoms*

Equalize the number of electrons exchanged between the two half-reactions:

$1 \times [6OH^- + Bi_2O_3(s) \rightarrow 2BiO_3^-(aq) + 3H_2O(l) + 4e^-]$

$2 \times [2e^- + H_2O(l) + ClO^-(aq) \rightarrow Cl^-(aq) + 2OH^-(aq)]$

$6OH^- + Bi_2O_3(s) \rightarrow 2BiO_3^-(aq) + 3H_2O(l) + 4e^-$

$4e^- + 2H_2O(l) + 2ClO^-(aq) \rightarrow 2Cl^-(aq) + 4OH^-(aq)$

$6OH^- + Bi_2O_3(s) \rightarrow 2BiO_3^-(aq) + 3H_2O(l) + \cancel{4e^-}$

$\cancel{4e^-} + 2H_2O(l) + 2ClO^-(aq) \rightarrow 2Cl^-(aq) + 4OH^-(aq)$

$\mathbf{2H_2O}(l) + \mathbf{6OH^-}(aq) + Bi_2O_3(s) + 2ClO^-(aq) \rightarrow 2BiO_3^-(aq) + 2Cl^-(aq) + \mathbf{3H_2O}(l) + \mathbf{4OH^-}(aq)$

Eliminate duplicate substances (bold type):

$2OH^-(aq) + Bi_2O_3(s) + 2ClO^-(aq) \rightarrow 2BiO_3^-(aq) + 2Cl^-(aq) + H_2O(l)$

14.81 $NH_4^+(aq) + NO_3^-(aq) \rightarrow N_2O(g) + 2H_2O(l)$

The oxidizing agent is NO_3^- (N is reduced from 5+ to 1+).
The reducing agent is NH_4^+ (N is oxidized from 3− to 1+).

3−   5+   1+

↓   ↓   ↓

$NH_4^+(aq) + NO_3^-(aq) \rightarrow N_2O(g) + H_2O(l)$ *determine oxidation numbers*

Balance oxidation half-reaction:

$2NH_4^+(aq) \rightarrow N_2O(g)$ *balance atoms being oxidized or reduced*

$2NH_4^+(aq) \rightarrow N_2O(g) + 8e^-$ *add e^- to balance oxidation numbers*

$2NH_4^+(aq) \rightarrow N_2O(g) + 10H^+(aq) + 8e^-$ *add H^+ to balance charge*

$H_2O(l) + 2NH_4^+(aq) \rightarrow N_2O(g) + 10H^+(aq) + 8e^-$ *add H_2O to balance atoms*

Balance reduction half-reaction:

$2NO_3^-(aq) \rightarrow N_2O(g)$ *balance atoms being oxidized or reduced*

$8e^- + 2NO_3^-(aq) \rightarrow N_2O(g)$ *add e^- to balance oxidation numbers*

$8e^- + 10H^+(aq) + 2NO_3^-(aq) \rightarrow N_2O(g)$ *add H^+ to balance charge*

$8e^- + 10H^+(aq) + 2NO_3^-(aq) \rightarrow N_2O(g) + 5H_2O(l)$ *add H_2O to balance atoms*

Equalize the number of electrons exchanged between the two half-reactions

$H_2O(l) + 2NH_4^+(aq)$ $\rightarrow N_2O(g) + 10H^+(aq) + \cancel{8e^-}$

$\cancel{8e^-} + 10H^+(aq) + 2NO_3^-(aq)$ $\rightarrow N_2O(g) + 5H_2O(l)$

$\overline{\textbf{H}_2\textbf{O}(l) + \textbf{10H}^+(aq) + 2NH_4^+(aq) + 2NO_3^-(aq) \rightarrow 2N_2O(g) + \textbf{5H}_2\textbf{O}(l) + \textbf{10H}^+(aq)}$

Eliminate species appearing on both sides of the equation (bold type) and divide coefficients by 2:

$NH_4^+(aq) + NO_3^-(aq) \rightarrow N_2O(g) + 2H_2O(l)$

14.83 The element which is highest on the activity series is the best reducing agent (Figure 14.22).
(a) Al; (b) Zn; (c) Mn; (d) Mg

14.85 Brass is an alloy of copper and zinc. Because sweat is generally acidic, it facilitates the oxidation of the copper from Cu to Cu^{2+} (remember acids taste sour and corrode metals). There are several copper compounds that are green (including $CuCO_3$).

14.87 $Cu(s) + H_2SO_4(aq) \rightarrow Cu^{2+}(aq) + SO_2(g)$
The copper half-reaction is balance simply by balancing charge:
$Cu(s) \rightarrow Cu^{2+}(aq) + 2e^-$

$H_2SO_4(aq) \rightarrow SO_2(g)$
$H_2SO_4(aq) + 2e^- \rightarrow SO_2(g)$ *add electrons*
$H_2SO_4(aq) + 2H^+(aq) + 2e^- \rightarrow SO_2(g)$ *add H^+*
$H_2SO_4(aq) + 2H^+(aq) + 2e^- (g) \rightarrow 2H_2O(l) + SO_2$ *add H_2O*
Then we add the two half-reactions and cancel the two electrons that appear on both sides of the equation.

$H_2SO_4(aq) + 2H^+(aq) + \cancel{2e^-}$ $\rightarrow 2H_2O(l) + SO_2(g)$

$Cu(s)$ $\rightarrow Cu^{2+}(aq) + \cancel{2e^-}$

$\overline{H_2SO_4(aq) + 2H^+(aq) + Cu(s)}$ $\rightarrow 2H_2O(l) + SO_2(g) + Cu^{2+}(aq)$

14.89 (a) The reducing agent is the substance that loses electrons. The oxidation state of Tl^+ increases to Tl^{3+}. This is represented by the oxidation half-reaction:
$Tl^+(aq) \rightarrow Tl^{3+}(aq) + 2e^-$
Since Tl^+ is giving up electrons, it is being oxidized and it is the reducing agent.
$Ce^{4+}(aq) + e^- \rightarrow Ce^{3+}(aq)$
Since Ce^{4+} is gaining electrons, it is being reduced and it is the oxidizing agent.

(b) The nitrogen in the reactant NO_3^- has an oxidation state of 5+, and in the product NO_2 the oxidation state of nitrogen is 4+. The nitrogen is being reduced from 5+ to 4+.
The reduction half-reaction is $NO_3^-(aq) + 2H^+(aq) + e^- \rightarrow NO_2(g) + H_2O(l)$.
Nitrate, NO_3^-, is the oxidizing agent.
The oxidation state of bromine in the Br– reactant is 1–, and in the Br_2 product the oxidation state of Br is zero. Bromine is being oxidized from 1– to 0.
The oxidation half-reaction is represented by $2Br^-(aq) \rightarrow Br_2(l) + 2e^-$.
Bromide ion, Br^-, is the reducing agent.

14.91 The activity series shows that copper is a more active metal than silver.
(a) The higher the metal on the activity series, the more easily it is oxidized, so copper is more easily oxidized the silver.
(b) The lower a metal on the activity series, the more easily its cation form is reduced back to elemental form. Since silver is lower than copper on the activity series, Ag^+ is more easily reduced than Cu^{2+}.
(c) Since Cu is more easily oxidized than Ag, and Ag^+ is more easily reduced than Cu^{2+}, then we would expect the following reaction to be spontaneous:

$$Cu(s) + 2Ag^+(aq) \rightarrow Cu^{2+}(aq) + 2Ag(s) \qquad \textit{balanced}$$

14.93 $2CH_3CH_2OH + 2Na \rightarrow 2Na^+CH_3CH_2O^- + H_2$
In this reaction Na is converted to Na^+. Its oxidation number increases from zero to 1+ as a result of losing an electron, so it is oxidized and is therefore the reducing agent. The carbons in ethanol, CH_3CH_2OH, and most of the hydrogens do not undergo a change in oxidation number but the hydrogen from the –OH group that converts to H_2 is reduced from an oxidation number of 1+ to zero. Because the hydrogen being reduced was in the ethanol reactant, then CH_3CH_2OH is the oxidizing agent.

14.95 $CH_4(g) + 2O_2(g) \rightarrow CO_2(g) + 2H_2O(g)$
(a) Carbon's oxidation number increases from 4– to 4+ so carbon is oxidized.
(b) Oxygen's oxidation number decreases from zero to 2– so oxygen is reduced.
(c) Carbon's oxidation number increases by 8 so for each carbon that is oxidized, 8 electrons are transferred to oxygen. Scaling up to mole scale and taking into account that there is one mole of carbon in every mole of CH_4, 8 moles of electrons are transferred for every mole of CH_4 that reacts.

14.97 The unbalanced equation is: $Br_2(aq) \rightarrow Br^-(aq) + BrO^-(aq)$. In this reaction, one bromine atom in Br_2 is oxidized while the other one is reduced.

Bromine, Br_2, is therefore the reactant in both half-reactions:

$$Br_2(aq) \rightarrow Br^-(aq) \qquad \text{and} \qquad Br_2(aq) \rightarrow BrO^-(aq)$$

Starting with the first half-reaction (reduction), we first balance the bromine atoms:

$$Br_2(aq) \rightarrow 2Br^-(aq)$$

In this half-reaction, each bromine atom's oxidation number changes from zero to 1–, and there are two bromine atoms reacting so it is a two-electron reduction:

$$2e^- + Br_2(aq) \rightarrow 2Br^-(aq)$$

There are no other elements, so this half-reaction is balanced. The number of bromine atoms and the charge is the same on both sides of the equation.

In the oxidation half-reaction, we must also start by balancing bromine atoms:

$$Br_2(aq) \rightarrow 2BrO^-(aq)$$

In this half-reaction, each bromine atom's oxidation number changes from zero to 1+, and there are two bromine atoms reacting so it is a two-electron oxidation:

$$Br_2(aq) \rightarrow 2BrO^-(aq) + 2e^-$$

Now we have to balance the charge. The total charge on the left is zero, and the total charge on the right is 4–. When in basic solution, this reaction occurs in the presence of hydroxide ions, so we can balance charge with OH^- ions:

$$4OH^-(aq) + Br_2(aq) \rightarrow 2BrO^-(aq) + 2e^-$$

Now that the charges are balanced, the last step is to balance the hydrogen and oxygen atoms by adding H_2O molecules. There are four H atoms and four O atoms on the left and two O atoms on the right, so there is an excess of four H atoms and two O atoms on the left. If we add two H_2O molecules to the right, the half-reaction is balanced:

$$4OH^-(aq) + Br_2(aq) \rightarrow 2BrO^-(aq) + 2e^- + 2H_2O(l)$$

Now that we have both half reactions balanced, we can combine them:

$$2e^- + Br_2(aq) \rightarrow 2Br^-(aq)$$
$$4OH^-(aq) + Br_2(aq) \rightarrow 2BrO^-(aq) + 2e^- + 2H_2O(l)$$

The number of electrons in both half-reactions are the same (electrons lost equal electrons gained), the electrons will cancel when we combine them:

$$\cancel{2e^-} + Br_2(aq) + 4OH^-(aq) + Br_2(aq) \rightarrow 2Br^-(aq) + 2BrO^-(aq) + \cancel{2e^-} + 2H_2O(l)$$

Nothing else cancels but we must combine the two $Br_2(aq)$:
$$2Br_2(aq) + 4OH^-(aq) \rightarrow 2Br^-(aq) + 2BrO^-(aq) + 2H_2O(l)$$

All coefficients can be divided by 2 to give to give the balanced equation:
$$Br_2(aq) + 2OH^-(aq) \rightarrow Br^-(aq) + BrO^-(aq) + H_2O(l)$$

14.99   In the activity series, the element higher in the series is more active and therefore more easily oxidized.

(a) The activity series shows that aluminum is more active than zinc, so aluminum is more easily oxidized than zinc. When these two half-reactions are combined in a voltaic cell, the aluminum half-reaction would occur at the anode and the zinc half-reaction would occur at the cathode:
Anode: $Al(s) \rightarrow Al^{3+}(aq) + 3e^-$; Cathode: $2e^- + Zn^{2+}(aq) \rightarrow Zn(s)$

(b) Chromium is more active than silver so the chromium half-reaction will occur at the anode and the silver half-reaction will occur at the cathode:
Anode: $Cr(s) \rightarrow Cr^{3+}(aq) + 3e^-$; Cathode: $e^- + Ag^+(aq) \rightarrow Ag(s)$

(c) Cadmium is more active than hydrogen so the cadmium half-reaction will occur at the anode and the hydrogen half-reaction will occur at the cathode:
Anode: $Cd(s) \rightarrow Cd^{2+}(aq) + 2e^-$; Cathode: $2e^- + 2H^+(aq) \rightarrow H_2(g)$

Chapter 15 – Nuclear Chemistry

15.1   (a) critical mass; (b) fission; (c) alpha particle; (d) radioactive; (e) gamma radiation; (f) nuclear bombardment; (g) nucleon; (h) radiation

15.3   Any substance that is composed of nuclei that spontaneously decay with the emission of radiation is radioactive.

15.5   Radioactivity produces photons (gamma rays) and particulate radiation (beta particles, positrons, neutrons, and alpha particles).

15.7   Of the elements with atomic numbers less than 84, technetium (Tc), promethium (Pm), and astatine (At) have no stable isotopes. No isotopes with atomic numbers equal to or greater than 84 are stable.

15.9   Dense materials are best at shielding gamma radiation. We can use lead, iron, or aluminum (metals) to contain the radiation. We could use materials such as concrete and water, but only in very thick layers.

15.11   Both nuclear decay reactions and nuclear bombardment reactions result in the production of radiation. In addition, new isotopes are produced by both types of reactions.

15.13   The particles and rays produced from nuclear decay include alpha particles, beta particles, positrons, and gamma rays. X-rays result from electron capture, but are not produced by a decaying nucleus.

15.15   When we balance nuclear reaction equations, we make certain that both nuclear charge (atomic number) and mass numbers are conserved.

15.17   We know that an alpha particle has an atomic number of 2 and mass number of 4 (Table 15.1). We can write the decay equation as:

$$^{226}_{88}\text{Ra} \rightarrow\ ^{A}_{Z}X +\ ^{4}_{2}\alpha$$

Because mass numbers are conserved, we can write:
$$226 = A + 4$$
$$A = 222$$
Atomic numbers are also conserved:
$$88 = Z + 2$$
$$Z = 86$$
From the periodic table, we find that the element with $Z = 86$ is Rn. The isotope produced by the decay reaction is $^{222}_{86}$Rn. The complete nuclear decay equation is:

$$^{226}_{88}\text{Ra} \rightarrow\ ^{222}_{86}\text{Rn} +\ ^{4}_{2}\alpha$$

15.19   We know that a beta particle (electron) has an atomic number of −1 and a mass number of 0 (Table 15.1). We can write the decay equation as:

$$^{99}_{42}\text{Mo} \rightarrow\ ^{A}_{Z}X +\ ^{0}_{-1}\beta^{-}$$

Because mass numbers are conserved, we can write:
$$99 = A + (0)$$
$$A = 99$$

Atomic numbers are also conserved:

$42 = Z + (-1)$

$Z = 43$

From the periodic table we find that the element with $Z = 43$ is Tc. The isotope produced by the decay reaction is $^{99}_{43}\text{Tc}$. The complete nuclear decay equation is:

$$^{99}_{42}\text{Mo} \rightarrow {}^{99}_{43}\text{Tc} + {}^{0}_{-1}\beta^-$$

15.21　We know that a positron has an atomic number of +1 and mass number of 0 (Table 15.1). We can write the decay equation as:

$$^{15}_{8}\text{O} \rightarrow {}^{A}_{Z}X + {}^{0}_{1}\beta^+$$

Because mass numbers are conserved, we can write:

$15 = A + (0)$

$A = 15$

Atomic numbers are also conserved:

$8 = Z + (+1)$

$Z = 7$

From the periodic table we find that the element with $Z = 7$ is N. The nuclide (isotope) produced by the decay reaction is $^{15}_{7}\text{N}$. The complete nuclear decay equation is:

$$^{15}_{8}\text{O} \rightarrow {}^{15}_{7}\text{N} + {}^{0}_{1}\beta^+$$

15.23　When a neutron is converted to a proton, the process is accompanied by emission of a beta particle (electron). We can apply the conservation rules to atomic particle conversions as well as to nuclear decay and nuclear bombardment equations. An electron has a mass number of zero and a charge of -1. We can summarize the overall conversion as follows: $^{1}_{0}\text{n} \rightarrow {}^{1}_{1}\text{p} + {}^{0}_{-1}\beta^-$

15.25　An electron has a mass number of 0 and an atomic number of -1 ($^{0}_{-1}\beta^-$). We depict the process of electron capture as shown below:

$$^{A}_{Z}X + {}^{0}_{-1}\beta^- \rightarrow {}^{7}_{3}\text{Li}$$

Because mass numbers are conserved, we can write:

$A + 0 = 7$

$A = 7$

Atomic numbers are also conserved:

$Z + (-1) = 3$

$Z = 4$

From the periodic table we find that the element with $Z = 4$ is Be. The reactant nuclide is $^{7}_{4}\text{Be}$. The complete nuclear equation is:

$$^{7}_{4}\text{Be} + {}^{0}_{-1}\beta^- \rightarrow {}^{7}_{3}\text{Li}$$

15.27　To complete the following reactions, we use the conservation rules to determine the mass number and atomic number of the unknown isotope. In each case, the sum of the atomic numbers of the reactants must equal the sum of the atomic numbers of the products. Also, the sum of the mass numbers of the reactants must equal the sum of the mass numbers of the products.

(a)　$^{238}_{92}\text{U} \rightarrow {}^{234}_{90}\text{Th} + {}^{4}_{2}\text{He}$

(b)　$^{234}_{90}\text{Th} \rightarrow {}^{234}_{91}\text{Pa} + {}^{0}_{-1}\beta^-$

(c) $^{234}_{92}\text{U} \rightarrow {}^{4}_{2}\alpha + {}^{230}_{90}\text{Th}$

(d) $^{210}_{83}\text{Bi} + {}^{0}_{-1}e^- \rightarrow {}^{210}_{82}\text{Pb}$

15.29 To complete the following reactions, we first determine which particles are reactants and which are products. All decay processes produce particles as products (alpha particles, beta particles (electrons), or positrons). Bombardment particles are reactants. Next, we write a general equation (see part (a) below) and determine the atomic number (Z) and mass number (A) of the unknown particle or isotope. The sum of the atomic numbers of the reactants must equal the sum of the atomic numbers of the products, and the sum of the mass numbers of the reactants must equal the sum of the mass numbers of the products.

(a) Alpha decay produces an alpha particle and a new isotope:

$$^{227}_{87}\text{Fr} \rightarrow {}^{4}_{2}\alpha + {}^{A}_{Z}\text{X}$$

From this we determine that:
$A = 227 - 4 = 223$
$Z = 87 - 2 = 85$
We determine the elemental symbol from Z and the periodic table.

$$^{227}_{87}\text{Fr} \rightarrow {}^{4}_{2}\alpha + {}^{223}_{85}\text{At}$$

(b) Positron emission produces a positron and a new isotope:

$$^{18}_{9}\text{F} \rightarrow {}^{0}_{1}\beta^+ + {}^{A}_{Z}X$$

$$^{18}_{9}\text{F} \rightarrow {}^{0}_{1}\beta^+ + {}^{18}_{8}\text{O}$$

(c) Beta emission increases the atomic number but does not affect the mass number.

$$^{14}_{6}\text{C} \rightarrow {}^{0}_{-1}\beta + {}^{14}_{7}\text{N}$$

15.31 In a nuclear bombardment reaction a particle accelerator is used to accelerate particles, or nuclides, to very high velocities. The particle beam is aimed at a target element. If the particles strike the target with sufficient energy to fuse a target element nucleus and beam particle, a newly created nucleus forms. The newly created nucleus usually spontaneously decays and is often identified by its decay products.

15.33 A neutron has a mass number of 1 and an atomic number of 0 (${}^{1}_{0}n$). We can write the bombardment reaction as:

$$^{250}_{98}\text{Cf} + {}^{11}_{5}\text{B} \rightarrow {}^{A}_{Z}X + 4{}^{1}_{0}n$$

Because mass numbers are conserved, we can write:
$250 + 11 = A + 4$
$A = 257$
Atomic numbers are also conserved:
$98 + 5 = Z + 0$
$Z = 103$
From the periodic table we find that the element with $Z = 103$ is Lr. The nuclide (isotope) that is produced by the bombardment reaction is $^{257}_{103}\text{Lr}$. The complete nuclear decay equation is:

$$^{250}_{98}\text{Cf} + {}^{11}_{5}\text{B} \rightarrow {}^{257}_{103}\text{Lr} + 4{}^{1}_{0}n$$

15.35   For each of these problems, first determine the mass and atomic number of the unknown substance. For example, in (a), we know that the atomic number of the unknown is 6 (i.e. $5 + 1 = 6 + 0$). Also, we can calculate the mass number as 11 ($10 + 2 = 11 + 1$). Since the atomic number is 6, this must be an isotope of carbon.

(a) $^{10}_{5}B + ^{2}_{1}H \rightarrow ^{11}_{6}C + ^{1}_{0}n$

(b) $^{209}_{83}Bi + ^{4}_{2}\alpha \rightarrow ^{211}_{85}At + 2^{1}_{0}n$

(c) $^{12}_{6}C + ^{1}_{1}p \rightarrow ^{13}_{7}N$

15.37   To solve these problems, you need to find the mass number and atomic number of the unknown element. We can do this because these two values are conserved. In (a) for example, we know the mass number of the unknown must be 96 ($96 + 1 = A + 1; A = 96$) and that the atomic number must be 43 ($42 + 1 = Z + 0; Z = 43$). All that remains is to look up the element symbol on the periodic table using the atomic number (Tc).

(a) $^{96}_{42}Mo + ^{1}_{1}p \rightarrow ^{96}_{43}Tc + ^{1}_{0}n$

(b) $^{209}_{83}Bi + ^{64}_{28}Ni \rightarrow ^{272}_{111}Rg + ^{1}_{0}n$

15.39   From Figure 15.4 we see that the neutron to proton ratio is approximately 1:1 (i.e. the slope of the line passing through those points is about 1 neutron for each proton).

15.41   When we plot the mass numbers of all the stable isotopes versus their atomic numbers, a pattern emerges (Figure 15.14). The band of stability is the outline of that pattern.

15.43   According to the band of stability on the plot of N versus Z (Figure 15.14), the ideal ratio of neutrons to protons is approximately 1:1 for isotopes with atomic numbers less than 20. Oxygen has 8 protons, so a stable isotope has 8 neutrons. This makes the expected mass number 16 (8p + 8n). The isotope $^{16}_{8}O$ should be the most stable.

15.45   We can use the neutron to proton ratios of nuclides to predict their modes of decay. For nuclides with atomic numbers of 20 or less: if $N/Z > 1$, neutrons are converted to protons through beta emission; if $N/Z < 1$, positron emission occurs. Electron capture could also occur, but is rare in the first 20 elements. For nuclides with atomic numbers equal to or greater than 84, alpha and beta emission are common. (a) $N/Z > 1$, beta emission; (b) $Z \geq 84$, alpha and beta decay; (c) $Z \geq 84$, alpha and beta decay; (d) $N/Z < 1$, positron emission; (e) $N/Z > 1$, beta emission

15.47   To determine the final product, we write Pa as the reactant and 5 alpha particles, 2 beta particles and an unknown nuclide as products. The number of steps that occur is not important because mass numbers are conserved.

$$^{234}_{91}Pa \rightarrow 5^{4}_{2}\alpha + 2^{0}_{-1}\beta^{-} + ^{A}_{Z}X$$

$234 = 5 \times 4 + 0 + A$ *mass number conservation calculation*
$A = 214$
$91 = 5 \times 2 + 2 \times (-1) + Z$ *atomic number conservation calculation*
$Z = 83$

From the periodic table we know that element 83 is bismuth. The isotope produced by the reaction is $^{214}_{83}Bi$. The overall decay equation is:

$$^{234}_{91}Pa \rightarrow 5^{4}_{2}\alpha + 2^{0}_{-1}\beta^{-} + ^{214}_{83}Bi$$

15.49   The half-life of a radioactive nuclide is the time it takes for half of the radioactive nuclei in a sample to decay to a different nuclide. Regardless of the number of radioactive nuclei present in a sample, each half-life has exactly the same duration.

15.51   Film, a Geiger-Müller counter, or a scintillation counter can be used to detect radiation.

15.53   Sixty hours represents four half-lives.

$$\text{Half-lives} = \frac{60 \text{ hrs}}{15 \text{ hrs / half-life}} = 4 \text{ half-lives}$$

After each half-life, the number of nuclides decreases by 1/2.

$$\text{Fraction of material remaining} = \frac{1}{2} \times \frac{1}{2} \times \frac{1}{2} \times \frac{1}{2} = \frac{1}{16} = 0.0625 \text{ or } 6.25\%$$

One-sixteenth (6.25%) of the original material will be present.

15.55   In each half-life, the number of nuclei would be reduced by half. This means after three half-lives, the number of nuclei is reduced to one-eighth ($\frac{1}{2} \times \frac{1}{2} \times \frac{1}{2} = 1/8$). Since one half-life is two weeks, it would be a total of 6 weeks to reach this value.

15.57   To reach a mass of 0.0125 mg, a 0.100 mg sample would have to decrease by a factor of 8. This amount of decrease occurs after a period of three half-lives:

$$\text{Fraction remaining} = \frac{1}{2} \times \frac{1}{2} \times \frac{1}{2} = \frac{1}{8} = 0.125$$

Because the half-life of the nuclide is 7.5 hours, its mass will drop from 0.100 mg to 0.0125 mg in 22.5 hours (three half-lives).

15.59   If the number of $^{14}_{6}C$ nuclides remaining in the grain is one half the number of $^{14}_{6}C$ nuclides present in living plants, then the grain is one half-life, or 5730 years old.

15.61   During the span of each half-life, one-half of the radioactive nuclides initially present decay. 6.25% of the nuclides remain after four half-lives. Twenty four hours represents four half-lives, so the half-life of this nuclide is 6 hours.

15.63   A battery powered by a radioisotope will output power as long as the radioisotope is present. As a result, radiation powered devices can last much longer than battery powered devices. The half-life of $^{238}_{92}Pu$ is approximately 87.7 years.

15.65   Doctors use ^{99m}Tc for medical diagnosis because it produces radiation that they can use to image tumors, and its short half-life ensures that it won't stay in the body for a long period of time.

15.67   Positron emission tomography (PET imaging) detects radioactive nuclides that emit positrons. Positron emission occurs in nuclides with N/Z ratios < 1. $^{11}_{6}C$ has a low N/Z ratio and emits positrons, whereas $^{14}_{6}C$ has a high N/Z ratio and does not emit positrons.

15.69   Radiation can 1) have no effect, 2) cause reparable damage, 3) cause damage leading to formation of mutant cells (cancer cells), or 4) kill cells (subsequently absorbed by the body).

15.71   This is a difficult question to answer. How you answer the question will depend on your view of our governmental system and your perception of the risk of radon and the benefit of regulations. Clearly, the best protection against unnecessary radon exposure is knowledge of what causes enhanced radon levels and what measures can be used to reduce radon exposure. The internet can be a good source of information if you use it carefully (i.e. http://www.epa.gov/iedweb00/radon/index.html).

15.73   You would most likely find the decay products of ^{214}Po. Polonium-214 is part of the uranium decay series (Figure 15.15). The isotope directly below ^{214}Po is $^{210}_{82}$Pb.

15.75   Fission occurs when unstable nuclei decay into to smaller nuclei and release other energetic products. Fusion results when smaller particles combine to make larger "heavier" nuclei.

15.77   To determine the number of neutrons emitted in this reaction we write a nuclear equation with the number of neutrons as the unknown.

$$^{235}_{92}U + {}^{1}_{0}n \rightarrow {}^{93}_{36}Kr + {}^{140}_{56}Ba + x\,{}^{1}_{0}n$$

From our rules of conservation, we see that three neutrons are required to balance the equation.

$$^{235}_{92}U + {}^{1}_{0}n \rightarrow {}^{93}_{36}Kr + {}^{140}_{56}Ba + 3\,{}^{1}_{0}n$$

15.79   In principle, for a chain reaction to occur the product of one step of the reaction must be a reactant in another step. If we stand dominoes up in a row and push the first domino over, it falls and pushes the next domino over, and the process continues until we run out of dominoes.

15.81   A fission reactor has fuel rods (containing the radioactive material), control rods (to absorb excess neutrons), a moderator (i.e. water, heavy water, or graphite), and energy converters (to convert water to steam which powers electric generators). Water is used in many ways in nuclear reactors. It is used to transfer heat energy to produce steam. It is also used to cool both the reactor and the steam generated by the reactor.

15.83   The heat from fission reactions is not used directly to heat water and create steam for safety reasons. The steam produced by directly heating the water with the heat generated by the nuclear reaction would be both highly radioactive and poisonous (water formed with deuterium rather than hydrogen is very toxic). If there were a leak of this steam, the released products would be very dangerous (Figure 15.24).

15.85   The waste from nuclear reactors contains many substances which are radioactive either directly or by contamination from the spent nuclear fuel.

15.87   Isotopes of hydrogen (H-1, H-2, H-3) and lithium are most likely to be used (Figure 15.26).

15.89   A mixture of deuterium, $^{2}_{1}$H, and tritium, $^{3}_{1}$H, are used in fusion reactors. The reaction process produces a helium isotope and a neutron:

$$^{2}_{1}H + {}^{3}_{1}H \rightarrow {}^{4}_{2}He + {}^{1}_{0}n$$

15.91   We begin by writing the nuclear equation with the product as an unknown:

$$^{1}_{1}H + {}^{2}_{1}H \rightarrow {}^{A}_{Z}X$$

Based on the rules of conservation, $A = 3$ and $Z = 2$. The product nucleus is $^{3}_{2}$He.

$$^{1}_{1}H + {}^{2}_{1}H \rightarrow {}^{3}_{2}He \qquad \textit{balanced equation}$$

15.93   To determine the age of a substance by radioactive dating, we determine the quantity of a particular radioisotope that is present in a sample of the substance. Then we compare that quantity with the quantity of that isotope that was initially present and use the half-life of the isotope to estimate the age of the substance. For example, if the radioisotope concentration is ¼ of its original concentration, we can say that the substance is two half-lives old.

15.95   The sample contains about 1/64th of the carbon-14 present in a modern sample. A decrease of this magnitude occurs after the passage of about 6 half-lives (i.e. $\left(1/2\right)^{6} = 1/64$), or approximately 34,000 years.

15.97 From the text (the discussion on fusion reactions) we find that there are three separate reactions that take place:

$$\ce{^1_1H + ^1_1H -> ^2_1H + ^0_1\beta^+}$$

$$\ce{^2_1H + ^1_1H -> ^3_2He}$$

$$\ce{^1_1H + ^3_2He -> ^4_2He + ^0_1\beta^+}$$

The overall reaction is:

$$\ce{4^1_1H -> ^4_2He + 2^0_1\beta^+}$$

15.99 We begin by writing the nuclear equation with the product as an unknown:

$$\ce{^9_4Be + ^4_2He -> ^{12}_6C + ^A_Z X}$$

Based on the rules of conservation, $A = 1$ and $Z = 0$. The additional particle is a neutron:

$$\ce{^9_4Be + ^4_2He -> ^{12}_6C + ^1_0n}$$

15.101 An isotope with a half-life of 138.4 days will go through approximately three half-lives in one year, after which time only $1/8^{th}$ of the original material will remain. You would be better off buying brushes every year (or as needed), rather than stockpiling a supply and having them lose their effectiveness while they sit on the shelf.

15.103 Even though scientists first recognized the existence of the element chlorine in 1774, chloride-containing compounds (e.g. salt) were used well before the details of their chemistry were known. Similarly, living things breathed oxygen long before Joseph Priestly discovered oxygen!

15.105 Start by writing an unbalanced equation for the nuclear reaction:

$$\ce{^{249}_{98}Cf + ^{15}_7N -> ^{260}_{105}Db + x^1_0n}$$

Next, balance the equation by conservation of atomic number and mass number:
$98 + 7 = 105 + x\,(0)$ (Note that atomic number is already conserved.)
$249 + 15 = 260 + x\,(1)$
$x = 249 + 15 - 260 = 4$

Thus, 4 neutrons are emitted, and the balanced equation is:

$$\ce{^{249}_{98}Cf + ^{15}_7N -> ^{260}_{105}Db + 4^1_0n}$$

$$\ce{^A_Z X + ^2_1H -> ^{97}_{43}Tc + 2^1_0n}$$

Use the conservation of mass number and conservation of atomic number to identify the unknown nuclide:
$A + 2 = 97 + 2\,(1)$
$A = 97 + 2 - 2 = 97$

$Z + 1 = 43 + 2\,(0)$
$Z = 43 - 1 = 42$

Element 42 is molybdenum, so the starting nuclide is $\ce{^{97}_{42}Mo}$.

15.107 The mass number decreases by four and the atomic number decreases by two.

15.109   Since the half-life is 12.5 years, the number of elapsed half-lives is:
100 years/12.5 years = 8 half-lives

During each half-life, the amount of tritium is reduced by ½. After 8 half-lives, the amount remaining is the original amount multiplied by $(½)^8$:

Amount of tritium left = 256 g × $(½)^8$ = 256 g × (1/256) = 1.00 g

Alternately, the starting amount can be halved eight times:
256 g → 128 g → 64.0 g → 32.0 g → 16.0 g → 8.00 g → 4.00 g → 2.00 g → 1.00 g

15.111   The isotope will become more stable by converting a neutron to a proton. This will be accompanied by emission of a beta particle.

15.113   Write an equation for the alpha-emission process:
$$^{218}_{85}At \rightarrow {}^{4}_{2}He + {}^{A}_{Z}X$$

Use the conservation conditions to determine the mass number and atomic number of the unknown nuclide:
$218 = 4 + A$
$A = 218 - 4 = 214$
$85 = 2 + Z$
$Z = 85 - 2 = 83$

Element 83 is bismuth, so the unknown nuclide is $^{214}_{83}Bi$.

Chapter 16 – Organic Chemistry

16.1   (a) isomer; (b) alcohol; (c) ester; (d) aldehyde; (e) cyclic hydrocarbon; (f) saturated hydrocarbon;
       (g) alkyl group; (h) cycloalkane; (i) aliphatic hydrocarbon; (j) amino acid; (k) organic chemistry;
       (l) unsaturated hydrocarbon

16.3   Draw the Lewis structure for each molecule and count the bonds. Remember, carbon has four bonds and each
       hydrogen can only make a single bond. In other words, if a carbon has only two hydrogens on it, it must have at
       least two other bonds. In these hydrocarbons, this means either a double bond or two single bonds to other
       carbons.

       (a) seven single bonds

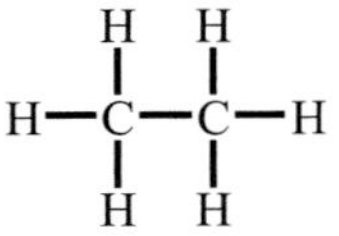

       (b) one double bond, four single bonds

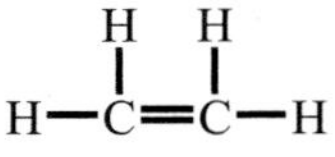

       (c) one triple bond, two single bonds

16.5   Table 16.1 summarizes the different types of functional groups. (a) amine; (b) ketone; (c) alcohol;
       (d) alkene; (e) carboxylic acid; (f) alkyne; (g) aldehyde; (h) ether; (i) ester

16.7   The molecular formula is determined by counting each type of atom in the molecule. For this molecule, the
       formula has 20 carbons, 28 hydrogens, and 1 oxygen. The molecular formula is $C_{20}H_{28}O$.
       The structural formula shows all the bonds and atoms found in the ball-and-stick model. The most common
       exceptions are hydrogens on different functional groups (i.e. methyl as CH_3— or H_3C—) because drawing all
       the bonds to all the hydrogens makes the drawing difficult to interpret.

       Structural formula

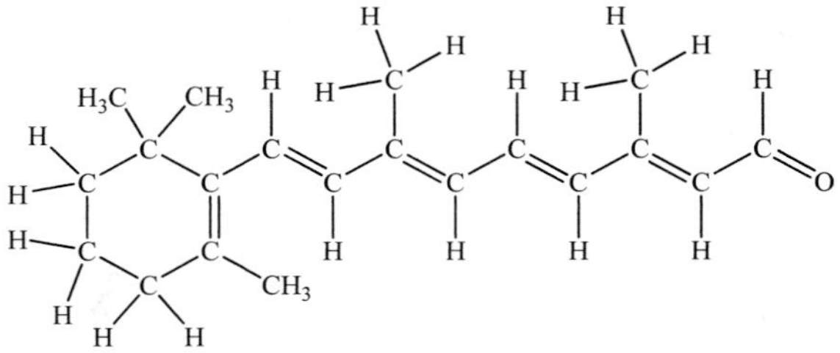

       In the line structure, hydrogens and carbons are not displayed. Notice that the CH_3— groups are represented by
       single lines. Hydrogens attached to functional groups are generally shown.

       Line structure

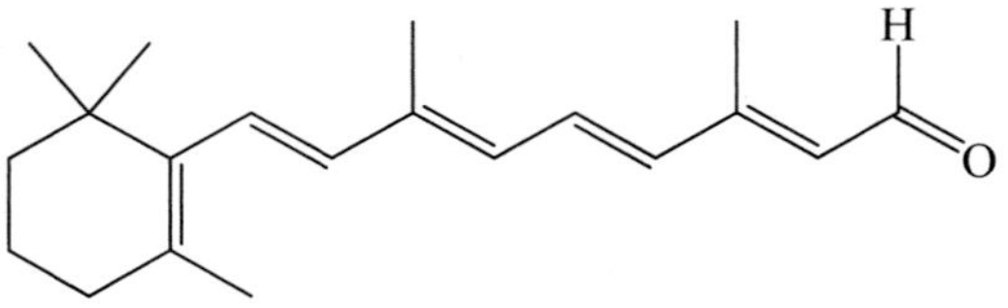

16.9   Remember each carbon must have four bonds. Bonds that are not shown on carbon are attached to hydrogens.
       Hydrogens attached to functional groups are usually shown. (a) $C_7H_{14}O$; (b) C_8H_{18}

16.11   The number of bonds that most atoms make in organic compounds is determined by the number of unpaired electrons in the Lewis dot symbol. This is generally why carbon makes four bonds and oxygen makes two bonds.

(a)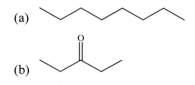

(b)

(c)

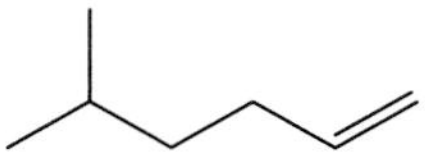

(d)

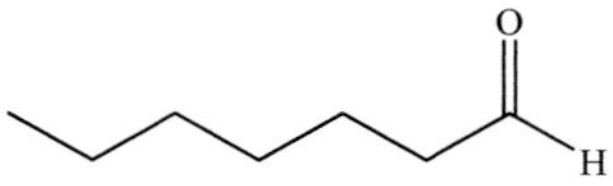

(e)

16.13   Structural formulas show all the bonds and atoms that are shown in the ball-and-stick model. Oxygen atoms are shown as red spheres in ball-and-stick models. When drawing line structures, remember that hydrogens attached to carbons are not shown. Carbons are indicated by "kinks" in the line structure.

(a)

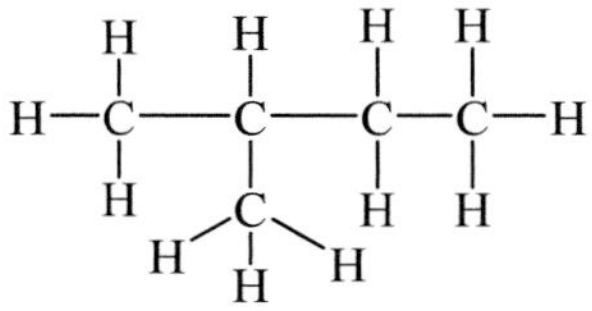

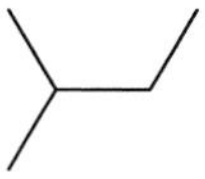

(b)

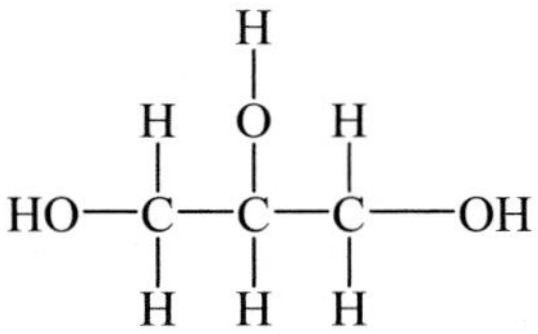

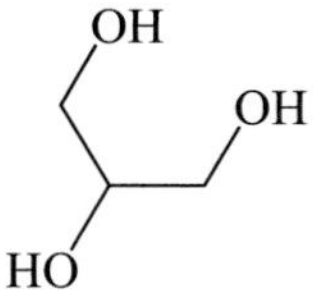

(c)

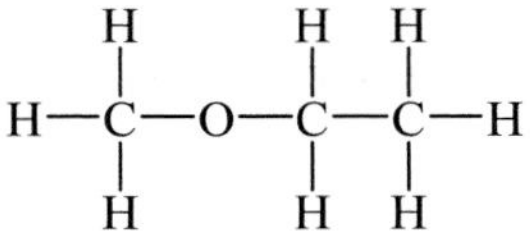

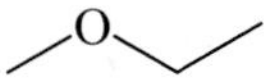

Molecules can also be represented by condensed structural formulas: (a) $(CH_3)_2CHCH_2CH_3$; (b) $(CH_2OH)_2CHOH$; (c) $CH_3OCH_2CH_3$.

16.15 Perspective drawings show atoms that are in one plane with solid lines. You could imagine that those atoms are in the plane of the paper. Wedge bonds are used to give the impression that the atom attached to the bond is pointing towards you (out of the surface of the paper) and a dashed wedge represents atoms pointed away from you (into the surface of the paper). In the methane molecule (a) two hydrogens are in the same plane with the carbon, but the other two are not.

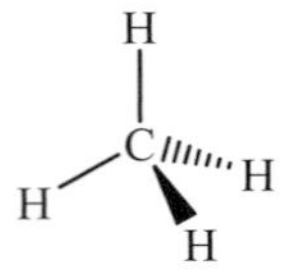

(a)

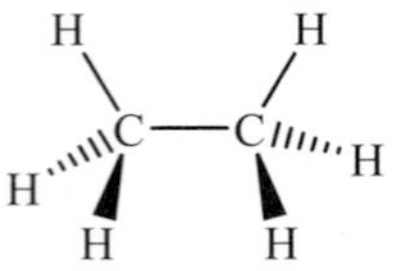

(b)

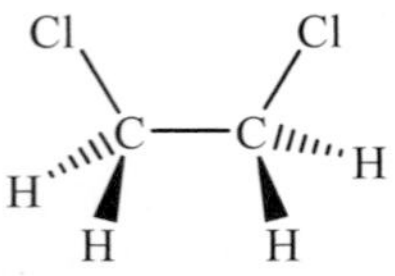

(c)

The chlorine can be in any of the positions occupied by hydrogen atoms. So there are numerous correct ways that the structure could be drawn:

(i.e. 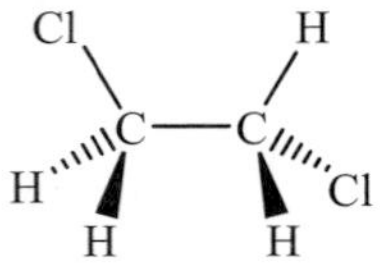).

16.17 An alkane hydrocarbon has only carbon-carbon single bonds, an alkene contains at least one carbon-carbon double bond, and an alkyne has at least one carbon-carbon triple bond.

16.19 Molecules that are unsaturated contain double or triple bonds. Saturated molecules follow the "2n+2" rule. That is, if a saturated molecule contains four carbons ($n = 4$) then it should have 10 hydrogen atoms (number of hydrogens = $2 \times 4 + 2 = 10$). So by looking at the line structure or by calculating the number of hydrogen atoms, we can determine if the molecule is saturated. (a) double bond, so it is unsaturated; (b) $n = 4$; would have 10 hydrogen atoms if saturated, so it is unsaturated; (c) $n = 5$, would have 12 hydrogen atoms if saturated, so it is unsaturated

16.21 Aromatic hydrocarbons are cyclic hydrocarbons that can be drawn with alternating double and single bonds between the carbon atoms. When the bonds are distributed in this way, the electrons are said to be delocalized.

16.23 The London dispersion forces must be stronger in butane (b.p. = $-0.5°C$) than in isobutane (2-methyl propane, (b.p. = $-11.6°C$)). London dispersion forces increase with the length of the molecule because the molecules have more opportunities to interact. Because the London forces are stronger in butane, it boils at a higher temperature.

16.25   In alkanes, the main intermolecular forces are London dispersion forces. London forces increase as the surface contact area between two molecules increases. This means that the intermolecular force increases with molecular weight (because the molecules get bigger) and with the length of the molecule. If two molecules have the same molecular weight, the molecule with the least amount of branching will have the stronger intermolecular attractions. (a) Propane, $CH_3CH_2CH_3$ has the highest molecular weight and is longer so it has the strongest intermolecular forces. (b) All three compounds have the same mass but n-hexane, $CH_3(CH_2)_4CH_3$, is the longest molecule so it has the strongest intermolecular forces.

16.27   By inspecting the chemical formulas you can determine that the compounds in this problem are all alkanes (each carbon is bonded to four other atoms). All alkane names end in the suffix –*ane*. To name straight chain alkanes count the number of carbons and use the prefixes listed in Table 16.3. The name can be written with or without the prefix *n*-. (a) ethane; (b) *n*-octane or octane; (c) propane; (d) *n*-nonane or nonane

16.29   Drawing isomers is much easier if you approach the problem systematically. First, draw the straight chain molecule. Next, shorten the chain by one carbon and place that carbon as a branch of the main chain. In this case there are two unique isomers with five-carbon chains. Notice that the molecule listed as a duplicate is the mirror image of one of the other isomers and is actually the same molecule flipped over. By shortening the main chain one more time, you have two branches to put on the main chain. For C_6H_{14} there are five unique isomers.

16.31   In naming branched-chain alkanes, first identify the parent chain. Number the main chain beginning at the end closest to a branch. It is helpful to convert the formula to a line drawing. The name of the branch is derived from the alkane name by changing the ending to –*yl*.
   (a)   The main chain is five carbons and the chain is numbered so that the methyl group is given the number 2. If the chain is numbered from the opposite side, the methyl group would be incorrectly numbered four.

2-methylpentane

(b)   2-methylbutane

(c)   2-methylbutane (same as previous structure)

16.33 When molecules contain more than one of the same type of branch, the prefixes *di-*, *tri-*, *tetra-*, etc., are used. The position of each group is indicated in the name of the compound. Number the longest (parent) chain so that the branches on the chain receive the lowest number.

(a) 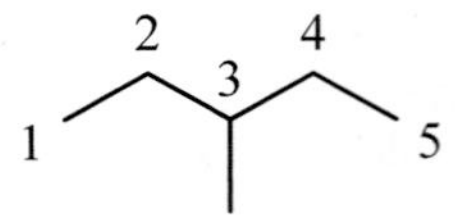 3-methylpentane

(b) 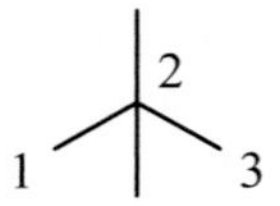 2,2-dimethylpropane

(c) 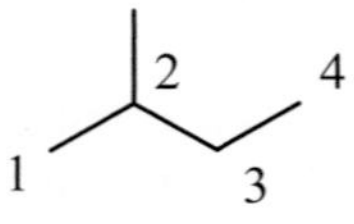 2-methylbutane

16.35 Start by writing the carbons for the parent chain. Next place the branches on the chain at the appropriate carbons. Finally add hydrogen atoms so that each carbon atom has four bonds.
(a) 2-methylpropane.

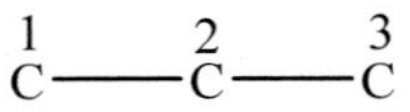 $\overset{1}{C}\text{---}\overset{2}{C}\text{---}\overset{3}{C}$	Write the parent chain: propane. The carbons are numbered for convenience. The numbers should not show up in your final structure.
$\overset{1}{C}\text{---}\underset{\underset{CH_3}{\|}}{\overset{2}{C}}\text{---}\overset{3}{C}$	Place the methyl group on carbon 2.
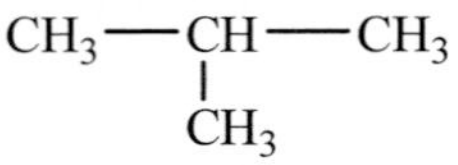 $CH_3\text{---}\underset{\underset{CH_3}{\|}}{CH}\text{---}CH_3$	Add the hydrogens necessary to complete the structure. The condensed structural formula is written from the structural formula: $CH_3CH(CH_3)CH_3$

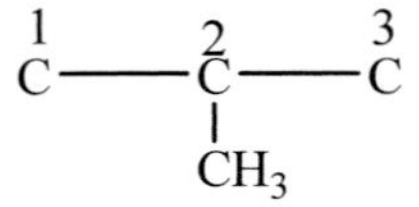

(b) 3,3-diethyl-2methypentane

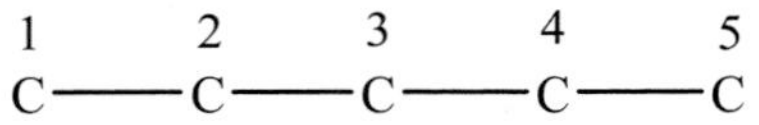 $\overset{1}{C}\text{---}\overset{2}{C}\text{---}\overset{3}{C}\text{---}\overset{4}{C}\text{---}\overset{5}{C}$	Write the parent chain: pentane. The carbons are numbered for convenience. The numbers should not show up in your final structure.
$\overset{1}{C}\text{---}\underset{\underset{CH_3}{\|}}{\overset{2}{C}}\text{---}\underset{\underset{CH_2}{\|}\underset{CH_3}{\|}}{\overset{\overset{CH_3}{\|}\overset{CH_2}{\|}}{\overset{3}{C}}}\text{---}\overset{4}{C}\text{---}\overset{5}{C}$	Place the ethyl groups at carbon 3 and a methyl group on carbon 2.
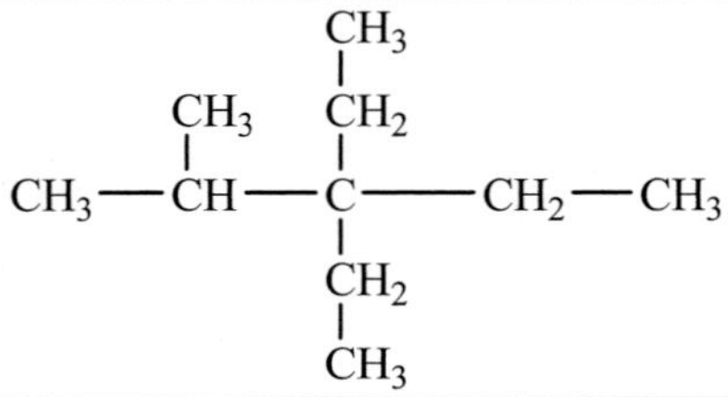 $CH_3\text{---}\underset{\underset{CH_3}{\|}}{CH}\text{---}\underset{\underset{CH_2}{\|}\underset{CH_3}{\|}}{\overset{\overset{CH_3}{\|}\overset{CH_2}{\|}}{C}}\text{---}CH_2\text{---}CH_3$	Add the hydrogens necessary to complete the structure. Condensed structural formula: $CH_3CH(CH_3)C(CH_2CH_3)_2CH_2CH_3$

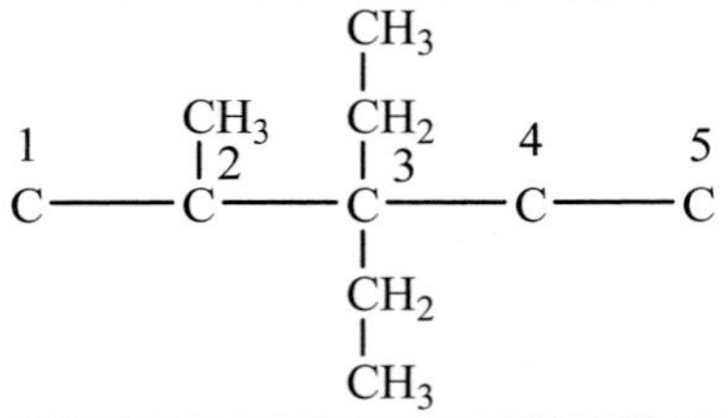

(c) 2,4-dimethyloctane

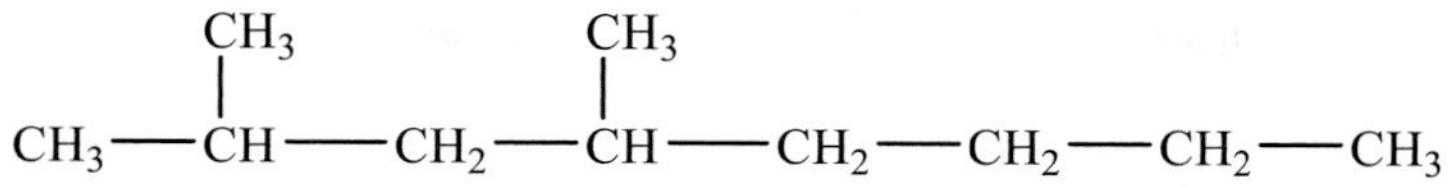

$$CH_3CH(CH_3)CH_2CH(CH_3)CH_2CH_2CH_2CH_3$$

Condensed structural formula: $CH_3CH(CH_3)CH_2CH(CH_3)CH_2CH_2CH_2CH_3$

(d) 3-ethyl-3-methylhexane

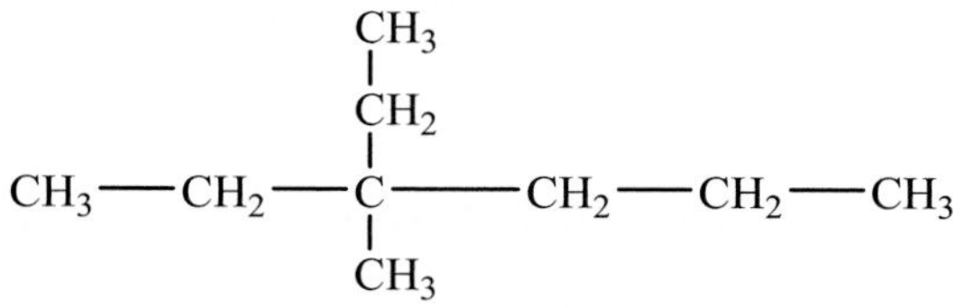

$$CH_3—CH_2—C(CH_2CH_3)(CH_3)—CH_2—CH_2—CH_3$$

Condensed structural formula: $CH_3CH_2C(CH_2CH_3)(CH_3)CH_2CH_2CH_3$

16.37 Alkane reactions are described in section 16.3.

(a) $CH_3CH_2CH_3 + Cl_2 \xrightarrow{\text{heat}} CH_3CHClCH_3 + HCl$

    or $CH_3CH_2CH_3 + Cl_2 \xrightarrow{\text{heat}} CH_3CH_2CH_2Cl + HCl$

(b) $(CH_3)_3CCH_2CH_2CH_3 + 11O_2 \xrightarrow{\text{heat}} 7CO_2 + 8H_2O$

(c) $CH_3CH_3 + Br_2 \xrightarrow{\text{light}} CH_3CH_2Br + HBr$

16.39 An alkene is a hydrocarbon containing one or more carbon-carbon double bonds.

16.41 For alkenes, number the longest chain starting at the end closest to the double bond.

(a)
$$\overset{1}{C}H_3\overset{2}{C}H_2\overset{3}{C}H=\overset{4}{C}H\overset{5}{C}H(CH_3)\overset{6}{C}H_2\overset{7}{C}H_3$$

The main chain is seven carbons long and the double bond follows the third carbon (3-heptene). The methyl group is attached to the fifth carbon (5-methyl): 5-methyl-3-heptene. Note that this structure could show cis-trans isomerism. If a structural formula or line diagram was given, the prefix *cis-* or *trans-* would be used (see (c) for an example).

(b)

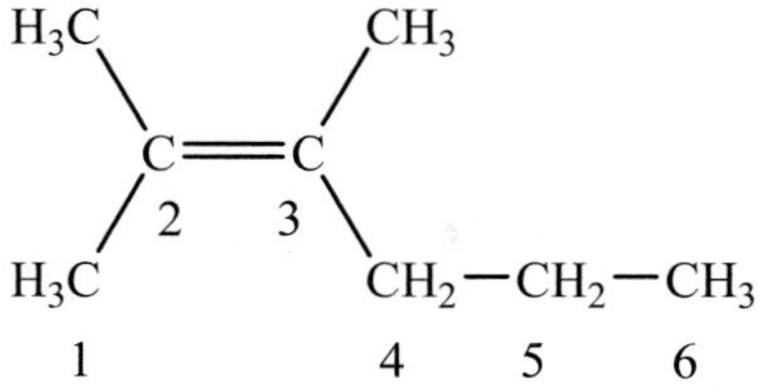

The longest chain is six carbons long and the double bond follows the second carbon (2-hexene). Methyl groups are attached to both the second and third carbons (2,3-dimethyl). The prefix *di-* is used to indicate that there are two methyl groups: 2,3-dimethyl-2-hexene.

(c)

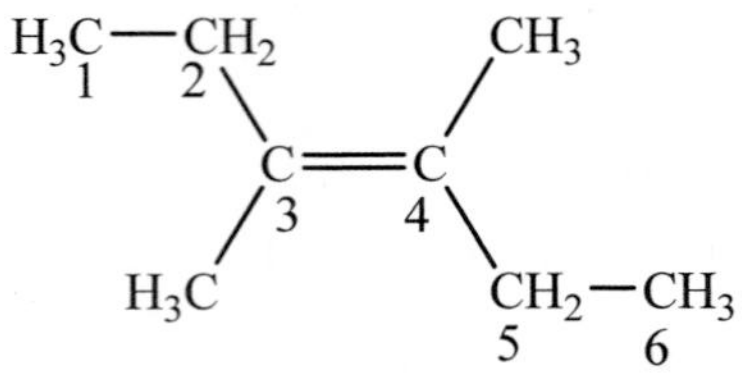

The longest chain is six carbons long and the double bond follows the third carbon (3-hexene). The methyl groups are on the third and fourth carbons (3,4-dimethyl). Unlike (b), this molecule shows cis-trans isomerism. Because the methyl groups are on opposite sides of the longest chain, the name contains the *trans-* prefix: *trans*-3,4-dimethyl-3-hexene.

16.43 In order to possess *cis-trans* isomerism, the groups attached to the double-bonded carbons must be different.
(a) CH_2CH_2 The groups (H's) are the same, no isomerism.
(b) $(CH_3)_2C=CHCH_3$ The methyl groups (on the left side of the double bond) are the same. No isomerism is present in this molecule.
(c) $CH_3CH_2C(CH_3)=C(CH_3)CH_2CH_3$ The double-bonded carbons have methyl and ethyl groups attached. Since the groups are different, this molecule has cis and trans isomers.

16.45 Line structures are most easily drawn by drawing the longest chain first. Keep in mind that you "zig-zag" the line for all structures except on the double bond of cis isomers (see (d)) and triple bonds (see (b)). Next place the double bonds and then the branches. To reduce confusion, always number the carbons from left to right. So, for example, if the longest chain is 1-butene, then the double bond will start on the far left and end on the second carbon (see (a)).

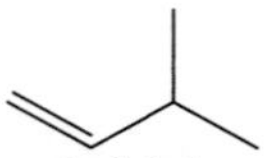

(a)   3-methyl-1-butene

(b)   propyne   The position of the triple bond was not given in the name because it is not necessary for drawing the correct structure.

(c)   *trans*-2-butene

(d)   *cis*-3-hexene

16.47 In addition reactions of alkenes, the atoms of the reactant are added on either side of the double bond.
(a) $CH_3CH=CHCH_3 + HBr \rightarrow CH_3CHBrCH_2CH_3$
(b) $CH_3CH_2CH=CHCH_3 + Br_2 \rightarrow CH_3CH_2CHBrCHBrCH_3$
(c)

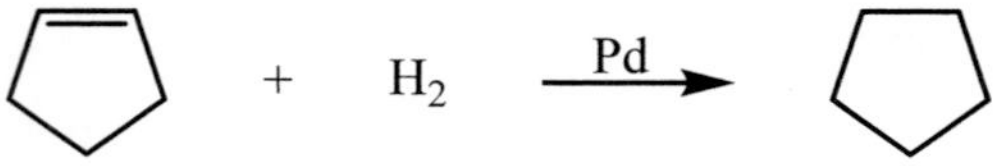

16.49   Aromatic hydrocarbons are cyclic and have alternating single and double bonds (delocalized electrons).

16.51   (a) When a benzene ring has two identical substituted groups, the ring is numbered so that the groups receive the lowest possible combination of numbers. The means that the bromo- groups receive the numbering 1,2 rather than 1,6.

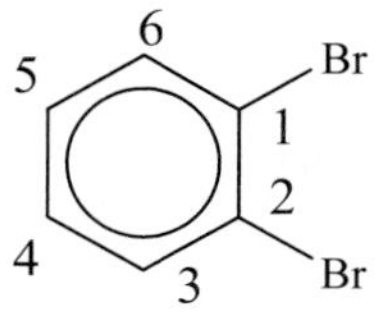

1,2-dibromobenzene

(b) For benzene rings which contain two different substituted groups, the ring is numbered so that the group which comes first alphabetically is given the lower number. Note that the group $-C_2H_5$ is an ethyl group ($-CH_2CH_3$).

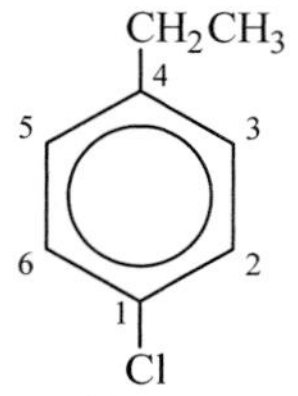

1-chloro-4-ethylbenzene

16.53   Begin by drawing a benzene ring and keep in mind that the carbons are numbered sequentially starting from any carbon on the ring.

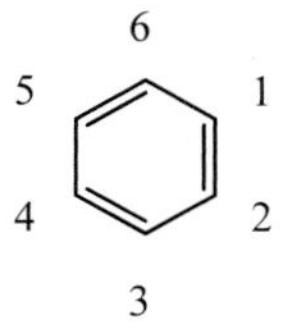

To finish drawing the structure, place the substituted groups on the appropriate carbon atoms.

(a)

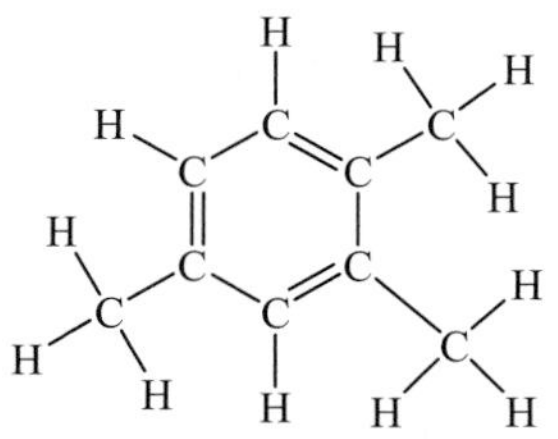

1,2,4-trimethylbenzene

(b)

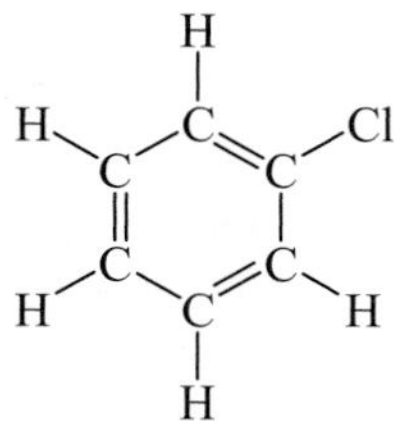

chlorobenzene

16.55 $C_6H_6 + Cl_2 \xrightarrow{FeCl_3} C_6H_5Cl + HCl$

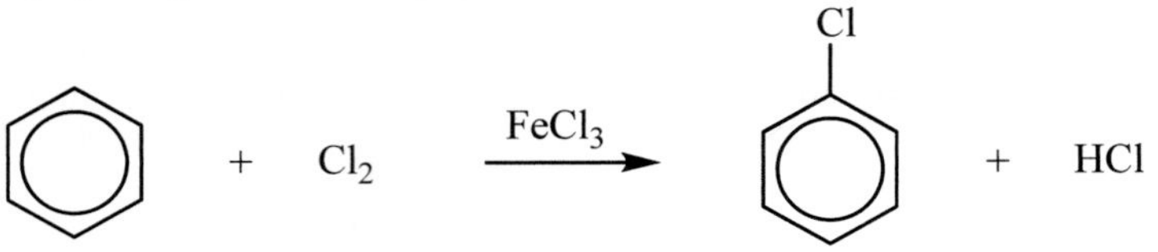

16.57   The functional group –OH is present in alcohols $\left[R^{\diagdown O \diagup} H \right]$. Both (b) and (c) are alcohols. Compounds in (a) and (d) are ethers and (e) is an aldehyde.

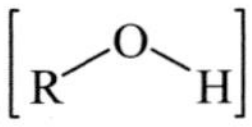

16.59   There are four isomers.

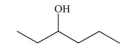

           1-butanol 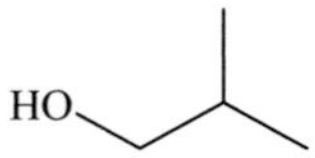   2-methyl-1-propanol

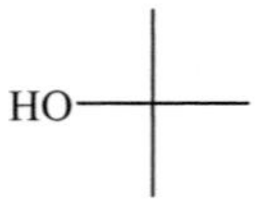

   2-methyl-2-propanol   2-butanol

16.61   Number the alkane chain first giving the alcohol the lowest number.

       $(CH_3)_2CHCH_2OH$
(a)   3   2 1   2-methyl-1-propanol;

       $CH_3CH(OH)CH_2CH_3$
(b) 1 2   3 4   2-butanol

16.63   Draw the longest chain and then place the substituted groups according to the positions designated by the name.

In (a) for example, first draw and number the hexane chain ($\overset{3}{\diagdown \diagup \diagdown \diagup \diagdown}$) and then place the alcohol functional group on the parent chain.

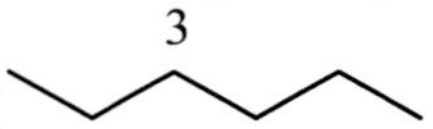

(a)   3-hexanol

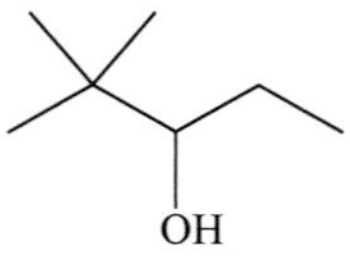

(b) 2,2-dimethyl-3-pentanol

16.65   The primary difference is that the intermolecular forces between ethanol (CH_3CH_2OH) molecules are hydrogen bonds while those between propane molecules are London dispersion forces. The stronger intermolecular forces between ethanol molecules means that a higher temperature (energy) is needed to raise the vapor pressure of ethanol to 1 atm.

16.67   Ethers contain the functional group –O– attached to two alkyl groups, $\left[R\diagdown^{\displaystyle O}\diagup R \right]$.

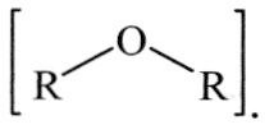

16.69   Only (a) CH_3OCH_3 is an ether. In an ether, the oxygen needs to be connected to carbons on either side. The compound in (d) initially looks like an ether, but the line drawing or structural formula reveals that it is a ketone

$$\text{(acetone, } H_3C\underset{}{\overset{\displaystyle O \atop \| \atop -C-}{\diagup\diagdown}}CH_3\text{).}$$

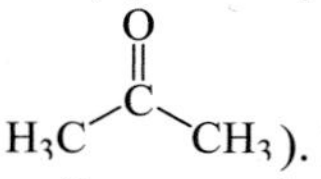

(acetone, $H_3C\diagdown C\diagup CH_3$). In condensed structural formulas, ketones are written COC and ethers are written as CH_2OCH_2. Compounds (b) and (c) are alcohols, (d) is a ketone, and (e) is an aldehyde.

16.71   The common names for ethers are derived from the lengths of the alkyl groups on either side of the oxygen. The names of the groups are written in alphabetical order.

(a)
　　　　　ethyl　　　propyl
　　　　$CH_3CH_2OCH_2CH_2CH_3$
　　　　ethyl propyl ether

(b)
　　　　methyl　　methyl
　　　　　CH_3OCH_3
　　　　dimethyl ether

16.73   The common names for ethers are derived from the lengths of the alkyl groups on either side of the oxygen.

(a) 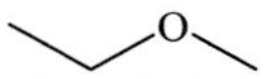   ethyl methyl ether

(b) 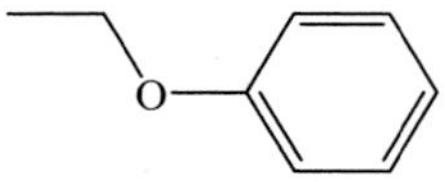   ethyl phenyl ether

16.75   Water is a hydrogen-bonding solvent, so substances that are hydrogen bonding or polar molecules will be more soluble in water. Ethers (CH_3OCH_3) do not directly hydrogen bond because the oxygen is attached to carbon atoms on either side. Alcohols are more water soluble because the alcohol group (–OH) can hydrogen bond.

16.77   An aldehyde has the functional group R–CHO, $\left[R\diagdown\overset{\displaystyle O \atop \| \atop -C-}{}\diagup H \right]$, where R can be an organic group or a hydrogen atom.

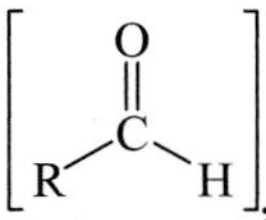

16.79   Structural formulas which contain – CHO are aldehydes, $\left[R\diagdown\overset{\displaystyle O \atop \| \atop -C-}{}\diagup H \right]$. Compounds (a), (c), and (d) are aldehydes. Compound (b) is a ketone.

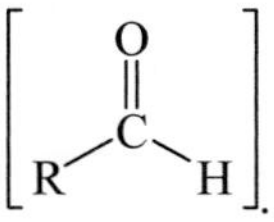

16.81   Both compounds are polar because of the C=O bond, but acetaldehyde or ethanal (CH_3CHO) has stronger London dispersion forces because it has a larger molecular mass. As a result, acetaldehyde has a higher boiling point.

16.83 A carboxylic acid contains the functional group R −COOH or R−CO$_2$H, $\left[\begin{array}{c} O \\ \| \\ R-C-O-H \end{array} \right]$, where R is an organic group.

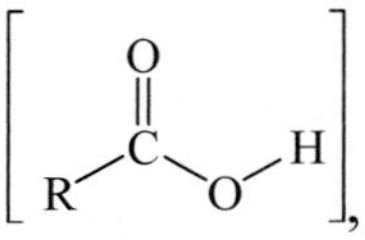

16.85 One route to the formation of esters is the condensation of a carboxylic acid and an alcohol. Methyl propionate (CH$_3$CH$_2$CO$_2$CH$_3$) can be formed from methanol (CH$_3$OH) and propionic acid (CH$_3$CH$_2$CO$_2$H).

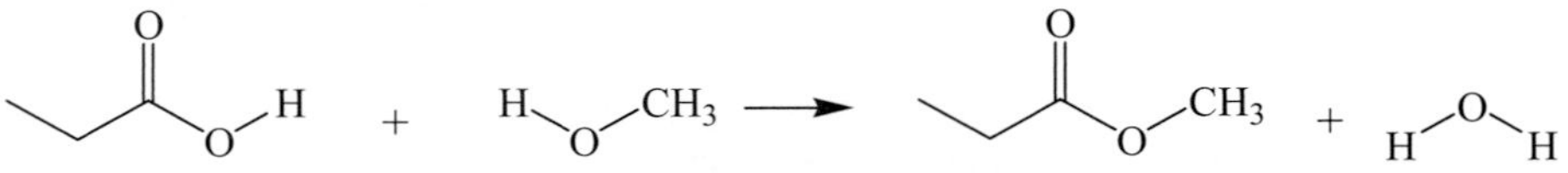

propionic acid   +   methanol   ⟶   methyl propionate   +   water

16.87 Saturated fatty acids are high molecular mass carboxylic acids which do not have double or triple bonds in the hydrocarbon chain. As a result, it is not possible to add any more hydrogen atoms to the hydrocarbon chain and they are said to be saturated with hydrogen.

16.89 Acetic acid (CH$_3$CO$_2$H) should have a higher boiling point than propanal (CH$_3$CH$_2$CHO) because acetic acid can hydrogen bond. Aldehydes are polar but do not hydrogen bond.

16.91 An amine is a derivative of ammonia (NH$_3$) with one or more of the H atoms replaced by organic groups.

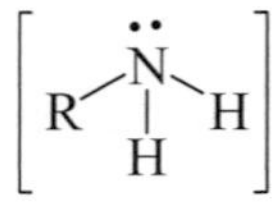

16.93 (a)

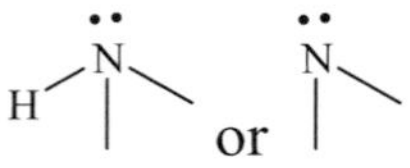

(b)

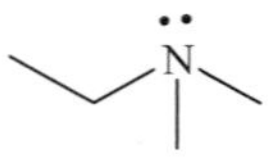

(c)

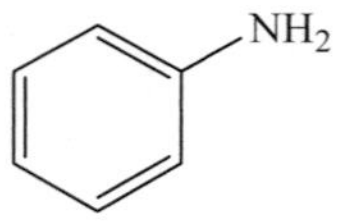

16.95 Methylamine (CH$_3$NH$_2$) can hydrogen bond so it has relatively strong intermolecular forces resulting in a higher boiling point than that of ethane (CH$_3$CH$_3$) which is nonpolar and cannot hydrogen bond.

16.97 Almost all amines can act as bases when reacted with HCl. The reaction is an acid base reaction to form an ammonium ion and the conjugate base of the acid.

(CH$_3$CH$_2$)$_2$NH   +   HCl(aq) ⟶ (CH$_3$CH$_2$)$_2$NH$_2^+$(aq) + Cl$^-$(aq)

16.99   A simple organization of hydrocarbons is given in the chart below. Cycloalkanes are considered saturated (i.e. they only contain single bonds). Also, many cycloalkenes are not aromatic because the electrons are not fully delocalized. Cycloalkynes (ring structures with triple bonds) are not common because the geometry of the alkynes (i.e. they are linear) does not fit well into small-ring structures and puts strain on the bonds in the ring.

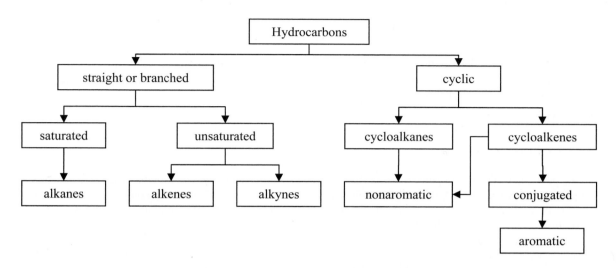

16.101   Alkenes and alkynes differ primarily in three ways: carbon-carbon bond type, bond angles, and number of hydrogen atoms in the structure.

	Alkanes	Alkenes	Alkynes
bond types	single	double bonds	triple bonds
bond angles	109.5°	120°	180°
general formula (straight or branched)	#H = 2n + 2	#H = 2n	#H = 2n − 2

#H = number of hydrogens
n = number of carbons. The formula assumes only one double or triple bond per structure.

16.103   Aldehydes and ketones have similar functional groups. A ketone has the general formula R–CO–R□,

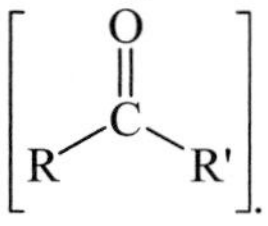

$$\left[\begin{array}{c} O \\ \parallel \\ R^{C}R' \end{array} \right]$$. The aldehyde functional group is R–CHO, $$\left[\begin{array}{c} O \\ \parallel \\ R^{C}H \end{array} \right]$$.

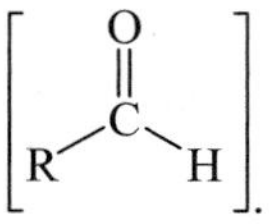

16.105   (a) Number the carbons so that the branches have the lowest possible numbers. Don't forget to number one of the methyl groups on the end of the structure.

CH₃CH(CH₃)CH₂CH(CH₃)₂
 5   4   3   2   1

2,4-dimethylpentane;

(b) Number the ring carbons so that the substituted groups get the lowest possible numbers.

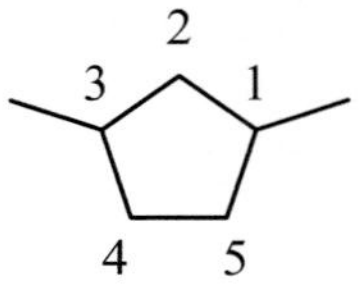

1,3-dimethylcyclopentane

(c) Number the carbons so that the branches have the lowest possible numbers. Don't forget that only one methyl group on each end of the molecule is part of the main chain, the other two are branches. This means there are four methyl branches in this molecule.

$(CH_3)_3C-C(CH_3)_3$
  1   2 3 4

2,2,3,3-tetramethylbutane

16.107   One possible test is to compare the solubility of each substance in an alcohol (such as ethanol). However, this test is not very specific. Addition of bromine, Br_2, which makes a reddish brown solution, would be a better alternative. Bromine will react with an alkene but not with an alcohol. If an alkene is present, the bromine solution will lose color as the bromine reacts with the alkene.

16.109   The trend in boiling points is directly related to the strength of the intermolecular forces in each substance. The weakest attractive forces are found in propane which is nonpolar. Dimethyl ether is polar, but ethanol is both polar and has hydrogen bonding. Because hydrogen bonds are so strong, ethanol has the highest boiling point.

16.111   To write the structural formulas, keep in mind that the suffix tells you what type of functional group is present in the molecule. In this case, you are writing structures for ketones (–one suffix) and aldehydes (–al suffix). Numbers either indicate the position of branches or, if the number is just prior to the name of the main chain, the position of the functional group on the main chain.

(a) CH_3COCH_3

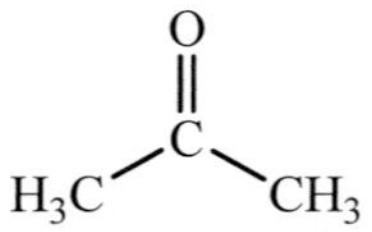

(b) $CH_3CH_2COCH_3$ or $CH_3COCH_2CH_3$ (both represent the same compound)

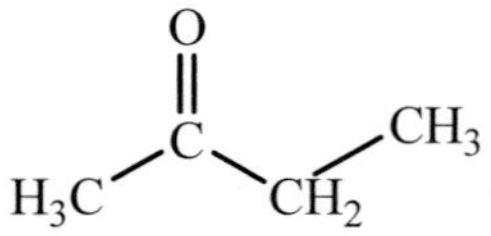

(c) CH_3CH_2CHO

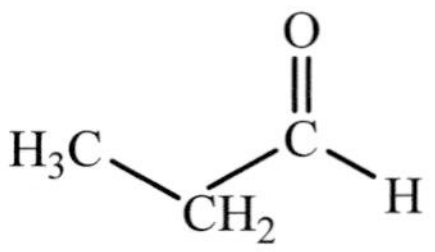

(d) CHOCHCHCH₃ (or CH₃CHCHCHO)

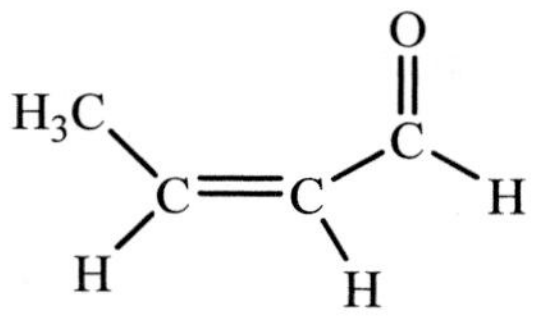

or

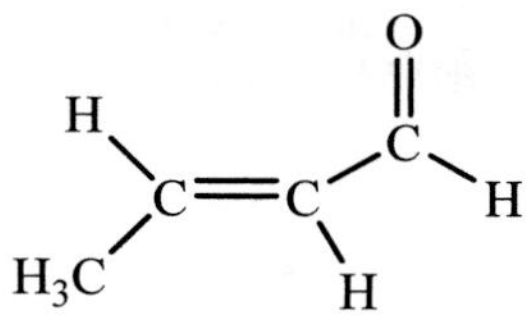

(e) CH₃CHO

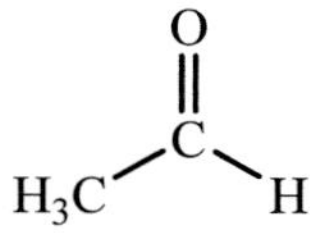

16.113 The names of carboxylic acids represent the number of carbons in the parent chain with the suffix –e replaced by -oic acid. Butanoic acid could be represented as:

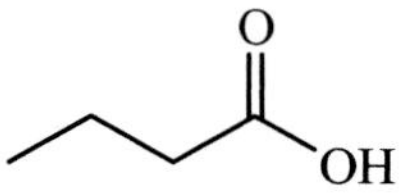

or $CH_3CH_2CH_2CO_2H$

The structures for esters show that they are essentially carboxylic acids in which the acidic hydrogen has been replaced by an alkyl group. The name of the group which replaced the acidic hydrogen is given first in the naming of the structure. Hexyl butanoate could be represented by

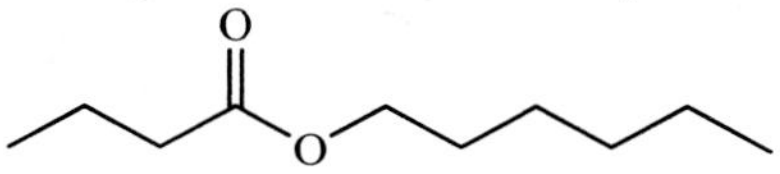

or $CH_3CH_2CH_2CO_2CH_2CH_2CH_2CH_2CH_2CH_3$

(a) Ethyl butanoate has a four carbon carboxylic acid and the acidic hydrogen has been replaced by an ethyl group: $CH_3CH_2CH_2CO_2CH_2CH_3$.

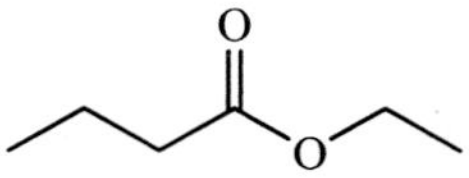

(b) Pentanoic acid is a five carbon carboxylic acid: $CH_3CH_2CH_2CH_2CO_2H$.

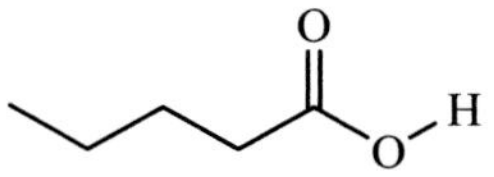

(c) Acetic acid is the common name for the two carbon carboxylic acid. The systematic name would be methyl ethanoate: $CH_3CO_2CH_3$.

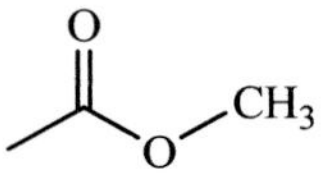

(d) Formic acid is the common name for a one carbon carboxylic acid: HCO_2H.

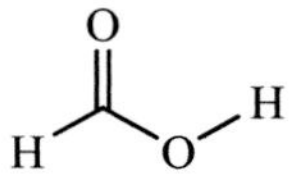

(e) Propanoic acid is a three carbon carboxylic acid: $CH_3CH_2CO_2H$.

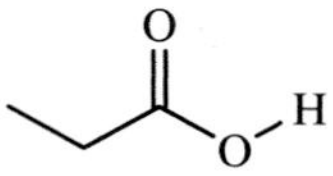

16.115 Common names for amines are derived by naming the alkyl groups attached to the nitrogen.
(a) dimethylamine; (b) ethyldimethylamine (the official IUPAC name is N,N-dimethyl-1-aminoethane, which is why we use the common name); (c) The compound $C_6H_5NH_2$ is usually called aniline, but is also called benzeneamine, phenylamine or aminobenzene (benzene with an amine group).

16.117 (a) $CH_3C(CH_3)_2CH_2CH(CH_3)_2$

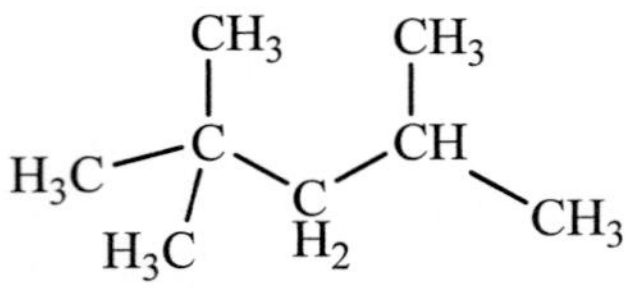

$$
\begin{array}{ccccc}
 & CH_3 & & CH_3 & \\
 & | & & | & \\
H_3C- & C & & CH & \\
 & / & \diagdown & & \diagdown \\
H_3C & & C & & CH_3 \\
 & & H_2 & &
\end{array}
$$

(b) $CH_3CHC(CH_2CH_3)_2$

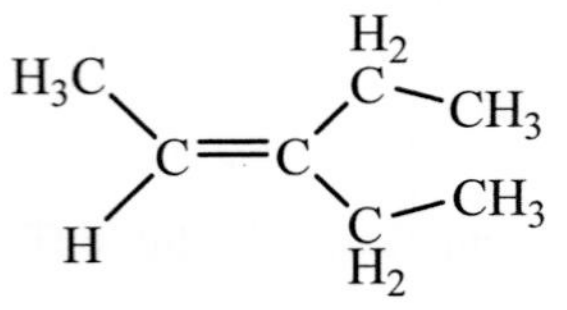

$$
\begin{array}{ccc}
H_3C & & H_2 \\
\diagdown & & C-CH_3 \\
 & C{=}C & \diagup \\
 & & \diagdown \\
\diagup & & C-CH_3 \\
H & & H_2
\end{array}
$$

(c) $CH_2CHCHCH_2$

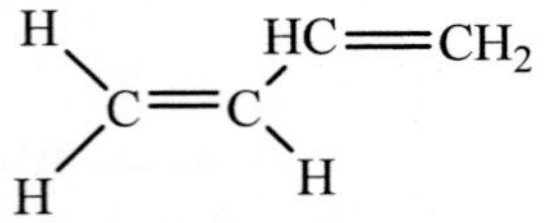

$$
\begin{array}{ccc}
H & & HC{=}CH_2 \\
\diagdown & & \diagup \\
 & C{=}C & \\
\diagup & & \diagdown \\
H & & H
\end{array}
$$

(d) C_4H_8;

$$
\begin{array}{ccc}
H_2C & \rule{1em}{0.4pt} & CH_2 \\
| & & | \\
H_2C & \rule{1em}{0.4pt} & CH_2
\end{array}
$$

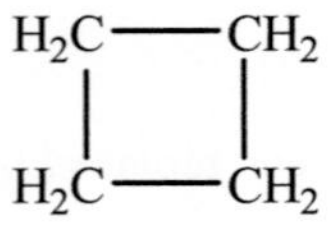

(e) C_6H_5Cl

$$
\begin{array}{ccccc}
 & & H & & \\
 & & | & & \\
H & & C & & Cl \\
\diagdown & & \diagdown & \diagup & \\
 & C & & C & \\
 & | & & | & \\
 & C & & C & \\
\diagup & & \diagdown & & \diagdown \\
H & & C & & H \\
 & & | & & \\
 & & H & &
\end{array}
$$

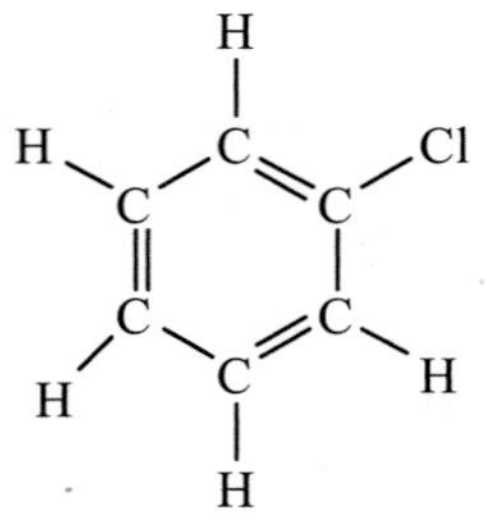

16.119   The following table is useful in determining the functional group type

Group name	Formula
alcohol	R–OH
aldehydes	R–CHO
alkane	No functional groups
alkene	C=C
alkynes	C≡C
amine	R-NH$_2$
carboxylic acids	–COOH
ethers	R–O–R'
ester	R–CO$_2$–R'
ketone	–CO–

(a) alkane; (b) ether; (c) carboxylic acid; (d) aldehyde; (e) ester; (f) amine; (g) ketone

16.121   Only (c) 2-butene and (d) 2-pentene can exist as *cis* and *trans* isomers. When the groups attached to one of the double-bonded carbons are the same, *cis-trans* isomerism cannot exist since the groups are interchangeable. For (a) propene and (b) 1-butene, there are two hydrogen atoms on the terminal carbon. Since these can be interchanged without changing the structure, *cis-trans* isomers do not exist. For 3-pentyne, alkynes do not form cis-trans isomers. For 2-butene, the methyl groups can either be together or on opposite sides of the double bond.

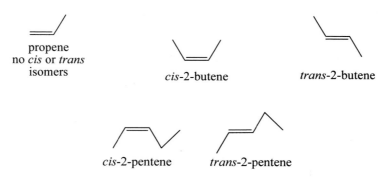

propene
no *cis* or *trans*
isomers

cis-2-butene

trans-2-butene

cis-2-pentene   *trans*-2-pentene

16.123 The most efficient way to do this is to number the parent chain on each as you would for naming the compound. Then make note of which have the same number of carbons (possible isomers) and whether they branch in different ways.

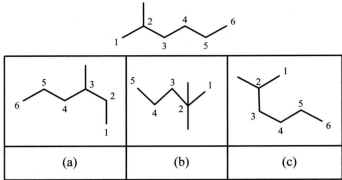

(a)	(b)	(c)

(a) has the same number of carbons (7) and is arranged differently (methyl group at C-3), so it is an isomer. (b) has the same number of carbons (7) and is arranged differently (two methyl groups at C-2), so it is an isomer. (c) is the same compound since it has the same structure.

16.125 Remember each carbon must have four bonds. Bonds that are not shown on carbon are attached to hydrogens. Hydrogens attached to functional groups are usually shown. (a) C_5H_9O; (b) $C_4H_{11}N$; (c) C_8H_{16}

16.127 The count of atoms in the formula indicate that the isomers could be alcohols or ethers. For the alcohols, start by drawing the possible alkane structures and then place the –OH, being careful not to duplicate any structures. For the possible ethers, we identify places in the alkane structures where an oxygen atom (which only forms two bonds) can be inserted. There are no possible ethers for the second alkane structure.

alkanes   alcohols   ethers

16.129

$$CH_3CH_2CH_2 \overset{\overset{\textstyle O}{\|}}{—C—} O—CH_2CH_2CH_3$$

16.131 Almost all amines can act as bases when reacted with HCl. The reaction is an acid base reaction to form an ion like ammonium and the conjugate base of the acid.

16.133 The most efficient way to do this is to number the parent chain on each as you would for naming the compound. Then make note of which have the same number of carbons (possible isomers) and whether they branch in different ways.

This compound is named 4-methyl-2-pentanol. Numbering begins along the chain closest to the functional group.

(a) This is the same compound, 4-methyl-2-pentanol.

(b) This is the same compound, 4-methyl-2-pentanol.

(c) This compound is an isomer with the name, 3-methyl-1-pentanol.

Chapter 17 – Biochemistry

17.1   (a) transfer RNA; (b) secondary structure; (c) nucleotide; (d) ribonucleic acid; (e) lipid; (f) primary structure; (g) deoxyribonucleic acid; (h) fatty acid; (i) active site; (j) carbohydrate; (k) peptide; (l) replication; (m) translation

17.3   Proteins play many important roles in the body. They are the catalysts for all chemical reactions in the body and thus regulate many cell process and reaction rates. They are found in the cell membranes and transport substances in and out of cells. They are also involved in the immune system and regulate the disease fighting ability of our bodies. They are the major structural component of muscles.

17.5   The general structures of all the alpha amino acids are given in Figure 17.4 and can be summarized by the structure:

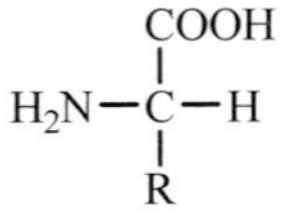

$$
\begin{array}{c}
\text{COOH} \\
| \\
\text{H}_2\text{N}-\text{C}-\text{H} \\
| \\
\text{R}
\end{array}
$$

17.7   The structures are found in Figure 17.4

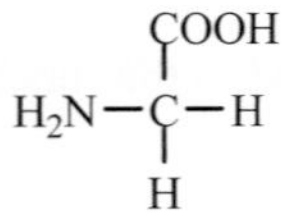 $\begin{array}{c}\text{COOH}\\|\\\text{H}_2\text{N}-\text{C}-\text{H}\\|\\\text{H}\end{array}$	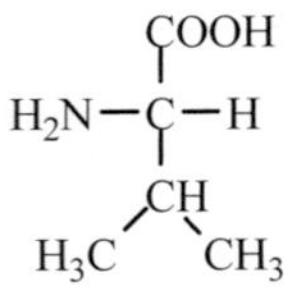 $\begin{array}{c}\text{COOH}\\|\\\text{H}_2\text{N}-\text{C}-\text{H}\\|\\\text{CH}\\ \text{H}_3\text{C}\quad\text{CH}_3\end{array}$	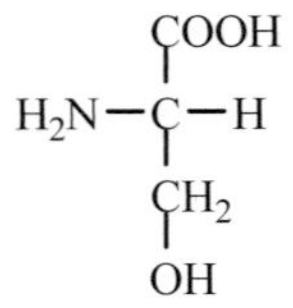 $\begin{array}{c}\text{COOH}\\|\\\text{H}_2\text{N}-\text{C}-\text{H}\\|\\\text{CH}_2\\|\\\text{OH}\end{array}$
Gly	Val	Ser

17.9   Both are aliphatic (contain only C and H) but alanine has a methyl group and valine has an isopropyl group.

17.11   (a) Ala-Gly   ; the methyl group is on the N-terminal amino acid;

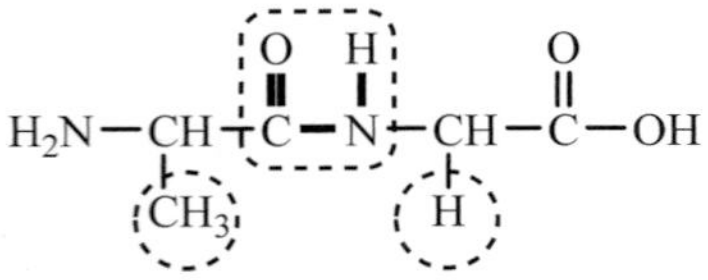

Gly-Ala   ; the methyl group is on the C-terminal amino acid; the (b) peptide bond is boxed and the (c) side chains are circled.

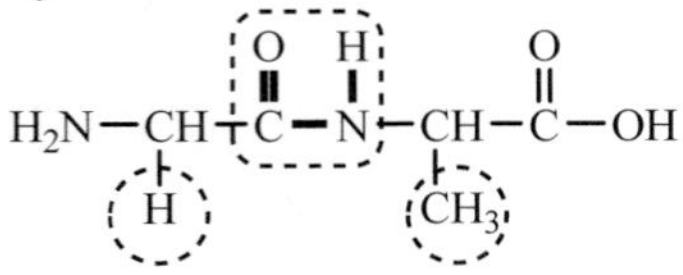

17.13   A polypeptide of alanine would simply have more than one alanine linked by peptide bonds. An example is shown below.

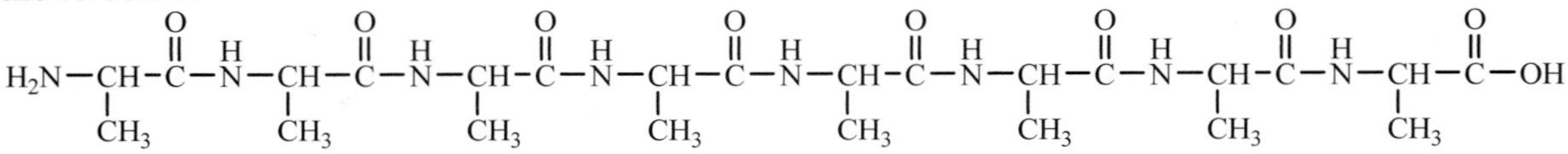

17.15   Proteins are large polypeptides with molar masses greater than about 5000 amu (often called Daltons). However, all are created by first linking two amino acids together in a peptide bond as shown below.

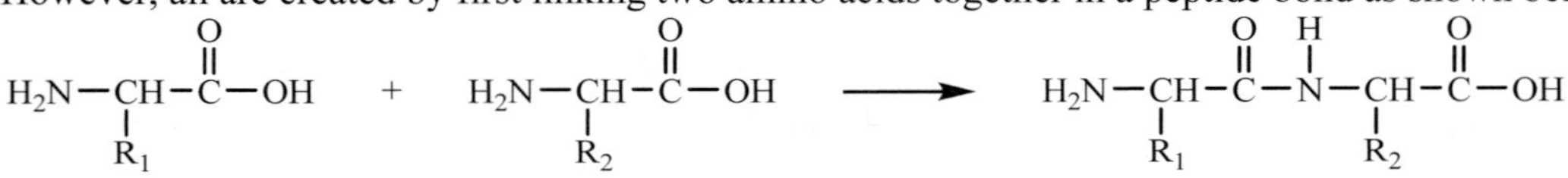

17.17 Either alanine or leucine could be on the N-terminal (amine end of the amino acid). The two structures that are possible are:

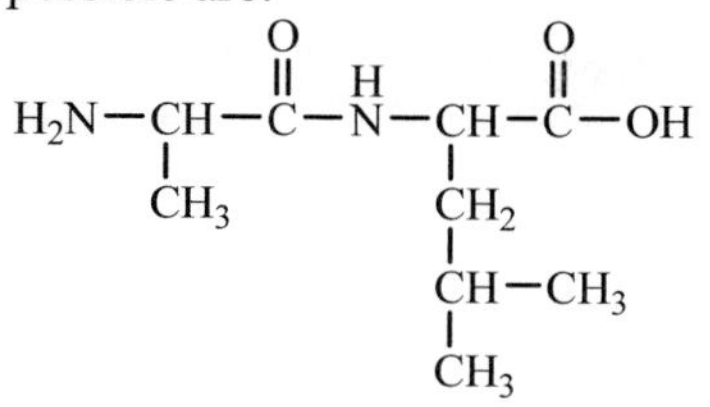

Ala-Leu 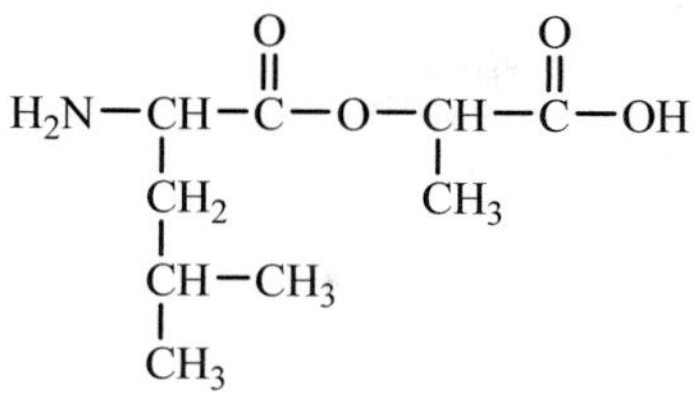   Leu-Ala

17.19 Like Problem 17.17 indicates, two different peptide bonds could form between alanine and valine. If alanine is the N-terminal amino acid, the following reaction would occur. Note that this is a condensation reaction because a water molecule is "removed" from the reactants.

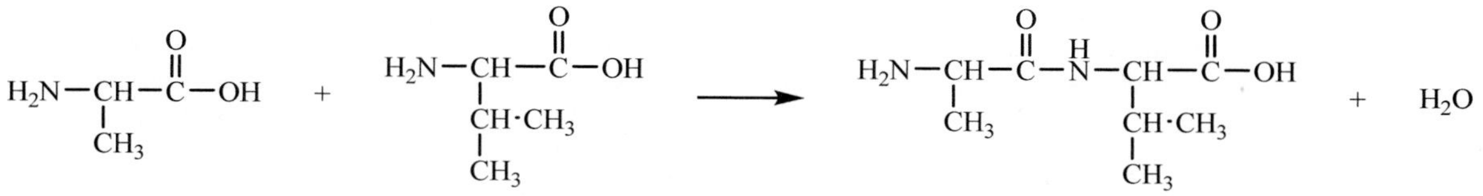

17.21 In Ala-Gly alanine is the N terminal amino acid and in Gly-Ala it is the C-terminal (carboxylic acid end) amino acid as shown below:

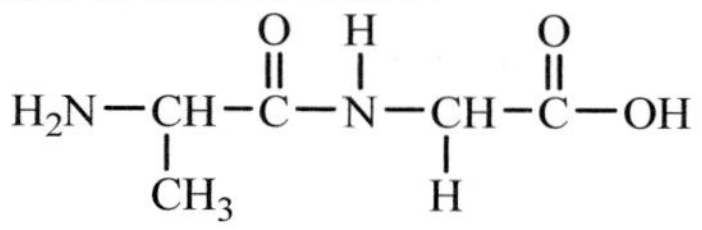

Ala-Gly 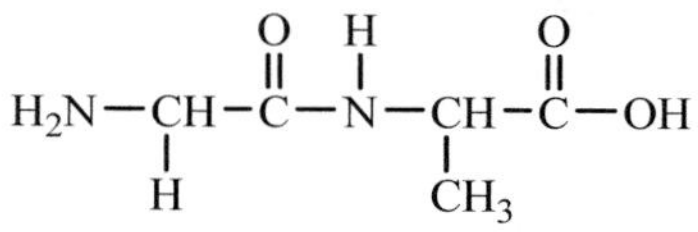   Gly-Ala

17.23 It is best to do this systematically. Consider all the permutations of the sequence 1-2-3:
1-2-3; 1-3-2; 2-1-3; 2-3-1; 3-1-2; 3-2-1
Note that 1, 2, and 3 each was first twice. The remaining numbers were simply switched in order. The same approach can be used to get the possible amino acid sequences:
Ser-Gly-Cys; Ser-Cys-Gly; Cys-Ser-Gly; Cys-Gly-Ser; Gly-Ser-Cys; Gly-Cys-Ser
There are a staggering number of possibilities if you consider that a protein might have several thousand amino acids in sequence!

17.25 See Table 17.1. Rice is low in certain amino acids (in particular Lys). To supplement their diet, vegetarians should consume food such as legumes. It is interesting to note that many cultures have developed diets which include beans and rice (i.e. Asian: rice and soybeans). http://www.fao.org/rice2004/en/f-sheet/factsheet3.pdf

17.27 Each amino acid in the tripeptide is formed when all the peptide bonds are hydrolyzed: Gly, Val, Ser.

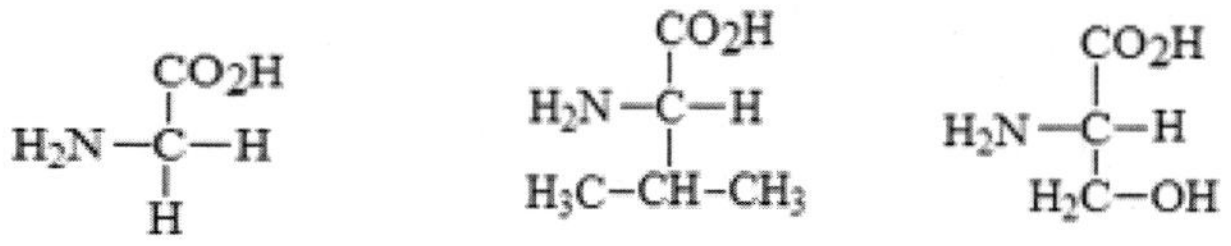

17.29 First identify the number of amino acids by locating the nitrogen atoms. Each peptide bond separates an amino acid. Also, the amine functional group is generally written as the first amino acid in the sequence. The R groups on the carbons between the peptide bonds (and the N and C terminal of the peptide) are $-H$, $-CH_3$, and $-CH(CH_3)_2$. These represent the amino acids glycine, alanine, and valine which would be obtained upon hydrolysis of the peptide. The abbreviated formula would be Gly-Ala-Val.

17.31 Ala-Pro-Phe-Gly

17.33 The primary structure is the amino acid sequence. The secondary structure results from intramolecular attractions such as hydrogen bonding and produces structural units such as the alpha helix and beta pleated sheets. The tertiary structure is the three-dimensional shape obtained from the folding of the peptide chain. Folding is caused by the intramolecular interactions of the secondary structural units. The quaternary structure is used to describe the shape of a protein when it is composed of more than one peptide chain. The quaternary structure is established through intermolecular interactions.

17.35 In myoglobin, a globular protein, the protein chain folds back on itself. Wool, a fibrous protein, is primarily composed of chains of alpha helices.

17.37 The three dimensional arrangement of the secondary structures caused by folding of the protein is the tertiary structure. The image shown is likely the tertiary structure of the G protein. It might also represent a quaternary structure if more than one peptide sequence is involved, but that is difficult to determine from a single image or further information about this particular protein.

17.39 Hydrogen bonding holds together the twisted-helical shape of the peptide chain. Repulsion of the amino acid R groups causes them to be on the outside of the helix.

17.41 Fibrous proteins have long rod like shapes. In globular proteins, the peptide chain folds back onto itself giving the protein a "glob"-like shape.

17.43 Denaturation disrupts the attractive forces which give rise to secondary, tertiary, and quaternary structures and causes the "unfolding" of the protein.

17.45 Ribonucleic acids and deoxyribonucleic acids; in deoxyribonucleic acids, the ribose unit lacks the hydroxide group at the 2 position.

17.47 A nucleic acid is composed of a chain of nucleotides.

17.49 The sugar is ribose and the base is adenine. Since the sugar is ribose, this would be a Ribonucleic Acid (RNA).

17.51 The shapes and sizes of these pairs (AT and CG) complement each other and provide for the maximum degree of hydrogen bonding. (Figure 17.21)

17.53 In DNA, the complimentary strand matches A to T and G to C. First write the given sequence and then, either above or below the original sequence, write the complimentary base:
TTAGCCAGTGCTA   Original strand
AATCGGTCACGAT   Complimentary strand

17.55   During transcription A, T, C, and G on DNA pair with U, A, G, and C in the mRNA. Write the DNA sequence and match the corresponding mRNA bases:
ACTAAGATC   Original DNA strand
UGAUUCUAG   Complimentary mRNA strand

17.57   Genetic information is encoded as complementary strands of nucleic acids. When genetic information is replicated it copies one strand using complementary base pairing of DNA nucleotides.

17.59   The complimentary sequence for UUU is AAA.

17.61   The AUG at the beginning sequence is the unique codon for Met and indicates the start of the protein synthesis. After that, break the sequence into three-base fragments. Each of these represents a different amino acid. The last six bases (CCC and CGA) are the codons for Pro and Arg, which also appear as the third and fifth codons, respectively. (See also Table 17.3.)
AUG   CUU   CCC   CAA   CGA   UUU   CCC   CGA
Met   Leu   Pro   Gln   Arg   Phe   Pro   Arg

17.63   There are several different structural formulas for glucose.

| Open Chain | β-glucose | α-glucose |

17.65   An example of a monosaccharide is glucose, $C_6H_{12}O_6$. Sucrose is a polysaccharide of glucose and fructose, $C_{12}H_{22}O_{11}$.

17.67   The −OH group on the first carbon can be either up or down. After that, if the OH is to the right on carbons 2, 3, and 4, in the linear molecule, it points down in the Hawthorn drawing (cyclic structure). This means that the first molecule, D-glucose, corresponds to the cyclic structure.

17.69   This is the cyclic form of a monosaccharide that is a six carbon sugar (hexose).

17.71   Sucrose yields glucose and fructose. Starch yields only glucose.

17.73   In starches, the main chain repeats with the glucose molecules maintaining the same orientation in the chains. Amylopectin also branches (which is not observed in cellulose).

amylose

In cellulose the, the glucose units invert after every ether bond (circled). Note the position of the –CH$_2$OH group.

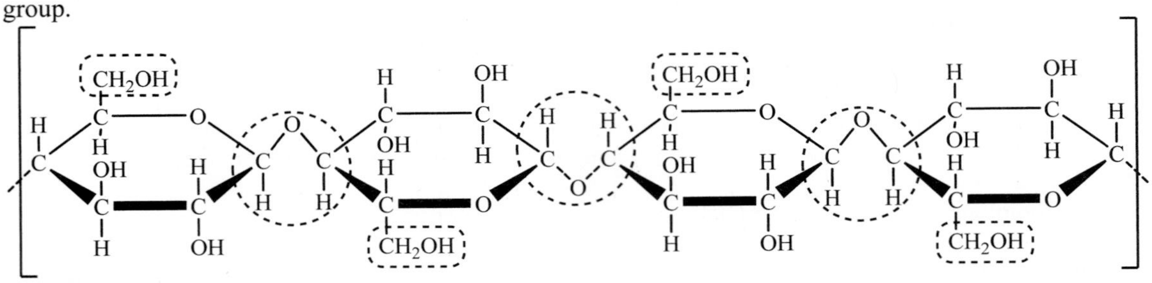

$$\text{[cellulose structure]}_{cell}$$

ulose

17.75   Before we can taste the sugars in starch, the glucose monomers must be cleaved from the terminal of the amylose or amylopectin. A receptor is believed to exist for sucrose sweetness, but sucrose is also considerably more soluble than the larger polysaccharides and would also be hydrolyzed more quickly.

17.77   The fatty acids in fat molecules tend to be completely or almost completely saturated (few or no C=C double bonds). The fatty acids in oil molecules are generally poly-unsaturated and contain several C=C double bonds on each fatty acid chain.

17.79   triacyl glycerol (tristearate)

$$\text{[triacylglycerol (tristearate) structure]}$$

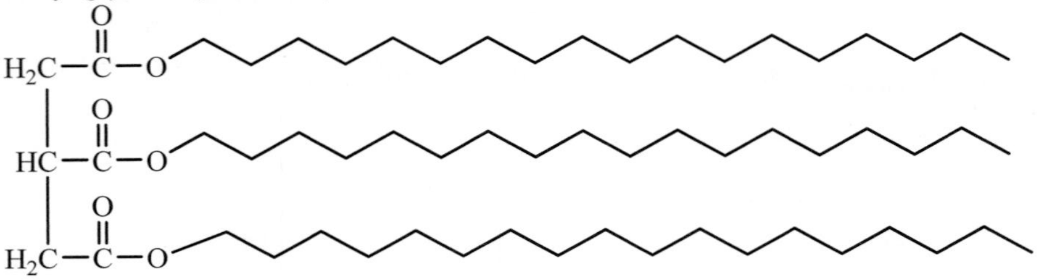

17.81   It would be a fat because it is saturated (no double bonds). The saturation causes the molecule to have a higher melting point and it would be a solid at room temperature.

17.83   The molecule is a phospholipid.

17.85   They are both lipids and all steroids are derived from cholesterol.

17.87   Ser-Ser-Ser has the following structure:

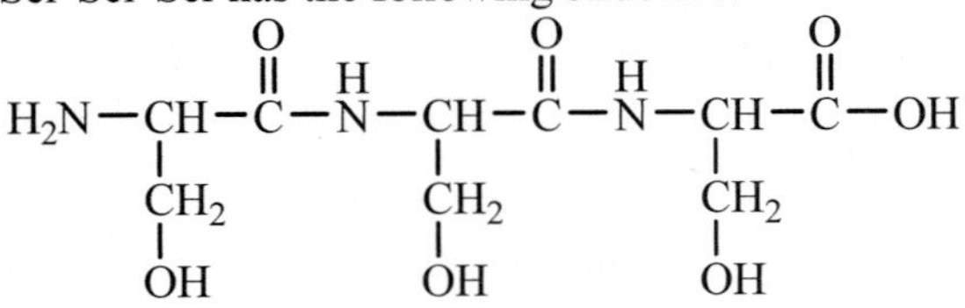

$$\text{[Ser-Ser-Ser tripeptide structure]}$$

17.89 Val-Cys-Tyr

H₂N—CH—C(=O)—N(H)—CH—C(=O)—N(H)—CH—C(=O)—OH with side chains: CH–CH₃/CH₃ (valine), CH₂–SH (cysteine), CH₂–(phenol ring with OH) (tyrosine)

17.91 Proteins perform many functions, one of the most important is to catalyze the biochemical reactions. They are also responsible for the machinery of the cell which regulates cellular processes such as transcription, translation, immune response, and energy production. Carbohydrates provide energy (i.e. glucose and metabolism) and play important structural roles in plants (cellulose).

17.93 DNA and RNA each have a sugar molecule, base, and a phosphate unit. The sugar in DNA is deoxyribose. The DNA contains all of the genetic information. RNA, derived from DNA through transcription, contains only information for a specific protein.

17.95 Genetic therapy can be used to produce novel substances from organisms. An example is human insulin production from bacteria. Plants can be bred to tolerate various chemical agents (i.e. herbicides) using this technique.

17.97 (a) valine; (b) methionine; (c) aspartic acid

17.99 valine   serine   methionine

Tripeptide structure: H₂N—C(H)—C(=O)—N(H)—C(H)—C(=O)—N(H)—C(H)—C(=O)—OH with side chains: H₃C—C(H)—CH₃ (valine), H₂C—OH (serine), H₂C—CH₂—S—CH₃ (methionine)

17.101 serine   valine   methionine

serine: H₂N—C(H)—C(=O)—OH with H₂C—OH side chain
valine: H—N(H)—C(H)—C(=O)—OH with H₃C—C(H)—CH₃ side chain
methionine: H—N(H)—C(H)—C(=O)—OH with H₂C—CH₂—S—CH₃ side chain

17.103 The sugar is ribose and the base is guanine. This nucleotide would be found in RNA.

17.105 There are no double bonds along the carbon chain of the fatty acid so it is saturated.

NOTES

NOTES

NOTES

NOTES

NOTES